METAL CLUSTERS

Wiley Series in Theoretical Chemistry

Series Editors

D. Clary, University College London, London, UK
A. Hinchliffe, UMIST, Manchester, UK
D. S. Urch, Queen Mary and Westfield College, London, UK
M. Springborg, Universität Konstanz, Germany

METAL CLUSTERS

Edited by

Walter Ekardt

Fritz-Haber-Institut der Max-Planck-Gesellschaft, Berlin,
Germany

JOHN WILEY & SONS, LTD

Chichester • New York • Weinheim • Brisbane • Singapore • Toronto

Chemistry Library

Other Wiley Editorial Offices

John Wiley & Sons, Inc., 605 Third Avenue,
New York, NY 10158-0012, USA

WILEY-VCH Verlag GmbH, Pappelallee 3,
D-69469 Weinheim, Germany

Jacaranda Wiley Ltd, 33 Park Road, Milton,
Queensland 4064, Australia

John Wiley & Sons (Asia) Pte Ltd, Clementi Loop #02-01,
Jin Xing Distripark, Singapore 129809

John Wiley & Sons (Canada) Ltd, 22 Worcester Road,
Rexdale, Ontario M9W 1L1, Canada

Library of Congress Cataloging-in-Publication Data

Metal clusters / edited by Walter Ekardt.
    p.   cm. — (Wiley series in theoretical chemistry)
  Includes bibliographical references and index.
  ISBN 0-471-98783-2 (hardback : alk. paper)
  1. Metal crystals.   I. Ekardt, Walter.   II. Series.
QD921. M465   1999
546′.3 — dc21   98–53083
                                                       CIP

British Library Cataloguing in Publication Data

A catalogue record for this book is available from the British Library

ISBN 0 471 98783 2

Typeset in 10/12pt Times by Laser Words, Madras, India
Printed and bound in Great Britain by Bookcraft (Bath) Ltd
This book is printed on acid-free paper responsibly manufactured from sustainable forestry,
in which at least two trees are planted for each one used for paper production

Contents

4 Dissociation, Fragmentation and Fission of Simple Metal Clusters **145**
Constantine Yannouleas, Uzi Landman and Robert N. Barnett

5 Optical and Thermal Properties of Sodium Clusters **181**
Hellmut Haberland

Contributors

Wanda Andreoni *IBM Research Division, Zurich Research Laboratory, 8803 Rüschlikon, Switzerland*

Pietro Ballone *Institut für Festkörperforschung Forschunhszentrum Jülich, 52425 Jülich, Germany*

Robert N. Barnett *School of Physics, Georgia Institute of Technology, Atlanta, Georgia 30332-0430, USA*

K. H. Bennemann *Institut für Theoretische Physik, Freie Universität Berlin, Arnimalle 14, D-14195 Berlin, Germany*

Vlasta Bonačić-Koutecký *Humboldt-Universität zu Berlin, Walther-Nernst-Institut für Physikalische und Theoretische Chemie, Bunsenstrasse 1, 10117 Berlin, Germany*

Walter Ekardt *Fritz-Haber-Institut der Max-Planck-Gesellschaft, Faradayweg 4-6, 14195 Berlin, Germany*

Piercarlo Fantucci *Dipartimento di Chimica, Inorganica, Metallorganica e Analitica, Centro CNR, Universita di Milano, Via Venezian 21, 20133 Milano, Italy*

Hellmut Haberland *Fakultät für Physik and Institute of Materials Research, Universität Freiburg, D-79104 Freiburg, Germany*

Jaroslav Koutecký *Freie Universität Berlin, Institut für Physikalische und Theoretische Chemie, Takustrasse 3, 14195 Berlin, Germany*

Uzi Landman *School of Physics, Georgia Institute of Technology, Atlanta, Georgia 30332-0430, USA*

J. M. Pacheco *Departamento de Física da Universidade, 3000 Coimbra, Portugal*

G. M. Pastor *Laboratoire de Physique Quantique, Unité Mixte de Recherche 5626 du CNRS, Université Paul Sabatier, 118 route de Narbonne, F-31062 Toulouse, France*

Jiří Pittner *Humboldt-Universität zu Berlin, Walther-Nernst-Institut für Physikalische und Theoretische Chemie, Bunsenstrasse 1, 10117 Berlin, Germany*

Detlef Reichardt *Humboldt-Universität zu Berlin, Walther-Nernst-Institut für Physikalische und Theoretische Chemie, Bunsenstrasse 1, 10117 Berlin, Germany*

P.-G. Reinhard *Institut für Theoretische Physik, Universität Erlangen, Staudstrasse 7, D-91058 Erlangen, Germany*

W.-D. Schöne *Fritz-Haber-Institut der Max-Planck-Gesellschaft, Faradayweg 4-6, 14195 Berlin, Germany*

E. Suraud *Laboratoire de Physique Quantique, Université Paul Sabatier, 118 route de Narbonne, F-31062, Toulouse Cedex, France*

Constantine Yannouleas *School of Physics, Georgia Institute of Technology, Atlanta, Georgia 30332-0430, USA*

Series Preface

Theoretical chemistry is one of the most rapidly advancing and exciting fields in the natural sciences today. This series is designed to show how the results of theoretical chemistry permeate and enlighten the whole of chemistry together with the multifarious applications of chemistry in modern technology. This is a series designed for those who are engaged in practical research, in teaching and for those who wish to learn about the role of theory in chemistry today. It will provide the foundation for all subjects which have their roots in the field of theoretical chemistry.

How does the materials scientist interpret the properties of a novel doped-fullerene superconductor or a solid-state semiconductor? How do we model a peptide and understand how it docks? How does an astrophysicist explain the components of the interstellar medium? Where does the industrial chemist turn when he wants to understand the catalytic properties of a zeolite or a surface layer? What is the meaning of 'far-from-equilibrium' and what is its significance in chemistry and in natural systems? How can we design the reaction pathway leading to the synthesis of a pharmaceutical compound? How does our modeling of intermolecular forces and potential energy surfaces yield a powerful understanding of natural systems at the molecular and ionic level? All these questions will be answered within our series which covers the broad range of endeavour referred to as 'theoretical chemistry'.

The aim of the series is to present the latest fundamental material for research chemists, lecturers and students across the breadth of the subject, reaching into the various applications of theoretical techniques and modeling. The series concentrates on teaching the fundamentals of chemical structure, symmetry, bonding, reactivity, reaction mechanism, solid-state chemistry and applications in molecular modeling. It will emphasize the transfer of theoretical ideas and results to practical situations so as to demonstrate the role of theory in the solution of chemical problems in the laboratory and in industry.

D. Clary, A. Hinchliffe, D. S. Urch and M. Springborg
June 1994

1 Application of the Jellium Model and its Refinements to the Study of the Electronic Properties of Metal Clusters

WALTER EKARDT

Fritz-Haber-Institut der Max-Planck-Gesellschaft, Berlin,

W.-D. SCHÖNE

Fritz-Haber-Institut der Max-Planck-Gesellschaft, Berlin

and

J. M. PACHECO

Departamento de Física da Universidade, Coimbra

1.1 INTRODUCTION

The jellium model, known from the study of the electronic properties of NFE (nearly free electron) metals [1] and from the understanding of the electronic properties of free metallic surfaces [2], is applied to the calculation of the electronic properties of metal clusters, mainly to the group Ia (alkaline metals) and to the group Ib metals (noble metals).

Metal Clusters. Edited by W. Ekardt

The model is shown to give results that are very often in *quantitative* agreement with experimental data and serve in most other cases as a good starting point for the calculation of the effects of the ionic structure, e.g. via pseudopotential perturbation theory.

Main experimental findings both for the ground state (magic numbers for the stability of clusters [3] and the existence of supershells [4]) and for excited states (the dominance of collective states in the photoabsorption of metal clusters Me_N with $N > 8$) were *predicted* [5] before their experimental confirmation. Recently we were able to explain the temperature dependence of the absorption of small metal clusters as observed by Haberland's group [6]. If the model is complemented by pseudopotential perturbation theory [7] the results obtained match *qualitatively* those obtained by demanding quantum-chemical methods (e.g. the photoabsorption spectra of Na_6). Further improvement of the model includes the removal of self-interaction effects, the so-called SIC [8–10] (a consequence of using the local density approximation (LDA) to general density functional theory (DFT)).

The development of the *super-atom model* for the description of electronic properties of metal clusters arose from the attempt to understand and interpret experimental data by W. Schulze and co-workers Figure 1.1 shows the absorption of small silver particles

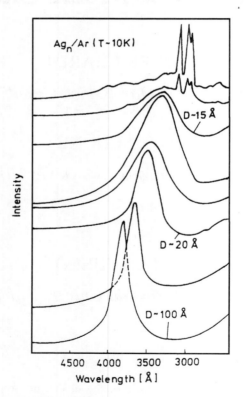

Figure 1.1 Optical absorption of small silver particles Ag_N embedded in argon at low temperatures, according to Ref. [11]. The huge absorption hump is a collective electronic oscillation localized at the interface Ag/Ar. This figure historically gave the impact for the development of the super-atom model for metal clusters. For large clusters a broad, damped peak is observed, whereas for small clusters the line is fragmented

(in arbitrary units) as a function of wavelength with the mean diameter of the clusters as parameter. As one can see, there is a pronounced blue-shift of the absorption hump as the cluster size decreases. If one tries to understand the observed absorption within *classical, macroscopic electrodynamics*, one has to look at the imaginary part of the dynamical polarizability of a small metal particle embedded in a dielectric host. The textbook solution for a spherical particle of radius R is [12]

$$\alpha(\omega) = R^3 \frac{\varepsilon(\omega) - \varepsilon_d(\omega)}{\varepsilon(\omega) + 2\varepsilon_d(\omega)}, \tag{1}$$

where $\varepsilon(\omega)$ and $\varepsilon_d(\omega)$ are the dielectric constants of bulk silver and the dielectric host, respectively. As in the case of bulk plasmons, defined in terms of the vanishing of $\varepsilon(\omega)$, collective interfacial excitations are characterized by the vanishing of the denominator, which means that there is an eigenoscillation *in the absence of external stimuli* [13]. Since both dielectric constants are size-independent, one sees immediately that classical, macroscopic electrodynamics does *not* work in this size regime. Therefore one has to resort to *microscopic or mesoscopic models*.

The first, most primitive, model is the infinite barrier model (IBM). Here the electronic motion is confined by a spherical potential hole with infinitely high barriers. Once the electronic wave functions (spherical Bessel functions) and eigenvalues are known, one can proceed and calculate the dynamic polarizability $\alpha(\omega)$. From this quantity the collective excitations are determined in a straightforward manner (see below). The theoretical prediction [50], shown in Figure 1.2, matches the experimental data (indicated by dots) rather well from very small to mesoscopic particle sizes. The result obtained shows that the IBM, which models the kinetic repulsion of the occupied 4d-shell of atomic Xe, works surprisingly well. This repulsion causes an *enhanced* electronic density, leading to the blue-shift of the surface-plasmon line.

For the description of *free* metal clusters, as observed and investigated in supersonic beams, the electrons relax (and tunnel) into the neighboring vacuum. In order to model

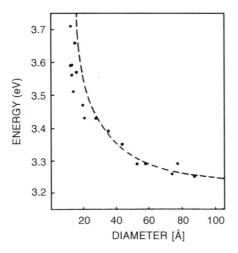

Figure 1.2   Crude theoretical interpretation of the experimental data of Figure 1.1. Here the electronic motion is confined by the IBM (infinite barrier model). For details see the original work by Ekardt *et al.* [50]. Reprinted by permission of Elsevier Science Publishers

this situation, it is useful to recall the series of papers by N. D. Lang and W. Kohn [14, 2], who applied the *self-consistent jellium model* to the study of the electronic properties of free metallic surfaces. A study along this line provides a first *qualitative* understanding of these properties, which is often even *quantitatively* correct (in the case of some alkali metals), and serves in all cases as a good starting point for refined models [15, 16]. As the defining property of a metal cluster is the very existence of its *surface*, the jellium model can be expected to produce good first-order results for clusters as well. These can be improved, if necessary, by pseudopotential perturbation theory to include the effects of the ionic structure (see below). Without any detailed calculation one can predict that the relaxation of the electronic cloud leads to a reduced density, and this in turn to a red-shift of the surface-plasmon line. Therefore, if this were the only active mechanism (as in the alkalines) and if there were *no* complications because of the dynamical coupling to d-electrons, as in Ag, one can immediately predict the experimentally observed red-shift of the surface-plasmon frequency of free alkaline metal clusters. In contrast, the surface-plasmon of neutral Ag-clusters in a molecular beam undergoes a *blue-shift* [17]. Most probably this is the effect of dynamical coupling to d-electrons. The frequency position of the plasmon of the s-electrons cannot be understood without taking into account their dynamical coupling to the d-electrons [13, 18]. For Ag *clusters*, there is the additional complication of the size-dependent hybridization between s and d electrons, respectively.

In this work we start with the primitive jellium model, as appropriate for alkaline metals. In the jellium model for metal clusters a fundamental input is the *size-dependent* ionic density. Fortunately, when one of us started this calculation in 1984 [3], some experimental data about the size dependence of the nearest-neighbor distance were available from EXAFS (extended X-ray absorption fine structure) measurements [19]. Except for fine details the size dependence is very weak. This means that in a first approximation the bulk density of the metal can be used as input for a cluster calculation. A second question is the size dependence of the *shape*. Since electron micrographs very often show a spherical shape, at least for the larger clusters, a spherical shape will be assumed for all cluster sizes. This means that for monovalent systems the radius R of the jellium cluster is determined by its bulk density n_+,

$$\frac{1}{n_+} = \frac{4\pi}{3} r_s^3 = \frac{1}{N} \frac{4\pi}{3} R^3, \tag{2}$$

where r_s is the Wigner–Seitz radius of the electron gas and N is the number of *delocalized* valence electrons within the cluster.

1.2   PROPERTIES OF THE GROUND STATE

The properties of metal clusters within the jellium model were first studied within the local density approximation (LDA) to the density functional theory (DFT) [3]. This means the following set of equations has to be solved self-consistently, starting from a proper initial density (for details see [3]). The total electronic density $\rho(\boldsymbol{r})$ obeys the subsidiary condition

$$\int \rho(\boldsymbol{r}) \, d\boldsymbol{r} = N, \tag{3}$$

where ρ is found by minimizing the total energy functional $E_{\text{tot}}[\rho]$ [20]:

$$E_{\text{tot}}[\rho] = \int d\boldsymbol{r}\, v_{\text{ex}}(\boldsymbol{r})\rho(\boldsymbol{r}) + G[\rho], \tag{4}$$

$$G[\rho] = E_{\text{kin}}[\rho] + E_{\text{es}}[\rho] + E_{\text{xc}}[\rho]. \tag{5}$$

E_{kin} describes the kinetic energy of the system of correlated electrons, E_{es} the classical Coulomb repulsion and E_{xc} the exchange–correlation energy. The traditional approximation to the unknown exchange–correlation part of the functional is the local density approximation [21]. The exact density, which determines the total energy in the ground state, is found by solving the Kohn–Sham equations [21]

$$\left(-\frac{\hbar^2}{2m}\Delta + v_{\text{eff}}[\rho(\boldsymbol{r})] \right)\psi_i(\boldsymbol{r}) = \varepsilon_i\psi_i(\boldsymbol{r}), \tag{6}$$

where the electronic density $\rho(\boldsymbol{r})$ is given by

$$\rho(\boldsymbol{r}) = 2\sum_{i=1}^{N} |\psi_i(\boldsymbol{r})|^2. \tag{7}$$

N is again the number of valence electrons and the factor of 2 accounts for the spin degeneracy. If the Gunnarsson–Lundqvist parameterization for the exchange and correlation potential is used [22], the effective potential v_{eff} is given (in Rydberg atomic units) by

$$v_{\text{eff}}(\boldsymbol{r}, \rho) = v_+^{\text{JBG}}(\boldsymbol{r}) + 2\int d\boldsymbol{r}'\, \frac{\rho(\boldsymbol{r}')}{|\boldsymbol{r} - \boldsymbol{r}'|} - \frac{1.222}{r_{\text{s}}(\boldsymbol{r})} - 0.0666\ln\left(1 + \frac{11.4}{r_{\text{s}}(\boldsymbol{r})}\right). \tag{8}$$

$v_+^{\text{JBG}}(\boldsymbol{r})$ is the external potential caused by the positive jellium background. The Wigner–Seitz radius $r_{\text{s}}(\boldsymbol{r})$ is obtained by

$$r_{\text{s}}(\boldsymbol{r}) = \left(\frac{4\pi\rho(\boldsymbol{r})}{3}\right)^{\frac{1}{3}}. \tag{9}$$

This set of equations has to be solved *self-consistently*, with some suitable initial density (for numerical details see [3]). This was done first for sodium as a function of the number of atoms N (which agrees with the number of delocalized electrons for monovalent metals). The set of equations was solved for $N = 2, 3, 4, \ldots$ up to 254. Figures 1.3 and 1.4 show typical charge densities and potentials for a small ($N = 20$) and a large ($N = 198$) particle number, respectively. The oscillatory electronic charge density is normalized to the constant background charge (the step-edge in the figures), and the continuous lines depict the occupied levels. The quantum numbers are those of a spherical potential. Hence the bottom level is 1s, followed by 1p, 1d and 2s (for $N = 20$) with the usual meaning of s ($l = 0$), p ($l = 1$) and d ($l = 2$) in terms of the angular momentum l. The level scheme transforms from a pronounced discrete structure, reminiscent of an atomic system, to the quasicontinuum of a mesoscopic system. Since level structure and nomenclature are like those of a very large atom this model was termed the *super-atom model* of metal clusters.

The first physical quantity of interest is the size dependence of the binding energy

$$\delta(N) = \frac{E(N) - NE(1)}{N}, \tag{10}$$

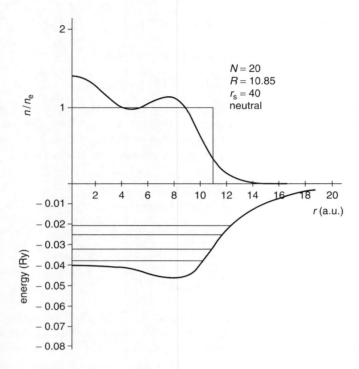

Figure 1.3 Electronic charge density, effective potential and occupied levels for Na_{20} as obtained within the jellium model. The electronic charge density is normalized to a constant ionic background (step-edge). For further explanation see text. Reproduced with permission from Reference [3]. Copyright 1984 by the American Physical Society

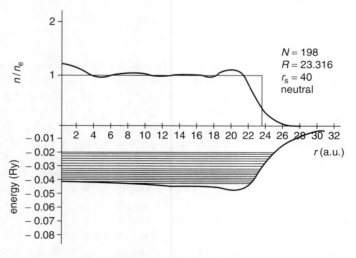

Figure 1.4 Same as Figure 1.3 but for Na_{198}. Comparing with Figure 1.3 one clearly sees how the system transforms from a discrete atomic-like level structure to the quasicontinuum of a mesoscopic system. Reproduced with permission from Reference [3]. Copyright 1984 by the American Physical Society

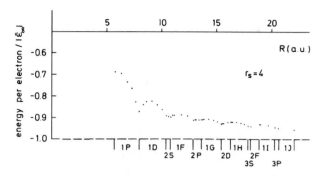

Figure 1.5 Total energy $E(N)/N$ for Na clusters. The pronounced dips at $N = 2, 8, 20, 40,$ 58 and 92 constitute so-called *magic numbers*, indicating especially stable cluster sizes. While small magic numbers coincide with shell-closings, the situation is different for larger numbers. For more explanation see text. Reproduced with permission from Reference [3]. Copyright 1984 by the American Physical Society

which requires the calculation of the total energy as a function of size N, with the Wigner–Seitz radius r_s as a parameter. Figure 1.5 shows the quantity $E(N)/N$ in units of its limiting value for $N \to \infty$ for the case of Na. The most remarkable property of this result is that the value at infinity is approached not in a monotonic but in a pronounced oscillatory fashion. In order to show the origin of this property more clearly, the bottom line indicates the symmetry of the top level, i.e. of the shell that is being filled. Looking at *small* particle numbers, one is tempted to conclude that these 'magic numbers' simply result from the closing of the spherical shell (hence $N = 2, 8, 20$, etc. are expected to be magic). That this conclusion is (at least *generally*) wrong can be seen by looking at *medium-size* numbers N. For instance, between $N = 58$ (closed 1g shell) and $N = 92$ (closed 3s shell) there are two more shell-closings (2d at $N = 68$ and 1h at $N = 90$), that do not produce magic numbers. Therefore one can state that the closing of a spherical shell is a necessary but not sufficient condition for the existence of magic numbers. By inspection of the single-particle level structure one easily recognizes that magic numbers are accompanied by especially large gaps between the highest occupied and the lowest unoccupied molecular orbital, the so-called HOMO–LUMO gaps. This property is genuine for the jellium model.

Two features of the jellium description of the super-atom model are experimentally confirmed:

(a)   The detailed geometrical structure of the ionic skeleton is of marginal importance [23].
(b)   The central property of the model, namely the quasifree motion of the delocalized electrons and especially their mutual correlation is *the* essence for the stability of metal clusters of this kind (all group Ia (alkali metals) and Ib (noble metals) and in addition a few divalent and three-valent metals.

These predictions made in February 1984 have been experimentally confirmed by pioneering experiments of the Berkeley group in June 1984 (see Figure 1.6) [24]. Whereas W. Knight in his early 1984 experiments focused on small mass numbers, these mass-abundance spectra have been afterwards extended to larger atomic numbers. S. Bjørnholm [23] was able to confirm that all magic numbers found for $N \le 1000$ are those of the jellium model (Figure 1.7). Notice the *absence* of magic numbers between $N = 58$

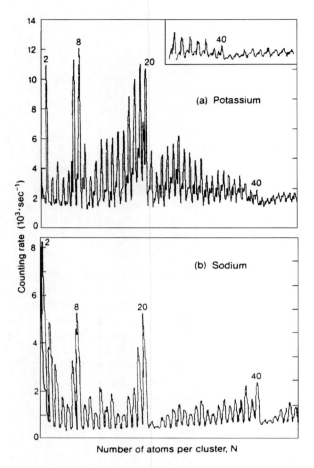

Figure 1.6 Early mass-abundance spectrum of the Berkeley group [27]. Magic numbers are clearly identified. Substructures between the main magic numbers were later sufficiently well explained within the deformed jellium model by Keith Clemenger [28] and Ekardt and Penzar [29]

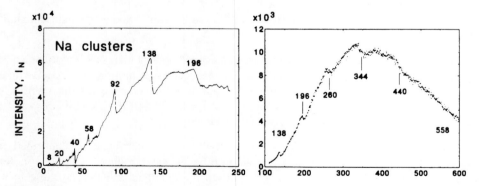

Figure 1.7 Experimental mass spectra by Bjørnholm *et al.* [23] for several hundreds of Na atoms. Note the *absence* of magic numbers between 58 and 92 and between 92 and 138. This can be considered as a direct confirmation of the jellium approximation. Reproduced with permission from Reference [23]. Copyright 1990 by the American Physical Society

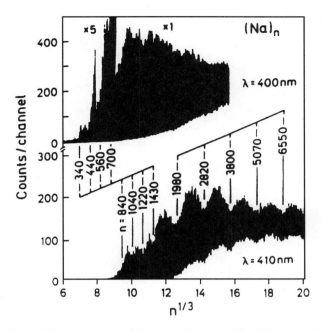

Figure 1.8 Experimental mass spectra of Na up to $N = 22\,000$ [25]. All magic numbers up to $N = 1500$ can be understood within the jellium model and it is only for still larger atomic numbers that the mass anomalies can be attributed to the formation of *ionic* shells. Reproduced by permission of Springer-Verlag from Reference [25]. Copyright by Springer-Verlag

and $N = 92$. In 1991 T. P. Martin [25] was able to take mass spectra for sodium clusters for up to 22 000 atoms, with the remarkable result that all magic numbers $N \leq 1500$ could be explained within the framework of the super-atom model. It is only for even larger N that mass anomalies (see Figure 1.8) could be related to *atomic rearrangements* [25].

The next physical quantities of interest are the size dependence of the ionization potential and of the electron affinity because these quantities can be related to the chemical reactivity of metal clusters. Within the DFT jellium model the size dependence of the ionization potential is easily obtained from two total energy calculations:

$$\Delta_{IP}[N] = E^+[N-1] - E^0[N]. \tag{11}$$

Here $E^+[N-1]$ denotes the total energy of a positively charged cluster with $N-1$ electrons and $E^0[N]$ the total energy of the neutral system. Since the total energy functional consists of various pieces (see Eqs (4) and (5)) this quantity can be written as follows:

$$\Delta_{IP}[N] = \Delta_{es} + \Delta_{rest}. \tag{12}$$

For $N \to \infty$ the first term tends to the electrostatic dipole barrier $v_{es}(+\infty) - v_{es}(-\infty)$, i.e. the difference between the electrostatic potentials far outside and deeply inside the metal, and the second term equals the chemical potential. In this limit Δ_{IP} is the work function of the solid [2].

Figure 1.9 shows the theoretical results for $\Delta_{IP}[N]$ and Δ_{es} for the case of Na together with the limiting values of the work function (3.02 eV) and the electrostatic surface barrier (0.98 eV), respectively (dashed lines). Whereas $\Delta_{IP}[N]$ shows large discontinuities

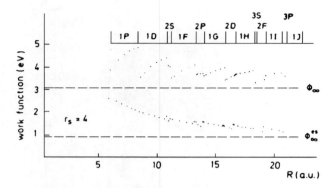

Figure 1.9 Ionization potential for Na clusters in the jellium model [3]. Notice that the total ionization potential does not approach the bulk value in a monotonic fashion but with a pronounced sawtooth behavior. In contrast, the electrostatic part of the total IP *is* monotonic. Here the two horizontal dashed lines represent, respectively, the total work function of the infinite half-space Φ_∞ and the electrostatic surface barrier Φ_∞^{es}. Reproduced with permission from Reference [3]. Copyright 1984 by the American Physical Society

at shell-closings, Δ_{es} is smooth and approaches the limiting value in a monotonic way. Δ_{es} corresponds to the additional amount of work required for ionizing the finite piece of metal and is approximately given by $e^2/(2R)$, the classical electrostatic self energy of one surface charge. This close agreement is related to the fact that also quantum-mechanically the remaining charge is confined mainly to the surface region, except for fine details originating from Friedel oscillations [3, 26].

Experimental data for the ionization potential (IP) are shown in Figure 1.10 [27]. The ionization potential shows large even–odd oscillations for *small* particle numbers, but no pronounced sawtooth patterns for medium-size numbers, in contrast to the theoretical predictions. This suggests that an important ingredient is lacking in the model. Indeed,

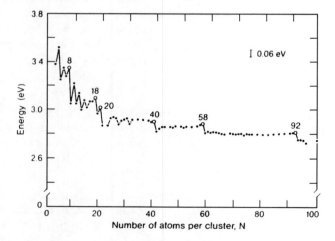

Figure 1.10 Experimental ionization potential (IP) of K clusters [27]. Note that the total IP shows strong discontinuities at magic numbers, whereas the predicted rise of the IP between two magic numbers is missing. As has been shown by Penzar and Ekardt [31] this can be reproduced within the self-consistent deformed jellium model. Reproduced by permission of Academic Press

noticing that at *spherical* shell-closing, the predicted *strong* discontinuities in the IP *are* experimentally confirmed, one realizes that for any *open* shell (all numbers except magic ones) the electronic charge density is non-spherical, a situation that can give rise to a Jahn−Teller distortion of the jellium nucleus. K. Clemenger [28] was the first to study spheroidal distortions of the jellium sphere by using the non-self-consistent Nilsson model of nuclear structure theory. His work was quickly followed by a complete self-consistent many-body calculation for the study of two-axial distortions of the jellium sphere [29]. Brack [30] was the first to study three-axial distortions of the jellium nucleus by using ideas from nuclear structure theory. If one includes these axial distortions one can successfully explain both the ionization potential and the electron affinity for open-shell clusters [30, 31]. Furthermore, fine details of the mass-abundance spectra are better understood.

In order to better understand the origin of the spheroidal distortion, let us recall the appearance of the electronic charge density for any particle number N,

$$\rho(\mathbf{r}) = 2 \sum_{n_l, l, m} |Y_{l,m}(\theta, \phi)|^2 |R_{n_l, l}(r)|^2. \tag{13}$$

Because of the relation

$$\sum_{m=-l}^{m=+l} |Y_{l,m}(\theta, \phi)|^2 = \frac{1}{4\pi}, \tag{14}$$

the total electronic charge density is *non-spherical* for all particle numbers except the magic numbers. According to general rules of quantum mechanics, a Jahn−Teller deformation of the external field is to be expected. This is efficiently achieved by introducing the distortion parameter δ describing the replacement of a sphere with radius R by a spheroid whose axes are given by

$$Z = \left[\frac{2+\delta}{2-\delta} \right]^{2/3} R,$$

$$X, Y = \left[\frac{2-\delta}{2+\delta} \right]^{1/3} R. \tag{15}$$

The introduction of δ makes the external potential axial symmetric. Consequently, the spherical symmetry of the wave functions is replaced with a spheroidal symmetry. As before, the radius R is determined by the particle number N and the Wigner−Seitz radius r_s as follows:

$$R = N^{1/3} r_s. \tag{16}$$

The total effective potential determining the electronic motion via the Kohn−Sham equations is expected to be spheroidal as well. Therefore all spherical shells n, l, m are expected to split into spheroidal subshells $|m|, p, k$. Here m is the preserved azimuthal quantum number. For time-reversal symmetry only its magnitude $|m|$ counts; p is the parity and k just enumerates the levels of a certain symmetry. The reduced spheroidal symmetry lifts the spherical degeneracy as depicted in Figure 1.11 for Na in the size-range of N from 3 to 18.

An immediate consequence of this spheroidal nature of the one-particle Kohn−Sham levels is the substructuring of the formerly spherical shells into spheroidal subshells. This is made evident in Figures 1.12 to 1.14. Figure 1.12 shows the analog to Figure 1.11

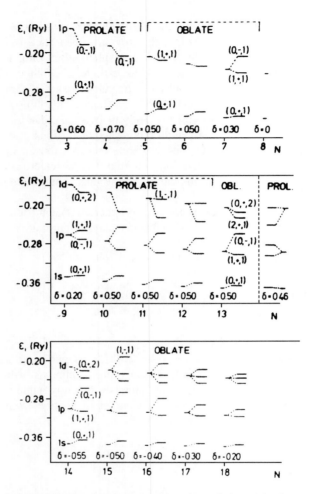

Figure 1.11 Evolution of the level structure as function of size N in the range $3 < N < 18$. In each case the deformation parameter δ, defined in Eq. (15), is given. For comparison, the spherical levels are also given. As one can see clearly, the spherical level structure is heavily perturbed in the midshell region. Note that in the case of $N = 18$ the cluster ends up with a spheroidal distortion (because of the interplay between 1d, and 2s levels). For $N = 13$, both prolate and oblate levels are given, because the two have a total energy, which is close (for the possibility of a three-axial distortion in this case, see [30]). Reproduced with permission from Reference [29]. Copyright 1988 by the American Physical Society

(spherical to spheroidal behavior). Figure 1.13 gives the second energy difference Δ_2 defined as:

$$\Delta_2(E) = E_{N+1} - (E_N + E_1) - (E_N - (E_{N-1} - E_1)). \qquad (17)$$

As has been shown [27], this quantity is decisive for the mass abundances observed experimentally in a supersonic beam experiment. Finally, we show in Figure 1.14 the ionization potentials obtained within the spheroidal jellium model. Note that there is a pronounced odd−even alternation, as in the abundances, and that this effect is purely electronic and *produced without regard to the spin of the electrons and also without any pairing effects*

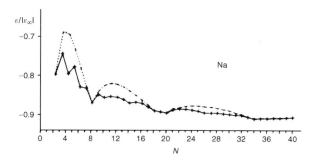

Figure 1.12 Energy per particle in units of $|\varepsilon_\infty|$. The solid line shows the data from the spheroidal jellium model, the dashed line the corresponding data from the spherical jellium model. The results are for $N < 41$. Reproduced with permission from Reference [29]. Copyright 1988 by the American Physical Society

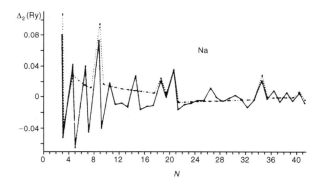

Figure 1.13 Second difference Δ_2 of the total energy as function of N. Solid line: distorted jellium; dashed line: spherical jellium. Reproduced with permission from Reference [29]. Copyright 1988 by the American Physical Society

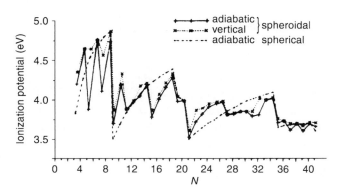

Figure 1.14 Ionization potential in various approximations. Note that the spherical sawtooth behavior is completely destroyed after allowing for spheroidal distortions. Originally it was believed that the strong odd−even alternation was related to the spin of the electrons. But, as shown first by the first author [29], it is a *kinematic* orbital effect. Reproduced with permission from Reference [29]. Copyright 1988 by the American Physical Society

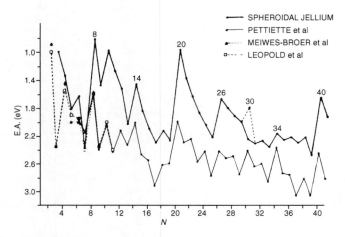

Figure 1.15 Comparison of the electron affinity within the spheroidal jellium model plus SIC and various experimental results (circles [44], triangles [45], squares [46]). For $N = 30$ theory predicts two isomers (prolate and oblate), which are nearly degenerate. But both do have different affinities and in the beam the signal will come from those clusters having the lower affinities (connected by dashed lines). Clearly, shell effects are very pronounced. Qualitatively theory and experiment agree rather well. In order to achieve *quantitative* agreement one has to introduce pseudopotential perturbation theory as sketched above. Reproduced by permission of Springer Verlag

as in nuclear structure theory. As explained in detail in [29] it is on the contrary an effect of the permanent change of level ordering and occupation as a function of N.

In addition we have calculated the electron affinities within the jellium model. But, as well known from experience in atomic physics, the LDA functional has to be corrected for so-called *self-interaction* effects (SIE) as originally proposed by Perdew and Zunger [8]. As explained in detail below, this leads to an orbital-dependent Kohn–Sham potential by the replacement [31]:

$$V_{\text{SIC}}^{\text{eff}} = V_{\text{LDA}}^{\text{eff}} - \delta_i. \tag{18}$$

Here i is the level in question and the functional δ_i is the self-interaction correction for the orbital i. We follow Perdew and Zunger [8] in writing

$$\delta_i = U_{\text{es}}[n_i] + U_{\text{xc}}^{\text{LDA}}[n_i]. \tag{19}$$

Here the first part is the electrostatic self-Coulomb interaction and the second part is the analog correction for the exchange–correlation part of the effective potential. Please note that both functionals depend on the *total* density n_i and *not* on the spin densities as in the case of Perdew and Zunger who corrected the LSDA (Local Spin Density Approximation). Results for Cu are reproduced in Figure 1.15 [31]. Though with this simple functional quantitative agreement with experimental data is not to be expected, the experimentally observed shell effects in the electron affinity are qualitatively well reproduced.

1.3   THE OPTICAL PROPERTIES

In the following we describe the optical properties of metal clusters within the jellium model. We begin by introducing the TDLDA (time-dependent local density approximation) to general TDDFT (time-dependent density functional theory) as was done *intuitively* by A. Zangwill and P. Soven in their seminal work [32]. The rigorous proof of this method was provided later by E. K. U. Gross and collaborators in a series of papers. For a rigorous derivation of this TDLDA (in this context sometimes called ADLDA (adiabatic local density approximation), which points to the range of validity of the TDLDA), the interested reader is referred to [33]. We continue with the intuitive introduction of the TDLDA.

We start by translating Fermi's golden rule concerning the photoabsorption into the language of density functional theory. If the jellium cluster is exposed to an external photon field $V_{\text{ext}}(r, t)$ an induced charge density is set up, which is given as

$$\rho_{\text{ind}}(r, \omega) = \int dr' \chi(r, r'; \omega) V_{\text{ext}}(r'; \omega). \tag{20}$$

In this equation χ is the *exact* dynamical density–density correlation function calculated with the inclusion of all many-body effects. As one can show [5], the exact χ is determined by solving the TDLDA integral equation

$$\chi(r, r'; \omega) = \chi^0(r, r'; \omega) + \int dr'' \, dr''' \chi^0(r, r''; \omega) K(r'', r''') \chi(r''', r'; \omega). \tag{21}$$

The so-called residual interaction $K(r, r')$ is defined (in Rydberg atomic units) as

$$K(r, r') = \frac{2}{|r - r'|} + \frac{d}{d\rho} V_{\text{xc}}[\rho(r)] \delta(r - r'). \tag{22}$$

The first part of the residual interaction K is the Coulomb potential established by the induced charge density $\rho_{\text{ind}}(r, \omega)$ and the other part is the exchange–correlation contribution to the induced effective field. Whereas Zangwill and Soven [32] introduced this part intuitively without exact proof for the existence of this functional (for time-dependent densities), Gross was able to show [33] the restrictions on $V_{\text{xc}}(r, t)$ and the assumptions to be made, to make this procedure an accurate one. To be brief, the procedure is valid if the system is at time t_0 in its non-degenerate ground state and if the external potential has a Taylor expansion in time around t_0. But the formulation of Gross is much more general; the interested reader should read the original papers by Gross.

Finally, the independent particle susceptibility $\chi_0(r, r'; \omega)$ is defined (see [5]) in terms of the one-particle Kohn–Sham eigenfunctions $\phi_i(r)$ and the retarded Green's function G of the Kohn–Sham Hamiltonian,

$$\chi^0(r, r'; \omega) = \sum_i^{\text{occ}} \phi_i^*(r)\phi_i(r')G(r, r'; \varepsilon_i + \hbar\omega) + \text{c.c.}(\omega \to -\omega). \tag{23}$$

Whereas the full density–density correlation function contains many-body effects (as collective excitations (as poles)), χ_0 does *not* contain these features. It is by construction an independent particle function. As such it is the response function to the total perturbing effective field,

$$\rho_{\text{ind}}(r, \omega) = \int dr' \chi^0(r, r'; \omega) V_{\text{eff}}(r', \omega), \tag{24}$$

with the effective potential defined as

$$V_{\text{eff}}(r, \omega) = V_{\text{ext}}(r, \omega) + \int dr' K(r, r') \rho_{\text{ind}}(r', \omega). \tag{25}$$

In Eq. (25) all many-body effects are implicitely contained in the effective field. The interesting point to note is that also in the presence of a *time-dependent field all* many-body effects can be stored in a local one-particle potential, to which the electrons respond as *independent* particles.

Once χ is determined we are ready to calculate the photoabsorption cross-section $\sigma(\omega)$ (for a detailed derivation of these formulas see [5]):

$$\sigma(\omega) = \frac{4\pi\omega}{c} \operatorname{Im}[\alpha(\omega)], \tag{26}$$

$$\alpha(\omega) = \int_0^\infty \mathrm{d}r \, r\alpha(r, \omega), \tag{27}$$

$$\alpha(r, \omega) = -\frac{4\pi}{3} r^2 \int_0^\infty \mathrm{d}r' \, r'^3 \chi_{l=1}(r, r'; \omega). \tag{28}$$

Here we have used spherical symmetry, which means that χ is diagonal in the angular momentum and that for the response to the photon field it is only the component with $l = 1$ we need to calculate. The last equation completes the formalism and we are ready to present the results for a number of jellium clusters.

We start with a typical result for jellium-cluster absorption in the range $100 \leq N \leq 200$, namely $N = 198$. In Figure 1.16 the dashed line gives $\operatorname{Im}[\alpha(\omega)]$. In comparison we show by a continuous line the absorption obtained within the Drude approximation to the dielectric constant $\varepsilon(\omega)$. Here the electrons are described as a system of damped 'oscillators' with eigenfrequency $\omega = 0$. For this reason there is no absorption feature beside

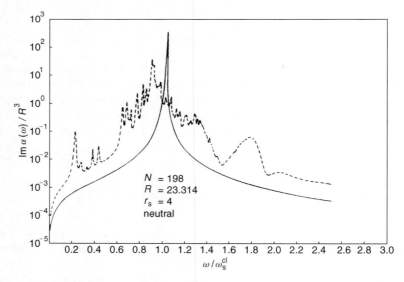

Figure 1.16 Imaginary part of the complex polarizability $\alpha(\omega)$ for an Na cluster with $N = 198$ in units of R^3. Effective single-pair excitations, as well as the surface plasmon and the volume plasmon, are clearly resolved. For comparison the result of the local Drude theory is also given. In this case there is only one mode of excitation, the classical surface-plasmon polariton or Mie-resonance at $\omega_p/\sqrt{3}$. Because the frequency is scaled with this frequency the Drude curve peaks trivially at 1. For more explanation see text. Reproduced with permission from Reference [5]. Copyright 1985 by the American Physical Society

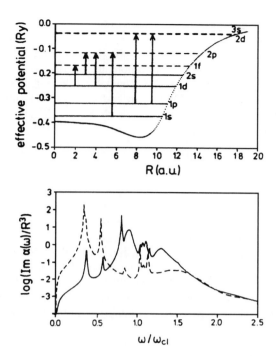

Figure 1.17 Dipole-allowed absorption in jellium Na$_{20}$ and its interpretation. The continuous line gives the result from the TDLDA. The nature of the double structure between 0.5 and 1.0 can be understood in two steps. First, the TDLDA is compared with LDA, the independent-particle response (dashed line). Each peak corresponds to one arrow in the upper part of the figure. After turning on the *interaction* among excited pairs, bare pairs are transformed into dressed pairs. Note that there is a one-to-one correspondence between the spikes in the two curves. As explained in the text there is another effect of this interaction, namely the formation of a collective surface mode at about 0.9. This feature has *no* counterpart in the dashed curve. Furthermore there is one more collective effect at about 1.2. For more explanation see text. Reproduced with permission from Ekardt, Pacheco and Schone, *Comments on Atomic and Molecular Physics*, **31**, 291 (1995). Copyright by OPA (Overseas Publishers Association) B.V

the collective surface plasmon–polariton. Quantum-mechanically there are a number of dipole-allowed single-particle transitions (see Figure 1.17). But, remarkably enough, these transitions are *very* weak. So the quantum corrections are weak in general and consist in the appearance of two kinds of transitions in addition to the classical Mie plasmon at $\omega_s = \omega_p/\sqrt{3}$, namely the various particle–hole transitions (the tiny spikes) and the broad hump at ω_p which is the precursor to the volume plasmon. The plasmon frequency ω_p is given by $\omega_p^2 = 4\pi n e^2/m$. For *smaller* particle numbers this picture changes gradually. This is demonstrated by Figure 1.17, showing the calculated absorption spectrum for Na$_{20}$ within the jellium model. In order to better identify the origin of the various spikes (single-particle transitions) we show in the upper panel the level structure of occupied (continuous line) and empty levels (dashed lines). The arrows mark allowed optical transitions in this potential. Clearly, there is a one-to-one correspondence between the arrows in the upper part and the spikes in the *dashed* curve, describing the absorption within the independent-particle approximation (i.e. *without* many-body effects). Those effects can be clearly identified by comparing the absorption at the TDLDA level (continuous line) with the dashed line. Upon taking into account many-body effects all naked particle–hole pairs are transformed to dressed ones. In addition to these effects there are two more features, one slightly below the classical Mie frequency and the second at about $1.2\,\omega_p$. The former is the surface plasmon and the latter is a complicated collective feature, already known from flat surfaces [51]. These two collective features gain oscillator strength at the cost of low-lying single pairs which are of reduced strength in the interacting system. In addition to these quantum effects, there is a *purely classical* effect, already known from the classical damped oscillator; if a damped oscillator is driven by an external field of frequency below its resonance frequency, it oscillates *in phase* with the external field. But if the external frequency is above resonance there is a phase shift by π.

This general statement applied to the problem in question means: the surface plasmon in the interacting system screens very efficiently the external electric field for all frequencies ω below ω_s. This leads to a further reduction of the oscillator strength of the single pair lines below the surface plasmon. This means the external field is *screened*. In contrast, at frequencies above ω_s the external field is *antiscreened*, which means enhanced! Indeed, for all frequencies above the surface-plasmon frequency the intensity of the independent particle lines (dashed line) is *enhanced*, a feature which is clearly seen in Figure 1.17. The remarkable feature of the absorption in Na_{20} is that there is not a single surface-plasmon line but a doublet. This behavior is called *Landau fragmentation* because it is the analog (for a discrete level structure) to Landau damping (coupling of the plasmon to a *continuous* single-particle spectrum). Remarkably enough, the jellium picture provides for Na_{20} a reliable, though oversimplifying, description of photoabsorption. The reason for this is discussed in the following section.

1.4   PSEUDOPOTENTIAL PERTURBATION THEORY

For a better understanding of the range of validity of the jellium model and in order to learn how it can be successively improved we discuss in the following the specific example of Na_8. A molecular dynamics study gives three different low-lying isomers (see Figure 1.18) with D_{2d} being the ground state within the Car–Parrinello pseudopotential plane-wave method [34]. Because Na has only very weak non-local components in the *ab initio* pseudopotentials we can use *local* pseudopotentials of the Heine–Abarenkov type [35]. As long as we are interested in the dynamical properties of the loosely bound valence electrons a pseudopotential description is perfectly adequate (for a tractable method for transition metals see below). With these restrictions, the cluster electrons are moving in an external potential of the type

$$v_{\mathrm{ext}} = \sum_{i=1}^{N} v_{\mathrm{ps}}(\boldsymbol{r} - \boldsymbol{R}_i). \tag{29}$$

Here the pseudopotentials v_{ps} are located and are spherically symmetric around the ionic sites $\boldsymbol{R}_i$. In order to better understand the performance of the jellium model each of these potentials is decomposed into its various angular parts with respect to an arbitrary cluster 'center', which we assume to be the center of mass (for a discussion of other 'centers' see [36]),

$$v_{\mathrm{ps}}(\boldsymbol{r} - \boldsymbol{R}_i) = \sum_{l=0}^{\infty} \sum_{m=-l}^{+l} \frac{\sqrt{4\pi}}{2l+1} \alpha_l(R_i, r) Y_{l,m}^{*}(\Theta_i, \Phi_i) Y_{l,m}(\theta, \phi). \tag{30}$$

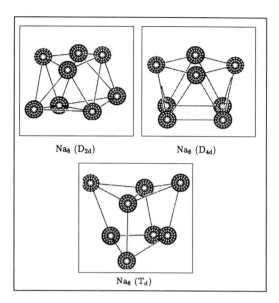

Na$_8$ (D$_{2d}$)   Na$_8$ (D$_{4d}$)

Na$_8$ (T$_d$)

Figure 1.18 Geometries of the three lowest isomers of Na$_8$ as obtained from Car–Parrinello calculations [34, 47]

Now we have to perform the total sum over all ionic sites in order to get the total potential. Hence we get the total spherical part of the potential as

$$v_0^{\text{ext}}(r, R) = \frac{1}{4\pi} \sum_i \alpha_{l=0}(R_i, r). \tag{31}$$

Please note that this part depends on (the modulus) of *all* ionic sites $\boldsymbol{R}_i$. The remaining part of the external potential contains geometrical information and effects finally the transition from the molecular structure (for few ionic numbers) to the solid (for a large number of ions). As we shall see in a number of examples the first, spherically symmetric part of the pseudopotential is *extremely well* represented by the jellium model and is by far the largest contribution to the total potential. Therefore the second part, containing the geometrical information, can be treated by perturbation theory.

To be specific we consider in what follows the example of Na$_8$. According to the Car–Parrinello method [34] there are three low-lying isomers (see Figure 1.18) with D$_{2d}$ being the ground state followed by two structures with a slightly higher total energy. Note in passing that quantum chemists, with their methods, find T$_d$ to be the ground-state symmetry. As we shall see below there exists the possibility via photoabsorption to determine the geometrical structure of a cluster. For this purpose we have to study the influence of the geometrical part of the total potential on the absorption. This is done by first investigating the spherical potential part of Na$_8$ in the D$_{2d}$ structure which is shown in the left-hand panel of Figure 1.19. Note that it is almost identical to the spherical jellium model potential discussed before. In this way we have a justification for its use in the case of the magic-number clusters $N = 8, 20, 40, 58, 92, 138$, etc. The two most important *nonspherical* potential parts are shown in the right-hand panel of Figure 1.19. Their influence is studied via perturbation theory. The result is as expected: very small changes in the single-particle energies, which are now split according to the point group of the

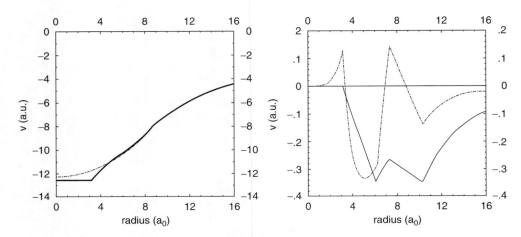

Figure 1.19 Various potential components resulting from Eq. (30) for the ground state of Na_8. The left-hand panel shows the spherical component (Eq. (31)) given by the continuous line. For comparison the spherical jellium potential (dashed line) is also given. As one can see the two are almost identical and much larger than the two most important nonspherical potential parts $v_{l=2,m=0}(r)$ (continuous line) and $v_{l=4,m=0}(r)$ (dashed line) displayed in the right-hand panel. Note that these two components are strongly fluctuating compared to the smooth spherical part. These potential parts make the optical response different in the three isomeric states. In the end they are responsible for the transition from the molecule with a point-group symmetry to the solid with a space-group symmetry — a problem that has not yet been solved

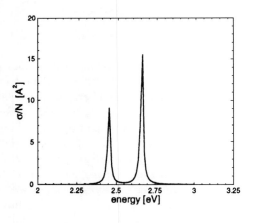

$Na_8(D_{2d})$

Figure 1.20 Optical absorption of Na_8 in its ground-state structure D_{2d}. This result from pseudopotential perturbation theory is explained in the main text. The spherical plasmon line at about 2.5 eV (see Figure 1.21) is split into two components which can be understood as follows. The moments of inertia of the structure D_{2d} point to a prolate spheroid within the jellium approximation to the distribution of ions. In such a system there are two collective excitations: one at higher frequencies (perpendicular to the axis of symmetry) and one for the motion along the axis of symmetry. Because the motion perpendicular is twofold degenerate its intensity is twice that of the low-frequency motion (with the cluster being statistically oriented in the beam (see [30])

cluster. More interesting is the effect of the geometry on the optical absorption. This effect is investigated and described in detail in [37] (essentially the TDLDA integral equation for $\chi(r, r'; \omega)$ is solved perturbatively). The result for Na_8 in its ground-state structure D_{2d} is shown in Figure 1.20. The result is unexpected and can hardly be understood within the jellium approximation to the external potential, namely the moments of inertia for the actual cluster structure point to a *prolate spheroid*. The spherical one-component surface plasmon is split into two components, with the high-frequency line about twice as intensive as the low-frequency line. This is because the fast oscillation perpendicular to the axis of symmetry is twofold degenerate. This result is unexpected because within the jellium approximation to the external potential the sphere is stable against deformations to a spheroid! The result is not only unexpected, but also annoying, because the *two*-peak structure seems to contradict the experimental findings of a broad one-peak hump centered around 2.5 eV. Before we can make a final decision about the relevance of our calculations, we present in the next figure absorption spectra of the other isomers. The upper part in Figure 1.21 shows the D_{4d}, whereas the lower part shows the T_d structure. Each is essentially a one-peak structure with the right position at about 2.5 eV. As we shall see later it is nevertheless the D_{2d} structure that is in agreement with the experiment — after taking into account the effects of temperature which we discuss next.

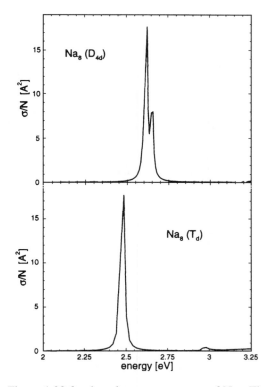

Figure 1.21   Same as Figure 1.20 for the other two structures of Na_8. The interpretation parallels the one for the D_{2d} symmetry. The absorption in Na_8 in the D_{4d} structure (upper panel) points to an *oblate* spheroid. Therefore the low-frequency component is twice as intensive as the high-frequency one. Finally, the T_d structure is almost spherical; therefore the absorption consists of *one* peak positioned at the spherical jellium line

1.5   THE EFFECT OF TEMPERATURE

Since most absorption spectra are taken at elevated temperatures T, the theoretical prediction should not be made for $T = 0$, but for $T > 0$. At higher temperatures the ions are oscillating around their equilibrium positions. For this reason the *measured* cross-section is an ensemble average over the canonical ensemble at temperature T,

$$\langle \sigma(\omega) \rangle = \frac{1}{Z} \sum_i \sigma(\omega; \boldsymbol{R}_i) e^{-\beta E[\boldsymbol{R}_i]}, \tag{32}$$

where Z is the partition function

$$Z = \sum_i e^{-\beta E[\boldsymbol{R}_i]}, \quad \beta = \frac{1}{kT}. \tag{33}$$

Here the canonical phase-space sampling is performed via the Monte Carlo method (for details see [6]). As the cluster in the beam is oriented statistically, we have to perform three different calculations,

$$\langle \sigma_{\mathrm{av}}(\omega) \rangle = \frac{1}{3} \sum_{i=x,y,z} \langle \sigma_i(\omega) \rangle. \tag{34}$$

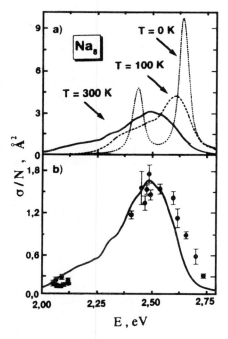

Figure 1.22   (a) Line shape of the photoabsorption cross-section of Na_8 at three different vibrational temperatures in its ground-state structure D_{2d} according to Eqs (32)–(34). A numerical damping of 0.02 eV has been assumed at $T = 0$ K. This curve is without ionic fluctuations; the other two contain ionic fluctuations at the corresponding temperatures 100 K and 300 K.
(b) Comparison with the experimental data of the Berkeley group [48]. The theoretical curve (solid line) has been renormalized in order to exhaust 55% of the Thomas–Reiche–Kuhn sum rule (f-sum rule) in accord with the experimental findings

Now we have a picture about how demanding temperature-dependent calculations are: we need at each T and $\omega 10^4$ Monte Carlo points and 10^2 frequencies in order to cover the experimental range. Furthermore we need the three different directions x, y and z. So we had to solve three million times the TDLDA integral equation with full inclusion of the ionic structure (via pseudopotential perturbation theory). Needless to say it seems almost impossible to perform calculations of this type for transition metals with the additional complication of the d-electrons!

The result of the Monte Carlo canonical sampling of the phase space is presented in Figure 1.22. At zero temperature, we reproduce the pronounced doublet, which seems to disagree with the experimental finding of a broad one-peak structure. With elevated temperature the two peaks merge into one and almost agree with the experimental data at 300 K, resolving a long-standing puzzle. The *observed* width of the surface-plasmon line is not related to dissipative processes or to Landau damping (coupling to particle–hole pairs), nor is it due to electron–phonon coupling. It is instead a line-broadening, due to the oscillations of the ions around their equilibrium positions.

We conclude with two more examples which demonstrate the power of pseudopotential perturbation theory. Figure 1.23 shows the geometry and absorption spectra of Na$_6$ which

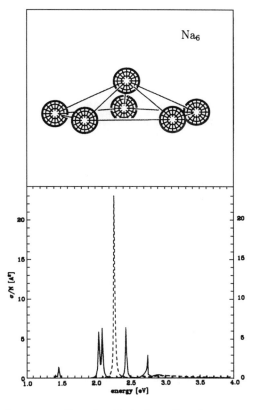

Figure 1.23 Geometry (upper panel) and optical absorption (lower panel) of Na$_6$ in its ground state C$_{5v}$ [34]. The spherical jellium line (dashed curve) disintegrates into various particle–hole lines (continuous curve) under the influence of the strongly fluctuating potential components shown in the next figure. The continuous curve agrees *qualitatively* with results obtained by quantum chemists [49]

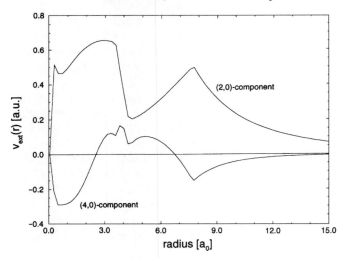

Figure 1.24   The most important nonspherical potential components of Na_6 in the geometry C_{5v}. Because these components are heavily fluctuating and *not* small, there is *no* collective peak at the TDLDA level

is an open-shell cluster (pentagonal pyramid). The dashed line is the result of the spherical jellium calculation; it shows one collective peak. Under the influence of the *strong and strongly fluctuating* nonspherical parts of the potential (shown in Figure 1.24) the efficient Coulomb coupling of (excited) particle–hole pairs is destroyed and the plasmon *disintegrates* into its constituent components. In the last example we show in Figure 1.25 the absorption spectra of Na_{90}. Because for such large clusters *ab initio* studies of the equilibrium structure do not exist we just built small crystal fractions of fcc, bcc, hcp and icosahedral symmetry and calculated the total energy within second-order pseudopotential perturbation theory (the first-order contribution is identical to zero for symmetry reasons). In the case of Na_{90} it is the hcp structure that is energetically favorable. Accurate experimental data on Na_{90} do not exist. First preliminary data confirm a peak on the high-energy side [38].

1.6   SUMMARY AND CONCLUSIONS

The alert reader will have realized that almost all examples given in this chapter are from Na. Of course this was on purpose: two of the most important conditions to be fulfilled for the excellent validity of the jellium model, namely (a) that the pseudopotential is local and (b) that the 'geometrical' parts of the ionic arrangement are weak, are best met in Na_N. In trying other elements we found only *one* other material that works comparably well — potassium. But the important point to note is that this simple model serves as a guideline for more complex cases. After electronic shells and plasmons have been found in Na, they have been found in almost all metal clusters. In order to get the same quantitative agreement as in the case of Na one has either to do all-electron calculations or to use non-local pseudopotentials, as has been done in the case of Li [39, 40]. But in these

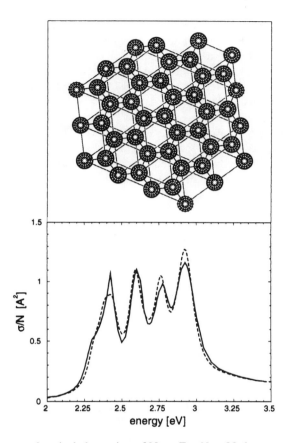

Figure 1.25   Geometry and optical absorption of Na$_{90}$. For $N = 90$ there are no determinations of the geometry at the *ab initio* level available. We have therefore simply calculated the total energy within pseudopotential perturbation theory of second order for various model clusters built as small crystal fractions of fcc, bcc, hcp and icosahedral type. For the case in question the hcp structure (see upper part of the figure) has the lowest total energy

calculations the geometrical part of the (non-local) pseudopotentials has not been studied. Only the spherical component has been modified and this results in just another position of the plasmon line and not in a qualitative change of the spectra. This is not sufficient to catch the full variety of metal-cluster absorption. For instance, the disintegration of plasmons in Na$_6$ is a new and qualitative aspect that can be obtained only in a model *beyond* the spherical average of the potential. This has never been done for *ab initio* pseudopotentials.

Another important point of complications is the existence of d-electrons. Though the noble metals, like the alkaline metals, show magic numbers [30], indicating that the d-electrons do not sufficiently strongly disturb the mutual correlation of the s-electrons (which is the root of the occurrence of shells), they do couple dynamically to the s-electrons, as is already known from the bulk crystals of the noble metals [13, 18]. For these cases we need a new and much more complicated theory. This is *ab initio* TDLDA [41, 42] which is currently being constructed. Here, we calculate the electronic structure *ab initio* at the all-electron level, within the FPLAPW method (full potential linearized augmented plane wave method; computer code WIEN95 [43]). In this way, the *dynamical* coupling between s- and d-electrons is taken into account from the very beginning. In the case of clusters an additional complication occurs, namely *hybridization* (e.g. in Ag_n the 4d- and 5s-electrons hybridize). This feature is hardly caught by any jellium-type model.

From a general point of view the example of the temperature-dependent absorption in Na_8 is of the utmost importance. Because the method TDLDA, which was introduced intuitively in the seminal work by Zangwill and Soven [32], has itself been doubted for several years. But as we know from the work of Gross's group this method is on a firm theoretical basis and should deliver exact results whenever the residual interaction is not frequency-dependent and is of the plasmon type. In the case of excitons it will fail, because here the screening in the Coulomb part of the residual interaction is essential. Therefore it is not surprising that all applications of the TDLDA to atoms, molecules and clusters and to solids are in a frequency region where no excitons are to be expected, i.e. the spectral region *above* the particle–hole part of the excitations but *not* that below it.

1.7   REFERENCES

[1] D. Pines and P. Nozières, *The Theory of Quantum Liquids*, W. A. Benjamin, New York, 1966.
[2] N. D. Lang, 'The density functional formalism and the electronic structure of metal surfaces', in *Solid State Physics, Vol. 28*, ed. H. Ehrenreich, F. Seitz and D. Turnbull (Academic Press, New York, 1973), p. 225.
[3] W. Ekardt, *Phys. Rev. B* **29**, 1558 (1984).
[4] H. Nishioka, K. Hansen and B. Mottelson, *Phys. Rev. B* **42**, 9377 (1990).
[5] W. Ekardt, *Phys. Rev. Lett.* **52**, 1925 (1984); *Phys. Rev. B* **31**, 6360 (1985).
[6] J. M. Pacheco and W.-D. Schöne, *Phys. Rev. Lett.* **79**, 4986 (1997).
[7] W.-D. Schöne, W. Ekardt and J. M. Pacheco, *Phys. Rev. B* **50**, 11079 (1994).
[8] A. Zunger and J. M. Perdew, *Phys. Rev. B* **23**, 5048 (1981).
[9] Z. Penzar and W. Ekardt, *Z. Phys. D* **17**, 69 (1990).
[10] J. M. Pacheco and W. Ekardt, *Ann. Physik* **1**, 254 (1992).
[11] W. Schulze, P. Frank, K.-P. Charlé and B. Tesche, *Berichte Bunsenges. Phys. Chem.* **88**, 263 (1984).
[12] L. D. Landau and M. Lifschitz, *Electrodynamics of Continuous Media*, Pergamon Press, Oxford, 1960.
[13] D. Pines, *Elementary Excitations in Solids*, Benjamin, New York, 1964.
[14] N. D. Lang and W. Kohn, *Phys. Rev. B* **1**, 4555 (1970); *Phys. Rev. B* **3**, 1215 (1971).
[15] A. G. Eguiluz, *Phys. Rev. B* **35**, 5473 (1987).
[16] R. N. Barnett, U. Landman and C. L. Cleveland, *Phys. Rev. B* **27**, 6834 (1983).
[17] K.-P. Charlé, L. König, S. Nepijko, I. Rabin, and W. Schulze, *Cryst. Res. Technol.* **33**, 7 (1998).
[18] H. Raether, *Plasmons*, Springer, Berlin, 1980.
[19] *EXAFS, Basic Principles, and Data Analysis*, Springer, Berlin, 1994.
[20] P. Hohenberg and W. Kohn, *Phys. Rev.* **136**, B864 (1964).
[21] W. Kohn and L. J. Sham, *Phys. Rev.* **140**, A1133 (1965).
[22] O. Gunnarsson and B. I. Lundqvist, *Phys. Rev. B* **13**, 4274 (1976).
[23] S. Bjørnholm, J. Borggreen, O. Echt, K. Hansen, J. Pedersen and M. P. Rasmussen, *Phys. Rev. Lett.* **65**, 13 (1990).
[24] W. D. Knight, K. Clemanger, W. A. de Heer, W.A. Saunders, M. Y. Choi and M. L. Cohen, *Phys. Rev. Lett.* **52**, 2141 (1984).
[25] T. P. Martin, T. Bergmann, H. Goehrlich and T. Lange, *Z. Phys. D* **19**, 25 (1991).

[26] M. Seidl, Die Ionisierungsenergie von Metallclustern, dissertation, Regensburg, 1989.

[27] W. A. de Heer, W. D. Knight, M. Y. Chou and M. L. Cohen, *Solid State Physics, Vol. 40*, ed. H. Ehrenreich, F. Seitz and D. Turnbull (Academic Press, New York, 1987), p. 93.

[28] K. Clemenger, *Phys. Rev. B* **32**, 1359 (1985).

[29] W. Ekardt and Z. Penzar, *Phys. Rev. B* **38**, 4273 (1988).

[30] M. Brack, *Rev. Mod. Phys.* **65**, 677 (1993).

[31] Z. Penzar and W. Ekardt, *Z. Phys. D* **17**, 69 (1990).

[32] A. Zangwill and P. Soven, *Phys. Rev. A* **71**, 1561 (1980).

[33] E. K. U. Gross, J. F. Dobson and M. Petersilka, *Topics in Current Chemistry* **81**, 82–172 (1996).

[34] U. Röthlisberger and W. Andreoni, *J. Chem. Phys.* **94**, 8129 (1991).

[35] V. Heine and I. V. Abarenkov, *Phil. Mag.* **9**, 451 (1964); I. V. Abarenkov and V. Heine, *Phil. Mag.* **12**, 529 (1965).

[36] M. Manninen, *Solid State Commun.* **59**, 281 (1986).

[37] W.-D. Schöne, Wiedereinführung der ionischen Struktur in das sphärische Jellium-Model, dissertation, Berlin, 1995.

[38] H. Haberland, priv. communication.

[39] Ll. Serra, G. B. Bachelet, N. V. Giai and E. Lipparini, *Phys. Rev. B* **48**, 14708 (1993).

[40] F. Alasia *et al.*, *Phys. Rev. B* **52**, 8488 (1995).

[41] M. Bandić, C. Ambrosch-Draxl and W. Ekardt, in preparation (1998).

[42] W.-D. Schöne and W. Ekardt, in preparation (1998).

[43] P. Blaha, K. Schwarz, Ph. Dufek and R. Augustyn, *WIEN95, A Full Potential Linearized Augmented Plane Wave Package for Calculating Crystal Properties, User's Guide*, Technical University Vienna, February 1995.

[44] C. L. Petiette, S. H. Yang, M. J. Craycraft, J. Conceicao, R. T. Laakasonen, O. Chesnovsky and R. E. Smalley, *J. Chem. Phys.*, **88**, 5377 (1988).

[45] K.-H. Meiwes-Broer, priv. communication; G. Gantef, M. Gausa, K.-H. Meiwes-Broer and H. O. Lutz, *Faraday Disc. R. Soc. Chem.* **86** (1988).

[46] D. G. Leopold, J. Ho and W. C. Lineberger, *J. Chem. Phys.* **86**, 1715 (1987).

[47] J. M. Pacheco, W. Ekardt and W.-D. Schöne, *Europhys. Lett.*, **34**, 13 (1996).

[48] K. Selby, *et al.*, Contribution to the Aix en Provence Conference *Meeting on Clusters*, July 1988; K. Selby, V. Kresin, J. M. Masui, M. Vollmer, W. A. de Heer, A. Scheidemann and W. Knight, *Phys. Rev. B* **43**, 4565 (1991).

[49] V. Bonačić-Koutecký, J. Pittner, C. Scheuch, M. F. Guest and J. Koutecký, *J. Chem. Phys.* **96**, 7938 (1992).

[50] W. Ekarat, D. B. Tran Thoai, P. Frank, and W. Schulze, *Solid State Commun.* **46**, 571, (1983).

[51] Ansgar Liebsch, *Electronic Excitations at Metal Surfaces*, Plenum Press, New York, 1997.

2 The Quantum-Chemical Approach

VLASTA BONAČIĆ-KOUTECKÝ, JIŘÍ PITTNER,
DETLEF REICHARDT

Humboldt-Universität zu Berlin

PIERCARLO FANTUCCI

Universita di Milano

and

JAROSLAV KOUTECKÝ

Freie Universität Berlin

2.0 ABBREVIATIONS

AIMD = *ab initio* molecular dynamics; B-LYP = Becke–Lee–Yang–Parr; CCSD = coupled cluster single double excitations; CVC = core–valence correlation; ECP = effective core potential DF = density functional; GDA = gradient corrected density approximation; MCLR = multiconfigurational linear response; MP2 = Møller–Plesset second-order; (MRD)CI = multi-reference double-excitation configuration interaction; RPA = random phase approximation; TD-MCSCF = time-dependent multiconfigurational self-consistent field; TD-SCF = time-dependent self-consistent field.

Metal Clusters. Edited by W. Ekardt
© 1999 John Wiley & Sons Ltd

2.1 INTRODUCTION

The importance of small atomic clusters has been fully recognized since it became evident that novel physical and chemical phenomena can be obtained by controlling the cluster size, shape and temperature [1]. In other words, the extrapolation from the bulk properties towards the atom [2] assuming relatively smooth behavior does not account for characteristics of the given cluster size. This has been confirmed by new areas of applications involving possible construction of quantum dot lasers or single-electron transistors which are still in a development stage [1, 3]. In spite of controversy about the success of constructing such devices it is obvious that by the addition of a single atom or a few atoms the color of emitted light will be changed and this will of course depend on the material used. Similarly, the rearrangements of atoms in sodium nanowires which can be influenced by elongating the wire (changing the temperature) seem to have a significant effect on the transport properties [4]. These nanowires assume molecular-type supported structures related to the known shapes of the gas-phase sodium clusters [5].

Overall, the need for *ab initio* quantum-chemical treatment of small pure and mixed metal clusters has been widely recognized since it allows for the precise determination of their structural, optical and dynamical properties [6–8]. In fact the *ab initio* studies of structures and stabilities of neutral and charged alkali metal clusters have been carried out prior to experiments which could have made use of these results [9–12]. Direct experimental evidence of the ground-state geometries of small metal clusters in the gas phase is still not available. Nevertheless the optical spectroscopic techniques [13–16] can probe the size-dependent structural properties via their excited states provided that the required theoretical information is available. In fact, the *ab initio* determination of optically allowed transitions and their intensities for the stable cluster structures has been established as a reliable predictive tool, particularly for simple metals [17–20] or mixed aggregates involving also combinations of different types of bonding [21]. A comparison of *ab initio* spectroscopic patterns with experimental findings allowed us to propose the assignment of the cluster structure to the measured features and to account for the characteristic of each cluster size and the type of atoms involved [17–21].

This comparison is strictly valid at zero temperature. Depletion spectra have been recorded first at relatively high temperatures [13, 14, 22], while low-temperature data [16] as well as spectra obtained by two-photon femtosecond spectroscopy [15] has more recently become available. However, the presence of energetically close-lying isomeric forms corresponding to local minima on the shallow ground-state energy surface is characteristic of small pure and mixed metallic clusters. Therefore, the consideration of the contributions of different isomeric forms might be necessary when comparison is made with experimental findings at nonzero temperature. Consequently, the information about the nature of line-broadening as a function of temperature is important [23]. One of the approximate ways to account for the influence of nuclear motion on the line shape of the absorption spectrum at the given temperature is the use of Monte Carlo-type sampling of the conformational space which allows us to generate the final spectrum as a sum of spectra calculated at each point [24]. In order to study the influence of temperature on structural properties, the ground-state *ab initio* molecular dynamic is particularly useful. It allows us to follow the behavior of different isomeric forms with increasing temperature including the mechanism of the isomerization and to determine the significance of individual isomeric contributions at the given temperature [25–29].

In this paper we will show that quantum chemistry provides suitable approaches to extracting the specific properties of small metallic and mixed nonstochiometric clusters which cannot be obtained by more approximate methods. This accuracy is needed for controlling the properties through size, shape and composition of the cluster. For this purpose the methodological aspects will be briefly sketched in Section 2.2. We will first address the methods used for the determination of cluster structures at zero temperature and outline the *ab initio* molecular dynamics method which we developed for the determination of temperature-dependent ground-state properties. Then, *ab initio* methods for calculation of excited states valid at $T = 0$ will be described.

In Section 2.3 the structural and optical properties of neutral and cationic Na_n clusters at $T = 0$ K as functions of size are presented and compared with experimental data recorded at low temperature. The temperature-dependent line-broadening will be illustrated by the example of Na_9^+, since in this case a comparison with experimental data at different temperatures is particularly instructive. In Section 2.4 the results of *ab initio* molecular dynamics (AIMD) studies on Li_9^+ will serve to show different temperature behavior of distinct types of structures as well as their isomerization mechanisms. The study of possible 'metal–insulator transitions' and 'segregation into metallic and ionic parts' in finite systems carried out on prototypes of nonstoichiometric alkali halide and alkali hydride clusters with single and multiple excess electrons is presented in Section 2.5. A comparison of structural and optical characteristics of Na_nF_m and Li_nH_m ($n > m$) series allows us to illustrate the influence of different bonding properties.

Conditions for collective, plasmon-type excitations will be addressed in Section 2.6. It will be shown that they can be fulfilled only for very large systems. Future prospects will be given in the concluding remarks.

2.2 COMPUTATIONAL METHODS

2.2.1 Gradient-based Methods for Determination of Cluster Structures at Zero Temperature

Gradient methods employed for geometry optimization are able to locate only the minimum nearest to the starting point and, therefore, they do not provide a guarantee to find the absolute minimum [30, 31]. Born–Oppenheimer (BO) surfaces of metal clusters are characterized by low curvature even in areas far from the stationary points. Therefore, high-accuracy minimization techniques based on second derivatives (Newton methods) with respect to nuclear displacements are better suited than analytical gradients involving only first derivatives, which are widely used particularly at the Hartree–Fock level. The former are computationally more demanding than the latter. At the correlated level of theory a geometry optimization employing second-order perturbation theory (MP2) is more suitable for applications involving directional bonds than to metal clusters with delocalized multi-center bonds. Therefore the density functional method with gradient corrections, GDF (e.g. as proposed by Becke [32] and by Lee, Yang and Parr [33] (B-LYP) for exchange and correlation, respectively) seems to be sufficiently reliable for determination of structures at a correlated level of theory. Great activity in proposing new parameterizations of functionals can be found in the literature. Finally, the gradient-based techniques in the framework of the coupled cluster method accounting for single and double excitations (CCSD) [34] have also become available and provide a good check for density functional methods in spite of their large computational demand. A full harmonic vibrational analysis allowing for identification of local minima is available in all the above-mentioned schemes.

Notice that in spite of considerable progress in developing gradient-based methods the identification of the stable isomers and particularly precise determination of their energy sequence on the flat energy surfaces of metal cluster is still a difficult task.

In the framework of our earlier studies we usually started from the HF optimized isomeric forms and recalculated their energy sequence at the large-scale configuration interaction (CI) level at which the excited states have also been determined. In the case where correlation effects strongly influenced the energy sequence of isomers we also carried out geometry optimization at the correlated level of theory either using GDF or CCSD techniques.

2.2.2 *Ab Initio* Molecular Dynamics for Determination of Structures and their Temperature Behavior

An alternative to the gradient-based search of geometries is *ab initio* molecular dynamics (AIMD) which couples quantum-mechanical treatment of electrons and forces acting on nuclei with classical equations of motions [35, 36]. The advantage of AIMD is that all local minima can in principle be accessed. Thus, the cluster geometries generated along the classical trajectories can be used in the search of isomeric forms on the energy potential surfaces. The accurate determination of the electronic energies and forces can be achieved at either the HF [37, 25–27] or the GDF [28, 29] level employing Gaussian basis sets. In addition, structural properties, dynamics and temperature behavior can be investigated.

Therefore we combined the idea of AIMD, which was initiated by Car and Parrinello [35] in connection with a density functional (DF) procedure employing a plane waves expansion, with the quantum-chemical calculations of energies and gradients in the framework of the HF [25–27] or GDF [28, 29] procedures employing the AO basis sets.

The time evolution of the atoms in a cluster is simulated by their classical trajectories using the Verlet algorithm (cf. Ref. [25]). According to this algorithm, the position and velocity of the nucleus $i(r_i, v_i)$ at time step $t_n = n\Delta t$, are obtained recurrently:

$$r_i^{(n+1)} = 2r_i^{(n)} - r_i^{(n-1)} + \frac{\Delta t^2}{m} F_i^{(n)}, \tag{1}$$

$$v_i^{(n)} = (r_i^{(n+1)} - r_i^{(n-1)})/(2\Delta t). \tag{2}$$

The force $F_i^{(n)}$ acting on nucleus i is related to the gradients of the total molecular energy computed at SCF (self-consistent field) (HF or DF) level. In our AIMD-HF or AIMD-GDF programs the ground-state electronic wave function and energy are computed at each time step (i.e. for each geometric configuration). The electronic energy obtained from an iterative HF or iterative Kohn–Sham procedure, contributes to the total potential energy and the forces acting on the nuclei are determined as corresponding derivatives.

Two important requirements for AIMD techniques have to be pointed out. First, the accuracy of calculated energies and gradients must be higher than usually needed for the geometry optimization. That is, the precise determination of forces is required for the conservation of the total energy, necessary to justify the semiclassical treatment. Second, sufficiently long trajectories have to be calculated in order to obtain meaningful information on dynamical properties. Both factors increase the computational demand. Therefore, different strategies had to be developed to speed up calculations. Both AIMD-HF and AIMD-GDF schemes are suitable for parallel processing. The efficiency of the programs is achieved by means of a full parallelization of the most numerically demanding algorithms, by keeping the integrals in the memory and thus eliminating input–output operations, and by optimizing the sequential parts [26, 29]. Applications are limited to small systems, however, e.g. in the case of AIMD-B-LYP, owing to time-consuming numerical integration required for the exchange–correlation contributions.

In spite of these limitations AIMD schemes represent a suitable tool to study not only structural features but also the influence of internal energy on the dynamics of metal clusters, allowing us to investigate mechanisms of isomerization processes with increasing temperature [25–29]. For this purpose the trajectories need to be calculated over a broad range of different fixed total energies. In the case when one assumes zero initial atomic velocities, the total energy of a cluster is varied by random distortions of its equilibrium geometry. The chosen initial conditions satisfy the requirement of zero linear and angular momenta of the clusters. The lengths of the simulations varied from 20 to 80 ps with a time step of 0.5 fs. Owing to the high-accuracy criteria for calculations of energies and derivatives, the conservation of the total cluster energy in the longest runs was better than 10^{-2} eV. A small split valence basis set with three s-functions and one p-function was used in our AIMD calculations [25–29]. Cluster geometries, which were generated along the trajectories at high internal energies were employed as initial coordinates in a gradient-based search for stable isomers, which were identified as local minima by carrying out a full harmonic vibrational analysis. In addition to the isomers already found by standard optimization techniques [6–12] the AIMD procedure made it possible to find other isomeric forms. The structures and energy sequence of the isomers of Li_n clusters obtained using a small AO basis were in agreement with the corresponding results obtained with considerably larger basis sets [10]. The long time average of the kinetic energy $\langle E_k \rangle$ over the entire length of a trajectory was used to estimate the temperature T of a cluster according to $T = 2\langle E_k \rangle/(3n - 6)k$, where n is the number of atoms and k is the Boltzmann constant [25–29]. The influence of internal energy (temperature) on the dynamics of an Li_9^+ cluster obtained in the framework of a gradient-corrected AIMD-DF approach will be presented in Section 2.4.

2.2.3 *Ab Initio* Methods for the Calculation of Excited States

The configuration interaction (CI) procedure is one of the commonly used methods for determination of electronically excited states [30]. Starting from a finite set ψ_j of orthonormal one-electron basis functions (which can be either Hartree–Fock (HF) or canonical multiconfigurational self-consistent field (MCSCF) orbitals) [30], a subset of all possible antisymmetrized products have to be constructed:

$$|\Phi_J\rangle = A(\psi_{j1}\sigma_1 \ldots \psi_{jN}\sigma_N), \tag{3}$$

where σ_i are the spin functions α, β of the ith electron. The many-electron functions $|\Phi_J >$ which are usually (partially) spin-adapted are called 'configurations'. In the framework of the CI procedure the time-independent Schrödinger equation is solved in the subspace spanned by the constructed configurations, which means that the lowest eigenvalues and the associated eigenvectors of the Hamiltonian matrix

$$(\langle \Phi_J|\hat{H}|\Phi_K\rangle) \tag{4}$$

have to be determined. Since in most cases it is impossible to carry out a full CI, i.e. to include all possible configurations that can be constructed within a given basis set, different strategies have been developed for choosing the 'most important' configurations. One of the most effective procedures is the multi-reference double excitation (MRD) CI method in which, first, a small set of important reference configurations $|\Phi_R\rangle$ ($R = 1, \ldots, n_R$) has to be chosen, then, a subset of those configurations that differ by at most two orbitals from one of the $|\Phi_R\rangle$ is constructed, and, finally, an energy-lowering criterion is used to include in (or discard from) the secular problem the specific configuration. For this purpose, small secular determinants of dimension n_{R+1} need to be solved. Various techniques have been developed to handle the Hamiltonian matrix which can easily reach large $(10^4 - 10^6)$ dimensions. The CI method is conceptually very simple, but computationally very demanding. In order to achieve accurate results, large CI spaces are needed and the selection of the reference configurations has to be done very carefully. When a large number of electrons have to be correlated or a large number of excited states have to be calculated, the CI becomes infeasible.

The linear response methods offer a viable alternative to the CI procedure [38]. A time-dependent (TD) perturbation theory (e.g. involving an oscillating electric field), combined with the SCF or MCSCF method is referred to as the TD-SCF (or random phase approximation, RPA) or the TD-MCSCF (or multiconfigurational linear response, MCLR), respectively. Let us consider the time development of the dipole moment (z-component for simplicity):

$$\langle\Psi(t)|\hat{\mu}_z|\Psi(t)\rangle = \langle\Psi^0|\hat{\mu}_z|\Psi^0\rangle + \varepsilon(\langle\Psi^0|\hat{\mu}_z|\Psi^1\rangle + \langle\Psi^1|\hat{\mu}_z|\Psi^0\rangle) + \cdots. \tag{5}$$

The first term on the right-hand side is the permanent dipole moment and the second one is the first-order change of the dipole moment:

$$\langle\Psi^0|\hat{\mu}_z|\Psi^1\rangle + \langle\Psi^1|\hat{\mu}_z|\Psi^0\rangle = \int_{-\infty}^{\infty} \langle\langle\hat{\mu}_z; \hat{\mu}_z\rangle\rangle_\omega e^{-i\omega t} \, d\omega. \tag{6}$$

The Fourier component of the first-order perturbed time-dependent wave function is called the 'linear response function':

$$\langle\langle\hat{\mu}_z; \hat{\mu}_z\rangle\rangle_\omega = \sum_{m>0} \left[\frac{\langle 0|\hat{\mu}_z|m\rangle\langle m|\hat{\mu}_z|0\rangle}{\omega - \omega_{m0}} - \frac{\langle 0|\hat{\mu}_z|m\rangle\langle m|\hat{\mu}_z|0\rangle}{\omega + \omega_{m0}} \right], \tag{7}$$

where $|0\rangle$ and $|m\rangle$ label the ground and excited states, respectively, and $\omega_{m0} = E_m - E_0$ are the corresponding transition energies. The transition energies and transition moments turn out to be the poles and residues of the linear response function. The dipole length oscillator strength f_e can be obtained from residues:

$$f_e(0 \rightarrow m) = \frac{2}{3}\omega_{m0} \sum_{j=x,y,z} \text{Res}_{\omega=\omega_{m0}} \langle\langle\mu_j; \mu_j\rangle\rangle_\omega. \tag{8}$$

Both transition energies and oscillator strengths are needed for determination of optically allowed absorption spectra. In the multi-configuration version of the linear response theory (MCLR) one constructs an approximation to the exact linear response function by exposing the optimized (MC) SCF wavefunction $|0\rangle$ to a time-dependent perturbation. In this case the time-dependent wave function assumes the form

$$|\Psi(t)\rangle = \exp[iK(t)] \exp[i\Lambda(t)]|0\rangle, \tag{9}$$

where the Hermitian operators

$$K(t) = \sum_{r>s} [\kappa_{rs}(t)E_{rs} + \kappa_{rs}^*(t)E_{sr}] \tag{10}$$

and

$$\Lambda(t) = \sum_{m>0} [\lambda_m(t)|m\rangle\langle 0| + \lambda_m^*(t)|0\rangle\langle m|] \tag{11}$$

generate arbitrary time-dependent rotations of internal orbitals and variations of the configuration mixing coefficients, respectively. (The E_{rs} are excitation operators and are state
transfer operators.) By inserting $|\Phi(t)\rangle$ in the Schrödinger equation, identifying the first-
order terms, and going over to the Fourier components, a generalized eigenvalue equation
is obtained:

$$(\omega\mathbf{S} - \mathbf{E})\chi = 0, \tag{12}$$

which allows us to determine the poles ω (transition energies) and the residues
(transition moments) of the linear response function in the MCLR approximation. The
response matrices $\mathbf{S}$ and $\mathbf{E}$ consist of submatrices which may in a self-explanatory
way be classified as orbital–orbital, orbital–configuration, configuration–orbital and
configuration–configuration parts (see Appendix).

In the MCLR approach the matrix $\mathbf{E}$ couples the time-dependent rotation of orbitals
and the mixing of configuration coefficients. In the framework of RPA which is the
simplest linear response procedure only the orbital–orbital parts of the matrices are
present. In contrast, the configuration–configuration parts are needed in the CI calculations. Consequently, the MCLR can be viewed as a compromise between the simple
RPA and computationally demanding CI. In the RPA method, only interactions between
single excitations are explicitly taken into account; double excitations are not treated on
the same footing, but are only allowed to interact with the ground-state configuration.
Consequently, energetically low-lying excited states in which single excitations play the
dominant role have a good chance of being adequately described by the RPA. The MCLR
approach allows us to include, in addition to single excitations, double and higher-order
excitations for a given number of 'active' electrons within a chosen active space of
orbitals. The accuracy of the results depends in this case on the set of active orbitals and
correspondingly on the number of configurations included. The enlargement of the size
of the active space can easily lead to demanding calculations.

With respect to the MRD-CI method, both RPA and MCLR offer the advantage that all
electronic transition energies (in a given energy interval) and the corresponding oscillator
strengths can be determined in a single calculation on the basis of a single optimized
ground-state wave function.

On one hand, the calculation of the optical transitions of small group I metal clusters
is demanding because a large number of close-lying excited states are required (e.g.
50 or more), but on the other hand only a relatively small number of these transitions
have appreciable intensities. This simplifies the methodological requirements since, for
example, only single excitations are dipole-allowed so that they are responsible for the
dominant transitions. Of course, the influence of higher excitations can be important, in
particular if several transitions share intensities over a larger energy interval.

A comparison of the results obtained from all three methods, RPA, MCLR and (MRD)
CI, has been carried out for a number of examples (small Na_n and Li_n clusters) [38]

for which, also, the experimental results are available [13, 14]. The RPA usually yields qualitatively correct global features of the spectrum, but in many cases is less suited for the assignment of the measured spectrum to a specific cluster structure than the MCLR and (MRD) CI methods, which are able to better reproduce fine features. This is in particular true for geometries of low symmetry. In contrast, the description is intrinsically easier for highly symmetrical structures since the selection rules for optically allowed transitions are more strict. According to our experience, it is possible to achieve an accuracy of 0.1 eV for the transition energies of the intense bands. This goal might be more difficult to reach if the total oscillator strength is spread over a large energy interval, thus requiring calculation of a larger number of excited states.

The choice of the AO basis set for the description of the excited states is also an important accuracy-determining factor. The AO basis sets are usually determined to yield an optimal energy of the atom, and are extended by polarization functions to account for effects due to the molecular environment. Flexible basis sets are necessary in particular for the description of excited states. Numerous AO bases for ground and excited states have been proposed and tested in the literature [30]. In many cases, however, it is reasonable to optimize spectroscopic orbitals for the excited states of interest since the choice of the basis set is closely related to the compromise between accuracy and feasibility. Fortunately, the description of spectroscopic properties of larger systems is sometimes less difficult than that of dimers since in the former case, at least at equilibrium geometries and for vertical spectra, there are several orbitals with adequate nodal properties to mimic high angular momentum functions. Therefore, we use calculations on dimers, which usually require high precision, as a test on the quality of the AO basis set, in connection with all-electron treatment of light atoms as well as with an effective core potential (ECP) approach for heavy atoms.

In the case of sodium a relatively small AO basis was designed for a one-electron valence ECP treatment which includes also corrections for core-valence correlation (CVC): the (4s 3p 1d) primitive Gaussian basis has been contracted to (3s 3p 1d) (cf. Ref. [20] and references therein). Contraction means that several primitive Gaussians are used as a linear combination with fixed coefficients. For example, three coefficients of an s-type Gaussian are varied instead of four. The ECP-CVC results for optically allowed transitions for Na_2 as well as for other small neutral and cationic clusters are very close to those obtained from all-electron calculations using considerably larger AO basis sets, thus confirming that at ECP level larger systems can be treated with acceptable accuracy. Therefore, not only for Na_n and Na_n^+ clusters but also for the calculation of Na_nF_m species we decided also to use the ECP for the F atom with an adequately designed AO basis set (cf. Ref. [21]). In the case of Li_nH_m the all-electron calculations are feasible with the basis set (14s 13p 1d) contracted to (7s 3p 1d) and (8s 2p) contracted to (5s 2p) for the Li and H atom, respectively [10, 21].

2.3 STRUCTURES AND ABSORPTION SPECTRA OF Na_n^+ AND Na_n CLUSTERS

2.3.1 Zero Temperature

Alkali metal clusters are formed by multi-center delocalized bonding rather than by directional two-center bonds as usual molecules, but they also assume well-defined structures [6].

Electron correlation effects contribute considerably to the stability of clusters. Binding energies per atom of neutral and charged species exhibit even-odd oscillations with larger values for clusters with an even number of valence electrons. Consequently the ground-state properties such as ionization potentials and electron affinities do not change smoothly with the cluster size [6].

Careful geometry optimization and *ab initio* MD techniques are essential tools for determination of local and global minima on the flat energy surfaces. In particular, the choice of approximation used at the correlated level of theory can be crucial for establishing the energy sequence of isomers of very similar stability. Therefore we carried out calculations of absorption patterns for all isomers with comparable stabilities, for a given cluster size. The comparison of the theoretical findings [20] with experimental data recorded at low temperature [16], usually allowed us to identify one isomer that is responsible for the measured features.

In order to demonstrate how the optical response (absorption spectra) of clusters can serve as fingerprints of their structural properties we compare in Figure 2.1 *ab initio* results [6, 20] and experimental findings recorded at low temperature [16] for different cluster sizes of Na_n^+. It is important to emphasize that our theoretical spectra reflecting structural properties have been obtained at the time when experimental depletion spectra could be recorded at high temperature only [13, 14, 22, 23], which were characterized by broad structureless bands. Consequently, *ab initio* predictions contained more information. In fact, several years later experimental spectra had to be completely revised due to the possibility of obtaining low-temperature, well-resolved spectra [16]. Calculated transition energies and oscillator strengths for the most stable isomers are therefore in good agreement with the most recent and refined experimental results. The correspondence between structures of the most stable isomers and spectral features at low temperature can be easily established as shown on Figure 2.1 [20], for all cluster sizes except for Na_7^+. In this case, the theoretical absorption spectrum of the second isomer agrees better with the experimental spectrum.

It is apparent from Figure 2.1 that locations of dominant transitions as well as spectroscopic patterns change with the cluster size. The cations larger than the tetramer assume three-dimensional structures. Except for Na_2^+ and Na_3^+, for all considered cluster sizes more than two intense transitions are present in the theoretical spectrum in the energy interval up to 4.0 eV. The sum of calculated oscillator strengths corresponds closely to the number of valence electrons for each cluster considered. The number and location of intense transitions are closely connected with the structural properties. This can be clearly demonstrated by comparing the calculated absorption spectra for different isomeric forms of the given cluster size. The wave functions of the excited states are dominated by few leading configurations, mostly single excitations (from two to a maximum of five) contributing usually with weights of 80% or more. Due to different values and signs of their coefficients, the oscillator strengths have negligible values for some transitions and large values for others. Such a behavior corresponds to interference patterns that are simular to those found in the molecular systems. It is obvious that the correlated wave function of the excited states are usually linear combinations of single (or higher-order) excitations reflecting many-electron features. In order to obtain collective excitations that are characteristic for surface plasmon or bulk material certain conditions have to be satisfied; these will be addressed in Section 2.6 and are not fulfilled for the considered cluster sizes.

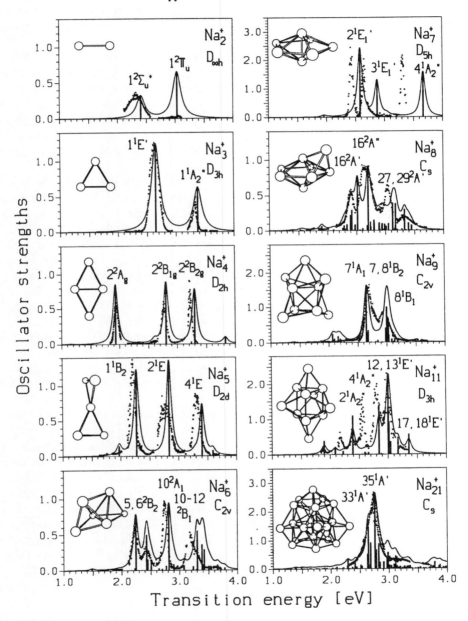

Figure 2.1 Comparison of photodepletion spectra of Na_n^+ ($n = 2-9$, 11, 21) recorded at low temperature (Ref. [16]) (dotted curves) and optically allowed transitions T_e (eV) and oscillator strength f_e (full lines) obtained from MCLR, CI or RPA calculations for the most stable structures of Na_n^+ (Ref. [20]). The electronic states to which intense transitions occur are labeled according to the irreducible representations of the point group. The Lorentzian broadening is indicated by the thin full line. (cf. J. Pittner, Dissertation, Humboldt University, Berlin (1996))

The structural and absorption properties of Na_n^+ presented in Figure 2.1 demonstrate clearly that each cluster size has specific features and no smooth behavior with increasing size can be depicted. This observation is also valid for the neutral Na_n clusters as shown in Figure 2 [6, 13, 17, 19]. A comparison of Figures 2.1 and 2.2 allows us to establish basic differences between optical and structural properties of neutral and cationic clusters.

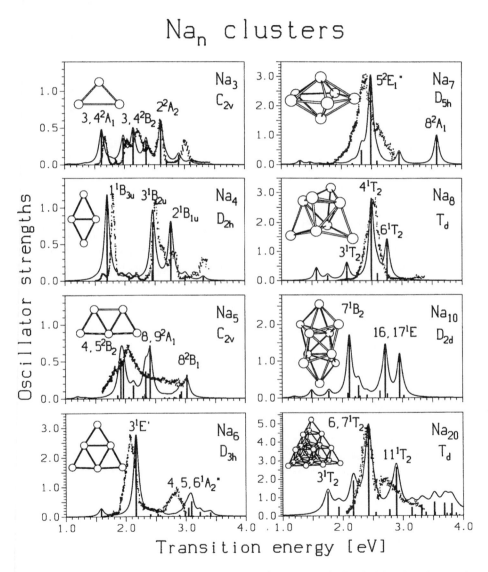

Figure 2.2 Comparison of photodepletion spectra of Na_n ($n = 3-8$, 10, 20) (dotted curves) and optically allowed transitions T_e (eV) and oscillator strength f_e (full lines) obtained from MCLR, CI or RPA calculations for the most stable structures of Na_n (Ref. [6]). The electronic states to which intense transitions occur are labeled according to the irreducible representations of the point group. The Lorentzian broadening is indicated by the thin full line. (cf. J. Pittner, Dissertation, Humboldt University, Berlin (1996))

Even when cationic and neutral forms share the same basic topology, as in the case of Na_4^+ and Na_4 with the rhombic structure, the positions of the most intense transitions are substantially red-shifted for the positively charged species. The transition from planar (2D) to three-dimensional (3D) structures takes place for smaller cluster sizes in the case of cations than for the neutral clusters (Na_5^+ as compared with Na_7). Distinct structural properties for the same size of neutral and cationic clusters such as for Na_6 and Na_6^+ or Na_8 and Na_8^+ give rise to very different absorption patterns. Similarly, clusters with the same number of valence electrons exhibit different optical response features (cf. Na_8 with Na_9^+ or Na_{20} with Na_{21}^+).

The agreement between calculated and experimentally measured optically allowed transitions for the most stable structures is very satisfactory. This means that the depletion spectra of neutral clusters, even if recorded at relatively high temperature [13], still reflect the structural properties; this aspect will be addressed separately in Section 2.4, when discussing the distinct temperature behavior of different isomers close in energy. It will be shown that the isomerization processes take place for different cluster sizes at distinct temperatures.

The influence of temperature on the spectroscopic patterns has been clearly shown for Na_n^+ clusters since the depletion spectra were recorded at different temperatures [16]. The change in the spectroscopic patterns with decreasing temperature is shown for the examples of Na_9^+ and Na_{11}^+ in Figure 2.3. For Na_9^+ the broad absorption band at $T \sim 560$ K splits into two bands with different intensities which become even narrower at ~35 K [16]. The calculated optical transitions and oscillator strengths for the most stable isomer with the C_{2v} form is in excellent agreement with the low-temperature spectrum [20] as can be seen from Figure 2.3. In other words, two groups of dominant transitions located at ~2.65 and ~3 eV allow for the straightforward association of the recorded spectral features at low temperature with the C_{2v} structure. In contrast, the calculated spectroscopic pattern for the C_{3v} structure of the isomer II, characterized by intense transitions all within the narrow energy interval 2.45–2.68, is not in agreement with the low-temperature recorded spectrum but it might contribute to the spectrum at ~560 K as it can be seen from Figure 2.3. Both isomers are very close in energy ($\Delta E = 0.03$ eV) but they assume distinct types of structures. The isomer I is formed by one atom capping the tetrahedral structure of Na_8 while the isomer II is a deformed section of the fcc lattice. Correspondingly the calculated spectroscopic patterns for the two isomers exhibit different features. This example illustrates clearly how the optical response of the clusters can serve as a fingerprint of their structural properties by comparing *ab initio* results at $T = 0$ K with low-temperature experimental spectra.

The right-hand side of Figure 2.3 demonstrates in the case of Na_{11}^+ once more that the recorded spectrum changes drastically by lowering the temperature: two broad bands at high temperature split into six narrow bands at $T \approx 35$ K [16]. The latter spectrum compares well with the calculated absorption pattern for the lowest-energy isomer of the Na_{11}^+ cluster which assumes a D_{3h} structure with the trigonal prism as a subunit being capped by one atom at each side [20]. All other less stable isomeric forms of Na_{11}^+ give rise to spectroscopic patterns that are not in agreement with the low-temperature spectrum (cf. Ref. [20]). The assignment of the D_{3h} structure to the recorded spectrum at $T \approx 35$ K with the six bands features confirms that the most stable isomer is by itself responsible for the rich molecular-like pattern.

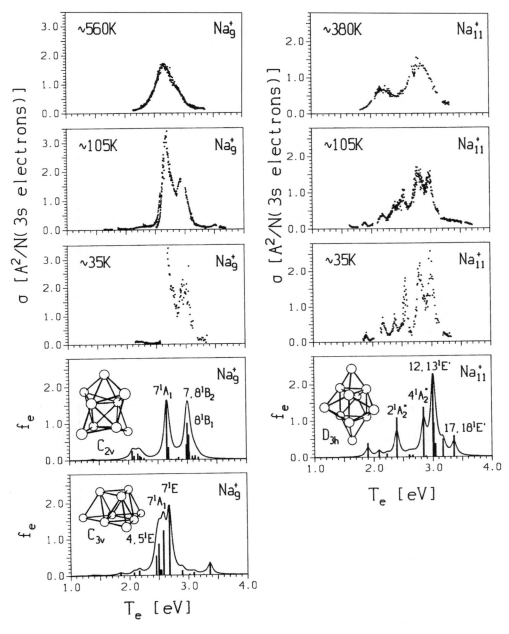

Figure 2.3 Comparison of photodepletion spectra (Ref. [16]) of Na_9^+ (left) and Na_{11}^+ (right) recorded at temperatures estimated to be $\sim$560, $\sim$150 and $\sim$35 K (dotted curves) with optically allowed transitions T_e in eV and oscillator strength f_e (full lines) for the two lowest-energy isomers of Na_9^+; (a) C_{2v} and (b) C_{3v} and for the lowest-energy structure of Na_{11}^+ (D_{3h}) obtained from CI calculations (Ref. [20]). The geometries were optimized at the SCF level. The electronic states to which intense transitions occur are labeled and Lorentzian broadening is indicated by the thin full line. (cf. J. Pittner, Dissertation, Humboldt University, Berlin (1996))

Figures 2.1–3 demonstrate that *ab initio* quantitative predictions of the ground and excited states properties permit not only the reproduction of the experimental findings, but also the identification of the most favorable structures which are not experimentally accessible.

2.3.2 Temperature Line-Broadening

The spectral behavior of the Na_9^+ cluster as a function of internal temperature has been investigated by means of *ab initio* calculations [24]. The idea is to follow the broadening of the spectral lines when the temperature increases, just by averaging the computed optical spectra over a set of distorted cluster geometries, using Boltzmann factors. The hypothesis that nuclear motion can lead to an important broadening of the spectral lines is supported by the evidence that the normal modes of a cluster such as Na_9^+ are characterized by low frequency values. In order to prove this point, calculations have been carried out for the C_{3v} and C_{2v} isomers of Na_9^+ using a contracted Gaussian basis of type (3s2p), the ECP-CVC Hamiltonian mentioned in Section 2.2 and the B-LYP density functional. These calculations serve also to confirm that the energy sequence of the two almost-degenerate isomers can dramatically depend on computational techniques. In fact, at the B-LYP level the C_{2v} structure is more stable than the C_{3v} one by 0.02 eV, while at the HF level the C_{3v} isomer is preferred by about 0.04 eV. Both isomers are characterized by low frequencies. In the case of the isomer C_{2v} the 21 B-LYP harmonic frequencies are distributed as follows: 4 are lower than 50 cm^{-1}, 17 lower than 100 cm^{-1} and only 4 are higher than 100 cm^{-1}. The zero-point vibrational energy is about 100 meV. Analogous results are valid for the C_{3v} isomer.

The analysis of the temperature dependence of the spectral line shape has been carried out employing an HF-RPA approach, using the same basis set and ECP approach. We considered that the Na_9^+ cluster had been in its low-energy isomers of C_{2v} and C_{3v} symmetry and their minimum-energy structures are subjected to small distortions induced by random displacements of randomly chosen atoms. If the distorted cluster is characterized by an energy higher than 0.30 eV above the ground-state energy (which corresponds approximately to a temperature value of 3500 K), then the random move is discarded. For accepted displacements, the RPA spectrum is computed in the energy interval 0–5 eV, which is large enough to include more than 85% of the states. The RPA spectrum together with the HF and MP2 (a 'byproduct' of the RPA calculation) ground-state energy values are stored for the following T-averaging procedure. The energy surface of Na_9^+ is sampled in about 50 000 points, of which more than 35 000 are characterized by an acceptable energy value. Once the random sampling is completed, the T-averaged spectrum is obtained using the (normalized) Boltzmann factors for each considered structure. We have verified that the averaging procedure gives results that are quite similar (from a graphical point of view) using either the HF or the MP2 energy values.

The T-averaged spectra are displayed in Figure 2.4 for temperature values ranging from 0 to 600 K. It is important to note that the fine structure is progressively lost with increasing temperature, while the most intense transition becomes increasingly broader. At $T = 600$ K, the spectrum is composed of a single dominant band centered around 2.6 eV, and a shoulder located at lower energy $T_e < 2.5$ eV (cf. Figure 2.4). In the recorded spectrum at high temperature the shoulder is observed at $T_e > 2.5$ eV (cf. Figure 2.3).

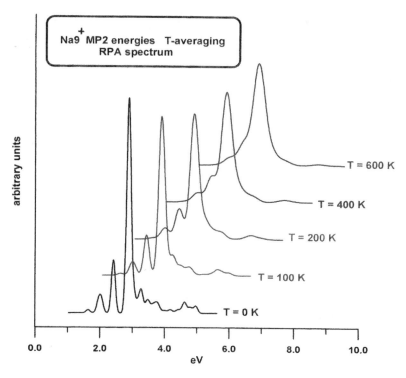

Figure 2.4 The temperature-averaged RPA spectra for the Na_9^+ cluster obtained by Monte Carlo-type sampling of the conformational space calculated for five different T-values

Unsatisfactory features of these particular calculations are due to the limitations of the HF-RPA procedure to produce quantitatively correct spectroscopic patterns, and to the very small basis set that we adopted in order to carry out evaluations of several thousands of HF ground-electronic-state wave functions and their associated RPA spectra. Such a quantitative inadequacy of the computational model, however, does not alter the physical meaning of the conclusion that can be drawn: the appearance of a single broad band in the optical spectrum of Na_9^+ is simply a thermal broadening effect, which, in general, is particularly effective in the presence of several low-lying isomers or in the case of cluster structures characterized by low-frequency vibrations, accompanied by large nuclear displacements due to the low curvature of the BO surface.

Another interesting T-averaged quantity is the set of principal moments of the inertia tensor, which in the case of the Na_9^+ cluster has been computed using a selection based on HF energies (atomic coordinates in A). At temperatures values equal to 1, 100, 400, 600 and 1000 K, the principal moments turn out to be equal to (4.07 4.07 7.12), (4.07 4.09 7.14), (3.70 3.81 7.21), (3.57 3.71 7.23) and (3.47 3.65 7.27), respectively. Clearly, as the temperature increases, the shape of the cluster loses the symmetry inherent in the form preferred at low temperature, and the three principal moments become progressively more distinct. The present calculations, therefore, do not support the idea that the shape of 'melted' clusters must necessarily be spheroidal.

2.4   STRUCTURAL AND TEMPERATURE BEHAVIOR OF METALLIC CLUSTERS

We have chosen to illustrate the influence of internal energy (temperature) on the dynamical properties of the Li_9^+ cluster. The simulations were initialized by randomly distorting equilibrium geometries of different isomers, since zero initial velocities were used (cf. Section 2.2.4 and Refs [25–29]). In previous work on structural properties, the centered antiprism (D_{4d}) has been identified as the most stable isomer at the HF level [9], a result confirmed also by an electron correlation treatment using large-scale CI [6, 10]. A further confirmation was obtained carrying out the geometry optimization using the B-LYP density functional. However, two other isomeric forms close in energy have been found by standard geometry optimization techniques as well as by AIMD schemes: the C_{2v} structure (which can be obtained by capping the T_d form of octamer or by bicapping a pentagonal bipyramid) and the C_{3v} form (a deformed section of the fcc lattice). They were identified as local minima by harmonic frequency analysis, both in the framework of the HF and the density-functional B-LYP procedure. At the HF level, the energy sequence is $D_{4d} < C_{2v} < C_{3v}$ with $\Delta E = 0.064$ and $\Delta E = 0.20$ eV with respect to the D_{4d} structure. The electron correlation effects in the B-LYP treatment increase the energy difference of these two isomers with respect to the isomer I ($\Delta E = 0.22$ and $\Delta E = 0.40$ eV) but the energy sequence remain unchanged [39].

We have chosen to discuss the temperature behavior of individual isomers of the Li_9^+ cluster for two reasons: (i) distinct types of structures characterize the isomers I, II and III, and (ii) the most stable isomers of Li_9^+ and Na_9^+ assume different types of structures: the C_{2v} form of isomer II of Li_9^+ becomes the most stable one for Na_9^+. Therefore distinct temperature behavior of Li_9^+ and Na_9^+ clusters is to be expected. Investigation of the temperature behavior of the neutral Na_n cluster has been carried out using the Car–Parrinello method [35] employing DF with gradient corrections and plane waves [40]. Notice that gradient corrections introduced in DF by Becke [32] for the exchange part of the functional, which improved the determination of binding energy and bond distances considerably, proved to be also very important for calculation of the energy sequences of the isomers.

In this section the results obtained from AIMD-B-LYP simulations by energizing three isomers of Li_9^+ clusters will be analyzed [39]. For this purpose three types of quantities will be used. One of them is the root-mean-square (RMS) bond-length fluctuation δ [25], which is calculated at the end of trajectories. A sharp increase of the δ value is known as the 'Lindemann criterion' for bulk melting, while in the context of finite-size clusters it can be taken as an indication of transition from solid-like to liquid-like state. We find it particularly instructive to analyze the trajectories in terms of 'atomic equivalence indexes',

$$\sigma_i(t) = \sum_j |\vec{r}_i(t) - \vec{r}_j(t)|, \tag{13}$$

where $\vec{r}_i(t)$ is the position of the atom i at time t [27]. The $\sigma_i(t)$ quantities account for all the structural information for a given atom, which depends on the position of all the surrounding atoms at the given time. The non-accidental degeneracies of the $\sigma_i(t)$ quantity reflect the symmetry of the given isomeric form. Structures that do not coincide with the configuration of the local minimum but belong to its basin of attraction are characterized by $\sigma_i(t)$ values that are 'similar' to those of the true isomer. Sharp variations in $\sigma_i(t)$ values along a trajectory indicate possible isomerization transitions. Moreover, $\sigma_i(t)$ curves allow the identification of the atomistic mechanism of an isomerization and to distinguish between pathways characterized by concerted motion of atoms from those along which exchange between atoms occurs. Also at relatively high internal energy (temperature) the analysis in terms of $\sigma_i(t)$ quantities is helpful, since one can easily identify the time intervals connected with different isomers. In contrast, when these trajectories are analyzed in terms of the δ quantity, the conclusion could be drawn that the cluster is in a liquid-like state.

Additional information about the dynamical behavior of a cluster associated with a local structure can be obtained from the power spectrum defined as

$$f(\omega) = \int_0^t \frac{\langle v(t)v(0) \rangle}{\langle v(0)v(0) \rangle} \cos(\omega t)\, dt \tag{14}$$

which is the Fourier transform of the velocity autocorrelation function defined as

$$\langle v(t)v(0) \rangle = \frac{1}{nn_t} \sum_{j=1}^{n_t} \sum_{i=1}^{n} v_i(t_{0j} + t)v_i(t_{0j}) \tag{15}$$

In Eq. (15), n_t is the number of time origins t_{0j} to be averaged over the trajectory.

The analysis of trajectories in terms of $\sigma_i(t)$ quantities, the power spectra, short-time-averaged kinetic energy per atom $\langle E_k/n \rangle$ and bond-length fluctuations δ are presented in Figures 2.5–9.

The plots of $\sigma_i(t)$ values at low internal energy (temperature) along trajectories initialized from the three isomers D_{4d}, C_{2v} and C_{3v} of the Li_9^+ cluster shown in Figure 2.5 serve as guidance for identification of groups of almost-degenerate $\sigma_i(t)$ values reflecting behavior of nearly equivalent atoms. For the same trajectories the calculated power spectra are shown on the right-hand side of Figure 2.5. The comparison of power spectra obtained by low-internal-energy distortion of all three isomers is instructive because different features can be identified. The positions of peaks correspond to frequencies obtained by harmonic vibrational analysis (vertical lines), whereas the intensities are dependent on the particular run. The power spectrum related to the D_{4d} is characterized by peaks located at ~ 100, 200–350 cm^{-1} and the particularly narrow one at ~ 400 cm^{-1}. This last is not present in the power spectra relating to the other two isomers with the C_{2v} and C_{3v} structures which are characterized by peaks spread in the interval between 100 and 300 cm^{-1}.

The difference in dynamical behavior of three isomers investigated at different cluster energies (temperature) can be clearly seen from Figures 2.6–9.

From the graphs of $\langle E_k/n \rangle$ resulting from trajectories generated by distorting the D_{4d} structure (isomer I) shown in Figure 2.6a, it is possible to see that the fluctuations are already present at $T = 182$ K and become larger with increasing temperature. This is even more evident from analysis of the dynamical behavior of the centered antiprism with increasing temperature in terms of $\sigma_i(t)$ quantities using the patterns of Figure 2.5 as shown on the right-hand side of Figure 2.6. The peripheral atoms of the centered antiprism (2–9) are highly mobile but no exchange occurs with the central atom (1) which is characterized by low σ_i, value, until $T \approx 472$ K. The sudden change of $\sigma_i(t)$

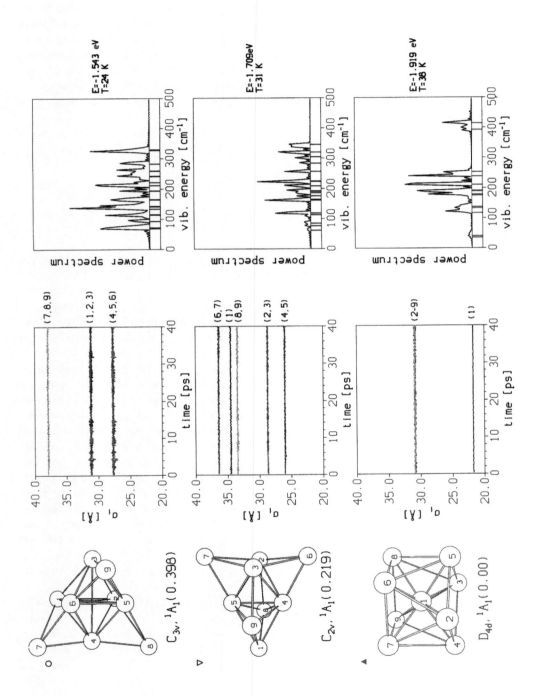

value for the central atom, indicating an exchange between one of the peripheral atoms and the central one, occurs first at $T \approx 513$ K. At this temperature (internal energy) the mechanism responsible for the isomerization can be followed, which is reflected in a sudden change of $\sigma_i(t)$, assuming values characteristic for the structures of isomers II and III (cf. Figure 2.5). This is in agreement with the results obtained by the gradient-based quenching procedure applied to the trajectory of the graph 2.6e ($T = 513$ K). The graph 2.6(b)e ($T = 513$ K) is particularly instructive since it shows that the liquid stage of Li_9^+ has not been reached yet at a temperature higher than 500 K.

The peripheral atoms of the centered antiprism perform a concerted type of motion leading to equivalent isomeric forms of I. The atomistic mechanism can be described in a simplified way as a concerted mutual rotation of two opposite square faces accompanied by relaxation of the interplane distance. For an idealized case of a cubo-octahedron, the atoms belonging to two faces (identified by the labels of Figure 2.5) collected in parenthesis (2345)/(6789) rearrange into an equivalent form, e.g. (2467)/(3589) by two successive rotations of one face with respect to the other by 45° in opposite directions. The first rotation, which would bring the D_{4d} structure into the D_{4h} one, has never been found, since the concerted motion of all eight atoms occurs, always avoiding the more symmetrical D_{4h} structure which lies in energy 0.15 eV above the D_{4d} one. The concerted motion defines a possible path of transformation between two equivalent forms along which a saddle point has been identified with the energy of 0.007 eV above the D_{4d} form. Notice that the exact determination of the saddle point is very difficult on such a flat energy surface. It should be noted that the concerted motion of peripheral atoms of the centered antiprism leading to equivalent isomeric forms is a relatively slow process at low excess of internal energy occurring on a time scale of a few picoseconds for the trajectory of graph 2.6a ($T = 182$ K).

A substantially different behavior was found when simulations were initialized by energizing an other two higher-energy isomers with C_{2v} and C_{3v} structures. The graphs of $\langle E_k/n \rangle$ and $\sigma_i(t)$ for the isomers II and III are shown in Figures 2.7 and 2.8.

Due to the compactness of the C_{2v} and C_{3v} structures, at low energy the fluctuations of atoms are considerably smaller than in the case of the centered antiprism (D_{4d}) but the isomerization process occurs at an energy (temperature) considerably lower than when initialized from the D_{4d} form (cf. Figures 2.6–8). Moreover, on the time scale of our simulations the isomerization processes take place from C_{2v} or C_{3v} to D_{4d} form while the inverse process never occurs (cf. Figures 2.7 and 2.8). In the case of dynamics initiated from the isomer C_{3v}, the isomerization occurs first into the basin of the C_{2v} isomer and than into the basin of the D_{4d} isomer (Figure 2.8). This suggests that the activation barriers that control the transition between C_{2v} and C_{3v} are lower than those between C_{2v} and D_{4d}

Figure 2.5 Short-time (0.25 ps) averaged 'atomic equivalence indexes' σ_i (A) as functions of time for trajectories of Li_9^+ cluster initialized from (i) a distorted D_{4d} structure (isomer I) with the fixed total energy of -1830.919 eV; (ii) a distorted C_{2v} structure (isomer II) with the total energy of -1830.709 eV; (iii) a distorted C_{3v} structure (isomer III) with the total energy -1830.543 eV. Nearly equivalent atoms are grouped in parenthesis on the right-hand side of windows in accordance with the numbering of atoms of the corresponding structures of the isomers drawn on the left. The relative energy sequence of the isomers obtained from the B-LYP procedure is given. Power spectra (evaluated for the corresponding trajectories of the left-hand side) obtained by Fourier transform of the velocity autocorrelation functions are shown on the right. Frequencies calculated by harmonic analysis of the given isomeric form are indicated by vertical lines [39]

48

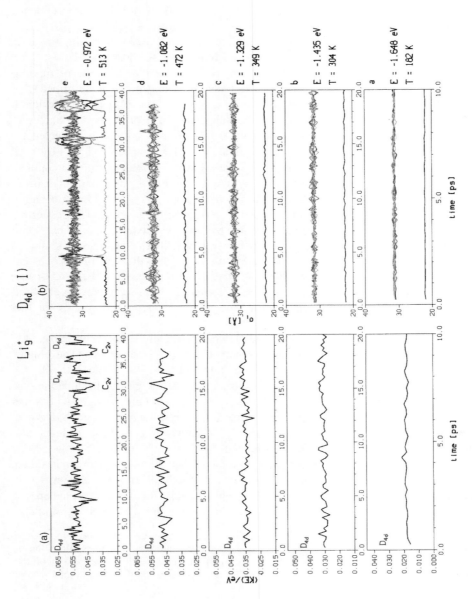

Figure 2.6 Short-time (0.25 ps) averaged kinetic energy per atom (KE) (a) and 'atomic equivalence indexes' $\sigma_i(t)$ (b) as functions of time for trajectories initialized by distorting the D_{4d} structures (isomer I) of the Li_9^+ cluster. For convenience, the corresponding total energies per atom (in eV), shifted by 1830, and the estimated temperature are also indicated on the right of the figure [39]

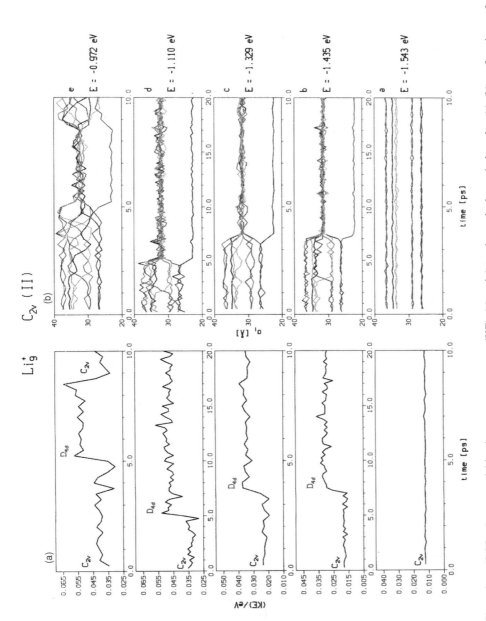

Figure 2.7 Short-time (0.25 ps) averaged kinetic energy per atom $\langle KE \rangle$ (a) and 'atomic equivalence indexes' $\sigma_i(t)$ (b) as functions of time for trajectories initialized by distorting the C_{2v} structures (isomer II) of the $\mathrm{Li}_9{}^+$ cluster. For convenience, the corresponding total energies per atom (in eV), shifted by 1830, are also indicated on the right of the figure [39]

50

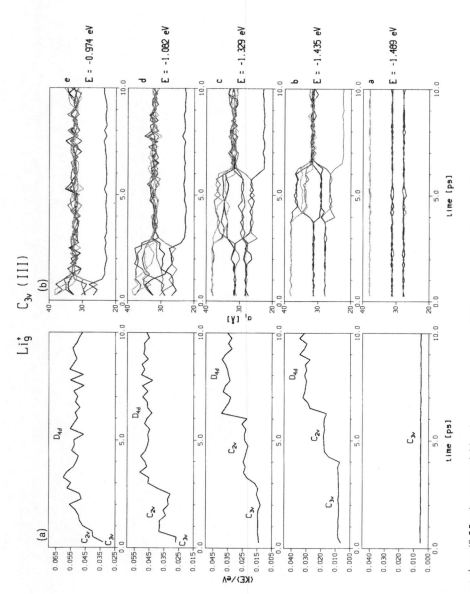

Figure 2.8 Short-time (0.25 ps) averaged kinetic energy per atom ⟨KE⟩ (a) and 'atomic equivalence indexes' $\sigma_i(t)$ (b) as functions of time for trajectories initialized by distorting the C_{3v} structures (isomer III) of the Li_9^+ cluster. For convenience, the corresponding total energies per atom (in eV), shifted by 1830, are also indicated on the right of the figure [39]

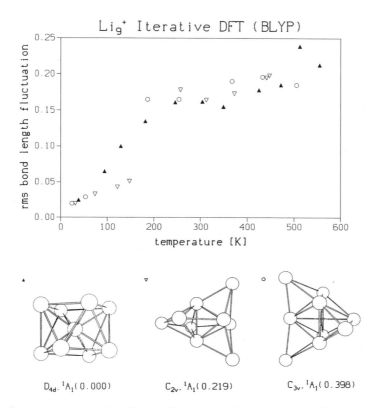

Figure 2.9 Root-mean-square bond-length fluctuations δ as a function of the total energy. The full and open symbols correspond to trajectories initialized from the D_{4d} and C_{2v}, C_{3v} structures of the isomers I and II, III, respectively, of $Li_9{}^+$ drawn with symmetry labels and energy sequence in eV [39]

or C_{3v} and D_{4d}. Notice that although the graphs of $\langle E_k/n \rangle$ exhibit pronounced branches which can be connected with individual isomers (cf. graphs 2.7(a) and 2.8(a)), the $\sigma_i(t)$ quantities provide the precise atomistic mechanism of isomerization (cf. graphs 2.7(b) and 2.8(b)).

The bond-length fluctuation δ plotted as a function of the cluster energy shown in Figure 2.9 exhibits two distinct types of energy-dependencies. One (full triangles) is associated with simulations initialized from the D_{4d} structure of the isomer I, the other (open symbols) has been obtained from trajectories started by distortion of the other two isomers with higher energies. We will refer to these two distinct classes of δ-dependencies as graphs (i) and (ii). At relatively low energy a sudden increase in δ value of the graph (i) is connected with a mobility of atoms of the centered antiprism which involves concerted motion of peripheral atoms according to analysis based on $\sigma_i(t)$ quantities. Notice that in spite of an increase in δ value at low excess energy for the isomer I (D_{4d}), the transition to liquid-like behavior occurs at much higher energy or temperature. In fact, the δ value for the graph (i) increases very slowly between 200 and 500 K and then a second abrupt increase occurs due to isomerizations involving the three isomers I, II and III.

In contrast, small values of δ are characteristic of the C_{2v} and C_{3v} structures at low energies indicating solid-like behavior and the abrupt change for small increase in energy is due to isomerizations to the D_{4d} form (cf. open circles and triangles along the graph (ii) in Figure 2.9).

In the range of $T \approx 200-500$ K the values of δ for simulations initialized from configurations related to the isomers C_{2v} and C_{3v} lie above the (almost constant) δ values obtained form trajectories initialized by distorting the D_{4d} form. The details may change in longer simulations.

The above analysis shows that the lowest-energy isomer of $Li_9{}^+$, the centered antiprism (D_{4d}), exhibits very different dynamical behavior with increasing energy (temperature) than the other two C_{2v} and C_{3v} isomers. All peripheral atoms of the D_{4d} structure perform concerted motion, but remain confined to the energy basin of the equilibrium structure and do not easily rearrange in other isomeric forms until considerably high temperature ($T \approx 500$ K) is reached. The opposite is the case for the C_{2v} and C_{3v} structures, which exhibit a solid-like behavior at low excess of internal energy but are already undergoing the isomerization process above the temperature of $T \approx 150$ K.

From investigating the dynamics of different isomeric forms of $Li_9{}^+$ the following conclusions can be drawn:

(i)   The averaged quantities such as bond-length fluctuation δ must be applied with caution to finite-size alkali-metal clusters. In particular, large increase in δ value does not necessarily indicate a solid-like to liquid-like transition.

(ii) The atomic equivalence indexes $\sigma_i(t)$ which contain inherently structural information are well suited to follow dynamical processes along a given trajectory. They proved to be very valuable for determining the atomistic mechanism of isomerization. It has been illustrated on the example of the centered antiprism of $Li_9{}^+$ that the σ_i quantities allow us to identify a transformation leading to equivalent isomeric forms at low excess energy as well as to follow multimodal behavior which is characteristic for isomerization among different geometric forms at higher excess energies.

(iii) Different energy (mean temperature) behavior of the individual isomers was identified. At low energy the structures with a central atom exhibit a particular kind of concerted mobility of the peripheral atoms coordinated with the central one, while in more compact forms derived from the tetrahedral structure of Li_8, atoms undergo only small fluctuations around equilibrium structures. At higher energies the latter forms undergo easier isomerization than the former ones. Therefore, in the case of $Na_9{}^+$ with the most stable isomer assuming the C_{2v} structure it is possible to estimate that the isomerization process occurs at lower temperature (between 200 and 300 K) than for $Li_9{}^+$.

(iv) The higher-energy isomers show a tendency toward mutual isomerization at lower excess energy (temperatures) due to their low activation barriers. Also activation energy that controls the transformation of higher-energy isomers to the most stable form is, of course, lower than that for the opposite process. This explains the tendency of the most stable forms to be present for relatively long times in trajectories characterized by moderate or even high excess energy, and suggests that the lowest-energy isomers might have the largest catchment areas.

From the above described results as well as from our AIMD studies of Li_8, Li_{10} and $Li_{11}{}^+$ clusters [27] it became clear that isomerization processes can take place for different cluster sizes at different temperature, dependent on the type of structure of the most stable isomer.

In other words, specific structural properties of cluster sizes investigated in the present work cannot be ignored at relatively high temperature in spite of a large mobility of the alkali metal atoms. This explains why the optical spectra recorded at moderate temperature still reflect structural properties of small clusters (cf. Figure 2.2).

2.5 STRUCTURES AND ABSORPTION SPECTRA OF NONSTOICHIOMETRIC ALKALI HALIDE AND ALKALI HYDRIDE CLUSTERS

The ground-state properties of nonstoichiometric $X_n Y_m$ clusters (X = Na, Li, K and Y = Cl, F) with single and multiple excess electrons have been extensively studied experimentally [41–43] and theoretically [44–46] since they are good candidates for possible 'metal–insulator transitions' and 'metallic–ionic' segregation in finite systems. Hydrogenation of lithium clusters has been also investigated [47, 48]. It is of interest to establish similarities and differences among properties of alkali halide and alkali hydride clusters, since both bulk materials have a common structure but the electron affinities of F and H atoms are very different (3.4 versus 0.75 eV). The question can be raised to what extent these differences are reflected in properties of small finite systems.

In addition to structures, stabilities and other ground-state properties we have determined optical response properties for both $Na_n F_m$ and $Li_n H_m$ series with one, two or more excess electrons [21]. The optically allowed transitions serve as suitable fingerprints for localization–delocalization of excess electrons as well as metallic–ionic segregation which cannot always be identified from the structural properties. This will be clearly demonstrated for each investigated series as well as by comparing structures and absorption spectra of sodium fluorides with those of lithium hydrides. The excited states of halide- and hydride-deficient clusters arise from the excitations of the one, two or more excess electrons being placed in a large energy gap between occupied and unoccupied molecular orbitals (MOs) in analogy with the valence and conduction bands in infinite systems. In fact, HOMO (highest occupied) or HOMO $- m$ (occupied with valence electrons) is well separated from other MOs involved in the ionic (or strongly polar) bonding by about 10.0 and 6.0 eV in halides and hydrides, respectively. In addition, k one-electron levels lie close in energy within ~ 1.0 eV above LUMO (lowest unoccupied). Therefore, the interaction between excitations from the HOMO to LUMO, HOMO to LUMO $+ k$, HOMO $- m$ to LUMO and HOMO $- m$ to LUMO $+ k$ levels give rise to low-energy optically allowed transitions in the visible–infrared regime with characteristic spectroscopic features connected with the structural and bonding properties. The experimental data are incomplete since the visible spectral region is more easily accessible than the infrared one, although the progress in the two — photon ionization technique allowed for extension of measurements towards lower energy values [43].

2.5.1 $Na_n F_m$ $(n > m)$

The absorption patterns involving transition energies and oscillator strengths obtained from the MCLR and CI procedures for the most stable structures of the $Na_n F_{n-1}$ $(n = 2-6)$ and $Na_n F_{n-2}$ $(n = 3-6)$ are shown in Figure 2.10 (cf. Ref. [21]).

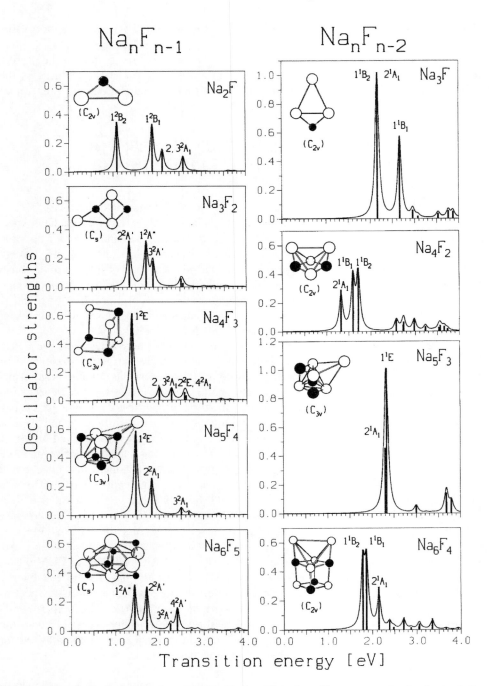

Figure 2.10 Optically allowed transitions T_e in eV and oscillator strength f_e obtained from the MCLR, CI or RPA calculations for the lowest-energy structures of Na_nF_{n-1} ($n = 2$–6) and Na_nF_{n-2} ($n = 3$–6). In the drawn structures full circles denote F atoms. The electronic states to which intense transitions occur are labeled according to the irreducible representations of the point group. Lorentzian line broadening is indicated by the thin full line. Compare Ref. [21] and J. Pittner, Dissertation, Humboldt University, Berlin (1996)

Two characteristic types of structures for Na_nF_{n-1} series can be deduced from Figure 2.10: the cuboid species with 'corner defects' and the structures with an atom attached to the ionic subunits. The prototypes of the former are the Na_4F_3 and Na_6F_5 clusters. In the second case it is necessary to distinguish between the planar (2D) and three-dimensional (3D) species. In Na_3F_2 and Na_2F one Na atom bridges a Na_2F_2 or an NaF subunit, while in the Na_5F_4 cluster one external Na atom is attached to the cuboid unit of Na_4F_4.

Concerning the ground-state properties of the Na_nF_{n-1} series ($n = 2$–6), the F-center Na_4F_3 cluster has the largest binding energy per atom. The calculated values of vertical and adiabatic ionization potentials (IP_v, IP_a) differ substantially, particularly for Na_2F, Na_3F_2 and Na_5F_4, indicating that a considerable geometrical rearrangement is induced by the ionization. However, variations in IP_v and IP_a values with the cluster size are relatively small (4.4–4.0 eV) and (3.7–3.2 eV) respectively. Two sets of experimental data [41, 42] are available and lie between calculated values of IP_v and IP_a in the energy interval (3.8–3.1 eV). This is considered to be relatively high with respect to the low values of IP measured for filled cuboids with a highly delocalized electron characterized as a diffused surfaces state such as $Na_{14}F_{13}$.

The common feature of the optical response of Na_nF_{n-1} clusters is the appearance of an intense transition in the infrared region below 1.5 eV regardless of whether the single excess electron is localized at the corner defect or at the external atom attached to the ionic subunit. It can be clearly seen from the left-hand side of Figure 2.10 that the locations of the dominant transitions for the C_{3v} structures of Na_4F_3 and Na_5F_4 almost coincide. The corner defect structure of Na_4F_3 has been assumed to represent a finite-size prototype of the surface F-center [41, 49]. The calculated spectroscopic pattern for Na_4F_3 is in good agreement with a two-photon ionization experiment that also accesses the infrared region [50]. Notice that the F-center absorption in NaF bulk is located at considerably higher energy (3.7 eV) than the dominant transition ($\sim$1.4 eV) in the infrared for Na_4F_3. The recorded spectrum of Na_5F_4 with the intense band at $\sim$1.5 eV and a weak feature above 1.8 eV [50] also confirms our findings.

From the optical response feature of the left-hand side of Figure 2.10 it is possible to conclude that classification according to the different characteristics expected for the F-center cuboid as compared with noncuboid clusters [41] has not been found in the optical response of small sodium fluoride clusters. This illustrates clearly that structural properties alone are not sufficient for classification of the cluster characteristics.

In contrast, the Na_nF_{n-2} series with two excess electrons allows for distinction of structural and optical response properties according to two classes: (i) the system with cuboid two-corner lattice defects structure (F-center), such as Na_4F_2 and Na_6F_4, give rise to intense transitions in the infrared and low-energy visible region; (ii) for clusters with structures in which segregation between ionic and 'metallic' parts occurs, such as Na_3F and Na_5F_3, the dominant transitions are located at considerably higher transition energies in the visible region [21].

These characteristic features for the two classes are shown on the right-hand side of Figure 2.10. The stable structure of Na_5F_3 is the smallest 3D prototype for 'metallic–ionic segregation'. The location of intense transitions at $\sim$2.3 eV coincides with an energy interval in which 3D structures of pure Na_n clusters give rise to dominant transitions. In contrast, the two excess electrons localized at two-corner-defect cubic structures of Na_4F_2 and Na_6F_4 are responsible for the dominant transitions located below 2.0 eV.

In fact, the optical response features of the 3D structure of Na_6F_4 that contains an F-center structure of Na_4F_2 as a subunit are just slightly blue-shifted with respect to the latter.

For small Na_nF_{n-2} clusters we have found two distinct optical response features that are closely connected with different type of structures in which delocalization of two excess electrons in a metallic subunit occurs or localization at F-center defects takes place.

2.5.2 Na_6F_3

The structural and optical properties of Na_6F_3 represent a very illustrative example of the joint characteristics of Na_5F_3 species capped by an additional Na atom. As can be seen from Figure 2.11, the spectroscopic pattern of Na_6F_3 is composed from intense transitions located close to 2.5 eV, which are typical for the metallic–ionic segregation of the Na_5F_3 cluster (cf. Figure 2.10), and those located below 2.0 eV, which arise due to the localization of an excess electron at the peripheral Na atom bound to the F atom. The calculated spectroscopic pattern is in good agreement with the resonant two-photon ionization spectrum [51] shown in Figure 2.11, allowing for the direct assignment of the C_s structure to the measured features.

2.5.3 Li_nH_{n-1} and Li_nH_{n-2}

In contrast to the alkali halides, the localization of one or two excess electrons in hydrogenated clusters is considerably weaker due to polar bonding even for structures with corner defects such as Li_4H_3 and Li_6H_5. This can be clearly seen from the left-hand side of Figure 2.12 (cf. Ref. [21]). The calculated spectroscopic patterns for the most stable structures of Li_nH_{n-1} clusters exhibit distinct overall features with respect to those found for Na_nF_{n-1} series even in the case of common structural properties such as for Na_4F_3 and Li_4H_3 or Na_5F_4 and Li_5F_4. For all Li_nH_{n-1} clusters the location of transitions with comparable intensities is distributed over a large energy interval so that the dominant transitions are considerably less pronounced than in the case of Na_nF_{n-1} series. For the 3D cuboid structure of Li_4H_3 with corner defect, the lowest intense transition is not in the infrared but in the visible regime with the relative oscillator strengths 3 to 2 with respect to the other two higher-energy transitions. These two facts reflect considerably weaker localization of the single excess electron at the corner defect in comparison with that of Na_4F_3 for which the dominant intense transition in the infrared regime is six times stronger than the other transitions in the visible energy interval. Similarly for the 3D structure of Li_5H_4 with external Li atom linked to the cuboid unit, the lowest energy transition is located in the infrared but due to the rich pattern in the visible region it is not so dominant as in the case of the Na_5F_4 cluster. The left-hand side of Figure 2.12 illustrates also that the stable structures of Li_2H, Li_3H_2 and Li_6H_5 give rise to rich spectroscopic patterns without transitions with large intensities. Notice also that Li_3H_2 and Li_6H_5 assume shapes considerably different from the corresponding alkali halide species.

In the case of the lower degree of hydrogenation, as in the Li_nH_{n-2} series we have found optical response properties which are common to 2D C_{2v} structures of Li_3H and Li_4H_2 and to 3D structures of Li_5H_3 and Li_6H_4: spectra of all clusters are characterized by a group of intense transitions in the visible region (2.1–2.6 eV) (cf. right-hand side of Figure 2.12). These features originate from the 'metallic' part attached to the

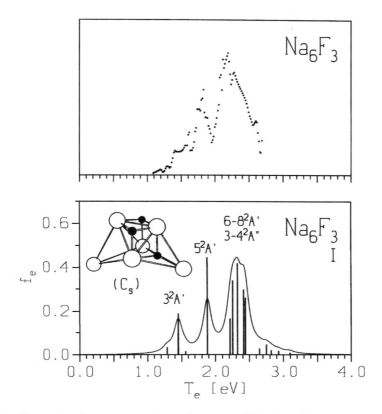

Figure 2.11 Comparison between experimental spectrum [51] (dotted line) and optically allowed transitions T_e in eV and oscillator strengths f_e obtained from the MR-CI calculations for the lowest-energy structure of the Na_6F_3 cluster [21]. For labels see Figure 2.10

ionic subunit although it seems artificial to consider the metalization process for clusters with two excess electrons. Notice that the calculated spectra of 2D and 3D structures of Li_nH_{n-2} species are very similar, which is not the case for pure Li_n clusters for which spectroscopic patterns drastically change by transition from 2D to 3D structures [18].

The global absorption features of Li_nH_{n-2} ($n = 3-6$) reflecting segregation between 'metallic' and ionic parts confirm again that the localization of two excess electrons is not present. This is clearly illustrated by comparing optical response for the most stable structures of Li_nH_{n-2} and Na_nF_{n-2} clusters (right-hand sides of Figures 2.10 and 2.12). The 3D structures of the 'F-center' type with characteristic intense infrared transitions are present only for halides and not for hydrides with the two excess electrons. Notice that in the case of Li_nH_{n-2} for each cluster size there are several isomers close in energy. However, different isomeric forms give rise to distinct spectroscopic patterns. Therefore, we assume that the assignment of the cluster structure to the measured feature will be relatively straightforward. The experimental optical response for Li_nH_{n-1} and Li_nH_{n-2} clusters data are not yet available.

In summary, we have shown that small sodium fluoride and lithium hydride clusters with one and two excess electrons exhibit their own characteristic absorption patterns,

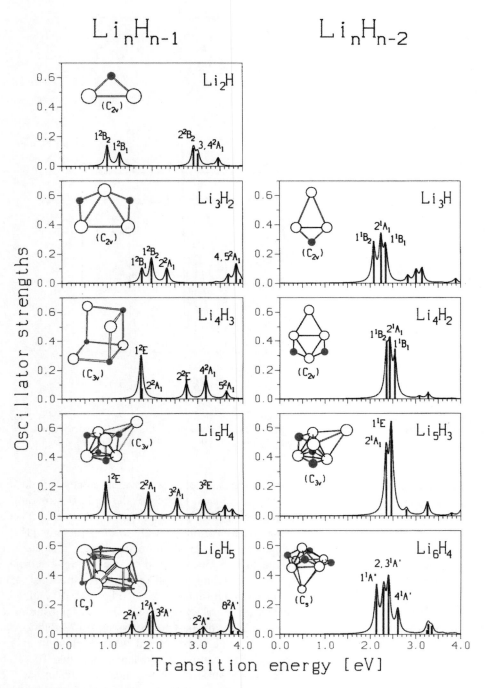

Figure 2.12 Optically allowed transitions T_e in eV and oscillator strength f_e obtained from the MCLR or CI calculations for the lowest-energy structures of $Li_n H_{n-1}$ ($n = 3-6$). Compare Ref. [21] and J. Pittner, Dissertation, Humboldt University, Berlin (1996). For labels see Figure 2.4

reflecting a strong interplay between structural and bonding properties, which are not accessible by extrapolation from their common bulk properties.

2.6 APPLICABILITY OF THE PLASMON THEORY TO CLUSTERS

The first depletion spectra obtained for neutral sodium clusters $N = 2-40$ were characterized by structureless broad features containing one or two bands. The results were interpreted in terms of collective resonances of valence electrons (plasmons) for all clusters larger than tetramers [2, 52–55]. The analogies between findings for metallic clusters and observations of giant dipole resonances in nuclei have attracted a large attention. Therefore the methods employed in nuclear physics, such as different versions of RPA in connection with the jellium model, have also been applied for studying the optical properties of small clusters. Another aspect was the onset of conductivity in metal–insulator transitions.

In contrast, the quantum-chemical approach has been suitable for the description of changes caused by the addition of each single atom influencing the structural and optical properties. Question has been raised as to whether a size regime can be found in which transition from optical properties of individual cluster size can go over to 'metallic' ones. Therefore we first present the theoretical background of the plasmon energy and examine under what conditions the plasmon-like excitations can be present in systems such as small clusters. The existence conditions for collective excitations are exactly formulated in the quantum theory of plasmons and the aim is to show under what assumptions they can be fulfilled for small finite systems.

2.6.1 Derivation of the Expression for the Plasmon Energy

In analogy to the plasmon theory [58, 59] we use as a starting point transition matrix elements of the commutator between the Hamiltonian and the one-electron 'replacement' operator $\hat{E}_j^k$ which correspond to the extended Brillouin theorem:

$$\langle P|[\hat{H}, \hat{E}_j^k]|0\rangle = (E_P - E_0)\langle P|\hat{E}_j^k|0\rangle, \tag{16}$$

where $|P\rangle$ is the many-electron function of plasmon and $|0\rangle$ is the ground state. The one-electron replacement operator is defined as

$$\hat{E}_j^k = \hat{a}_k^+ \hat{a}_j. \tag{17}$$

The Hamiltonian in the Born–Oppenheimer approximation takes the following form:

$$\hat{H} = \left\{ \left[h_{ss} + \tfrac{1}{2}(s, u|s, u)_A \hat{n}_u \right] \hat{n}_s + \left[h_{st} + (u, t|u, s)_A \hat{n}_u + \tfrac{1}{4}(t, v|s, u)_A \hat{E}_u^v \right] \hat{E}_s^t \right\}_{\neq}, \tag{18}$$

where $\hat{n}_s = \hat{E}_s^s$ is the number operator. Both one-electron integrals h_{st} and two-electron integrals

$$(t, v|s, u)_A = (t, v|s, u) - (t, v|u, s) \tag{19}$$

are defined for spin orbitals. The notation $\{\}_{\neq}$ denotes summation over all indices that are different from each other. The form of Hamiltonian given by Eq. (18) has the advantage that all operators commute. We follow the solid state approach in which the operator equations are used instead of the wave function ones. Therefore the commutator of Eq. (16) is written in the form that makes it possible to analyze the approximations and assumptions:

$$[\hat{H}, \hat{E}_j^k] = \{(\hat{f}_k - \hat{f}_j)\hat{E}_j^k + (\hat{n}_j - \hat{n}_k)\hat{A}_{jk} + \hat{E}_j^t\hat{B}_{jk}^t - \hat{E}_s^k\hat{C}_{jk}^s + \hat{D}_{jk}\}_{\neq, j, k}, \qquad (20)$$

where

$$\hat{f}_s = h_{ss} + \{(s, u|s, u)_A\hat{n}_u + (s, v|s, u)_A\hat{E}_u^v\}_{\neq s;\ s=j,k},$$

$$\hat{A}_{jk} = h_{jk} + \{(j, u|k, u)_A\hat{n}_u + (j, v|k, u)_A\hat{E}_u^v\}_{\neq j, k},$$

$$\hat{B}_{jk}^t = h_{kt} + \left\{(t, u|k, u)_A\hat{n}_u + \tfrac{1}{2}(t, v|k, u)_A\hat{E}_u^v\right\}_{\neq j, k, t},$$

$$\hat{C}_{jk}^s = h_{sj} + \left\{(j, u|s, u)_A\hat{n}_u + \tfrac{1}{2}(j, v|s, u)_A\hat{E}_u^v\right\}_{\neq j, k, s},$$

$$\hat{D}_{jk} = \{\hat{n}_k\hat{E}_j^v(v, j|k, j)_A - \hat{n}_j\hat{E}_u^k(j, k|u, k)\}_{\neq j, k}. \qquad (21)$$

Equation (20) has been derived using general rules for products of the replacement operators valid for fermions and does not contain any approximations.

Let us consider the Sawada model [58, 59] which corresponds to generalized random phase approximation and allows for clear definitions and interpretation. Space $\mathcal{R}$ represents the complete one-electron basis of spin-orbitals which is divided into two subspaces and it is assumed that only transitions between them are allowed:

$$\mathcal{R} = \mathcal{K} \cup \overline{\mathcal{K}}, \quad \mathcal{K} \cap \overline{\mathcal{K}} = \emptyset. \qquad (22)$$

Therefore, operators $\hat{B}_{jk}^t$ and $\hat{C}_{jk}^s$ define in Eqs (21) reduce to the one-electron integrals h_{kt} and h_{sj}, respectively. Moreover they vanish if the indices belong to different irreducible representations. Therefore it follows that

$$h_{st} = \delta_{st}h_{ss}. \qquad (23)$$

If the assumptions given by Eqs (22) and (23) are introduced into Eqs (20) and (21), the generalized Brillouin theorem given by Eq. (16) can be written in the form of two equations for $|j\rangle, |l\rangle \in \mathcal{K}$ and $|k\rangle, |m\rangle \in \overline{\mathcal{K}}$ corresponding to one-electron transitions from $\mathcal{K}$ to $\overline{\mathcal{K}}$ and vice versa:

$$(\Delta E - \Delta\varepsilon_{jk})\langle P|\hat{E}_j^k|0\rangle$$

$$= (\overline{n}_j - \overline{n}_k)\left\{\sum_{\substack{u\in\mathcal{K}\\v\in\overline{\mathcal{K}}}}(j, v|ku)_A\langle P|\hat{E}_u^v|0\rangle + \sum_{\substack{u\in\mathcal{K}\\v\in\overline{\mathcal{K}}}}(j, u|k, v)_A\langle P|\hat{E}_v^u|0\rangle\right\},$$

$$(\Delta E + \Delta\varepsilon_{lm})\langle P|\hat{E}_m^l|0\rangle$$

$$= -(\overline{n}_l - \overline{n}_m)\left\{\sum_{\substack{u\in\mathcal{K}\\v\in\overline{\mathcal{K}}}}(m, v|l, u)_A\langle P|\hat{E}_u^v|0\rangle + \sum_{\substack{u\in\mathcal{K}\\v\in\overline{\mathcal{K}}}}(m, u|l, v)_A\langle P|\hat{E}_v^u|0\rangle\right\}, \qquad (24)$$

with

$$\Delta E = E_P - E_0,$$

$$\Delta\varepsilon_{jk} = \varepsilon_k - \varepsilon_j = h_{kk} - h_{jj} + \sum_u [(k, u|k, u)_A - (j, u|j, u)_A]\hat{n}_u, \qquad (25)$$

where E_P is the plasmon energy and $\overline{n}_i$ are expectation values of number operators.

If it is assumed that the space $\mathcal{K}$ is almost full (in analogy to the Fermi sea) and that the space $\overline{\mathcal{K}}$ is almost empty:

$$\overline{n}_j \doteq \overline{n}_l \doteq 1; \quad \overline{n}_k \doteq \overline{n}_m \doteq 0, \qquad (26)$$

Eqs (24) take the form of the random phase relations [60]:

$$(\hbar\omega - \Delta\varepsilon_{jk})X_{kj} = \sum_{\substack{u\in\mathcal{K}\\v\in\overline{\mathcal{K}}}}(k, u|j, v)_A X_{vu} + \sum_{\substack{u\in\mathcal{K}\\v\in\overline{\mathcal{K}}}}(k, v|j, u)_A Y_{vu},$$

$$(\hbar\omega + \Delta\varepsilon_{lm})Y_{ml} = -\sum_{\substack{u\in\mathcal{K}\\v\in\overline{\mathcal{K}}}}(m, u|l, v)_A X_{vu} - \sum_{\substack{u\in\mathcal{K}\\v\in\overline{\mathcal{K}}}}(m, v|l, u)_A Y_{vu}. \qquad (27)$$

Notice that Eqs (27) have been obtained introducing only approximations for the one-electron integrals (Eq. (23)) and occupation numbers (Eq. (26)). For definition of integrals see Eq. (19). Notice that the positive and negative parts are called 'direct' and 'exchange' contributions. X_{ij} and Y_{lm} are coefficients next to the excitation operators in linear response methods.

Let us now introduce the further assumptions valid in the free-electron theory of metals which are due to the translational symmetry:

$$(\vec{a}, \vec{b}|\vec{c}, \vec{d}) = \delta_{(\vec{a}+\vec{b}),(\vec{c}+\vec{d})}V(|\vec{c} - \vec{a}|), \qquad (28)$$

where $\vec{a}$ is a pseudoimpulse vector labeling an IREP of the translational subgroup. Notice that the symmetry conditions for two-electron integrals given by Eq. (28) have the same form as in the zero-differential overlap (ZDO) MO theory, used in the semiempirical quantum chemistry methods.

Introducing the assumptions (28) in Eqs (24) one obtains the expressions for the Brillouin theorem Eq. (16) for the free-electron gas or ZDO-MO for the IREP labeled by the vector $\vec{q}$:

$$(\Delta E(\vec{q}) - \Delta\varepsilon_{\vec{j}(\vec{j}+\vec{q})})\langle P|\hat{E}_{\vec{j}}^{(\vec{j}+\vec{q})}|0\rangle = (\overline{n}_{\vec{j}} - \overline{n}_{(\vec{j}+\vec{q})})\left\{ V(|\vec{q}|)\langle P|\hat{\varrho}_{\vec{q}}|0\rangle - \sum_{\vec{u}\in\mathcal{K}} V(|\vec{j} - \vec{u}|)\right.$$

$$\left. \times\langle P|\hat{E}_{\vec{u}}^{(\vec{u}+\vec{q})}|0\rangle - \sum_{-\vec{u}\in\mathcal{K}} V(|\vec{j} + \vec{u} + \vec{q}|)\langle P|\hat{E}_{-(\vec{u}+\vec{q})}^{-\vec{u}}|0\rangle\right\},$$

$$(\Delta E(\vec{q}) + \Delta\varepsilon_{\vec{j},(\vec{j}+\vec{q})})\langle P|\hat{E}_{-(\vec{j}+\vec{q})}^{-\vec{j}}|0\rangle = -(\overline{n}_{\vec{j}} - \overline{n}_{(\vec{j}+\vec{q})})\left\{ V(|\vec{q}|)\langle P|\hat{\varrho}_{\vec{q}}|0\rangle - \sum_{\vec{u}\in\mathcal{K}} V(|\vec{j} + \vec{u} + \vec{q}|)\right.$$

$$\left. \times\langle P|\hat{E}_{\vec{u}}^{(\vec{u}+\vec{q})}|0\rangle - \sum_{-\vec{u}\in\mathcal{K}} V(|\vec{j} - \vec{u}|)\langle P|\hat{E}_{-(\vec{u}+\vec{q})}^{-\vec{u}}|0\rangle)\right\},$$

$$(29)$$

where $\vec{j}$, $-\vec{j}$, $\vec{u}$ and $-\vec{u}$ belong to $\mathcal{K}$ while $(\vec{j}+\vec{q})$, $-(\vec{j}+\vec{q})$, $(\vec{u}+\vec{q})$ and $-(\vec{u}+\vec{q})$ belong to $\overline{\mathcal{K}}$. Equations (29) take a particularly simple form if the following assumptions are valid for the majority of integrals:

$$V(|\vec{j}+\vec{u}+\vec{q}|) \ll V(|\vec{q}|); \quad V(|\vec{j}-\vec{u}|) \ll V(|\vec{q}|). \tag{30}$$

This is the case for free-electron gas with a high density of one-electron states which is fulfilled for large metallic multidimensional systems. Then, the Eqs (29) reduce to:

$$\langle P|\hat{E}_{\vec{j}}^{(\vec{j}+\vec{q})}|0\rangle = V(|\vec{q}|)G_{\vec{q}}^{+}(\vec{j})\langle P|\hat{\varrho}_{\vec{q}}|0\rangle,$$

$$\langle P|\hat{E}_{-(\vec{j}+\vec{q})}^{-\vec{j}}|0\rangle = -V(|\vec{q}|)G_{\vec{q}}^{-}(\vec{j})\langle P|\hat{\varrho}_{\vec{q}}|0\rangle, \tag{31}$$

with

$$G_{\vec{q}}^{\pm}(\vec{j}) = \frac{\overline{n}_{\pm\vec{j}} - \overline{n}_{\pm(\vec{j}+\vec{q})}}{\Delta E(\vec{q}) \pm \Delta\varepsilon_{\vec{j},(\vec{j}+\vec{q})}}$$

and

$$\hat{\varrho}_{\vec{q}} = \sum_{\pm\vec{u}\in\mathcal{K}} \sum_{\pm(\vec{u}+\vec{q})\in\overline{\mathcal{K}}} (\hat{E}_{\vec{u}}^{(\vec{u}+\vec{q})} + \hat{E}_{-(\vec{u}+\vec{q})}^{-\vec{u}}), \tag{32}$$

where $\hat{\varrho}_{\vec{q}}$ is the $\vec{q}$ component of the density operator.

By summing Eqs (31) for all $\vec{j}$ and $-\vec{j}$ one obtains the following expression:

$$\langle P|\hat{\varrho}_{\vec{q}}|0\rangle = 2V(|\vec{q}|)\left\{ \sum_{\substack{\pm\vec{j}\in\mathcal{K}\\ \pm(\vec{j}+\vec{q})\in\overline{\mathcal{K}}}} \frac{(\overline{n}_{\vec{j}} - \overline{n}_{(\vec{j}+\vec{q})})\Delta\varepsilon_{\vec{j},(\vec{j}+\vec{q})}}{[\Delta E^2 - (\Delta\varepsilon_{\vec{j},(\vec{j}+\vec{q})})^2]} \right\} \langle P|\hat{\varrho}_{\vec{q}}|0\rangle. \tag{33}$$

Notice that the expression $\langle P|\hat{\varrho}_{\vec{q}}|0\rangle$ on the left-hand side of Eq. (33) describes the reaction of the electronic system on the excitation caused by the Coulombic interaction (see the integral $V(|\vec{q}|)$). On the other hand, the matrix element on the right-hand side of Eq. (33) characterizes only the primary excitation process. Therefore Eq. (33) describes the resonance. The characteristic property of this energy equation is presence of the poles for the one-electron excitation energies $\Delta\varepsilon_{\vec{j},(\vec{j}+\vec{q})}$. If it exists the solution $E_P(\vec{q})$ (cf. Eq. (25)), which is much larger than any $\Delta\varepsilon_{\vec{j},(\vec{j}+\vec{q})}$, can be considered as the energy of a plasmon. If such a collective state exists it results from a constructive interaction of many electron–hole pair excitations with small individual energy contributions.

2.6.2 Two Models of Periodic Systems

In order to illustrate how Eq. (33) can be connected with the customary plasmon theory we find it particularly instructive to consider two cases:

(i) In the free-electron model the two-electron integral takes the form:

$$V(|\vec{q}|) = \frac{4\pi\hbar^2 e^2}{\Omega|\vec{q}|^2}, \tag{34}$$

where Ω is the volume of the system and the difference between one-electron integrals is

$$\Delta\varepsilon_q(\vec{j}) = \Delta_{\vec{q}}h(\vec{j}) = h_{(\vec{j}+\vec{q}),(\vec{j}+\vec{q})} - h_{\vec{j},\vec{j}}$$

$$= (\vec{j}\cdot\vec{q})/m + |\vec{q}|^2/2m. \tag{35}$$

Consequently the energy equation (33) takes exactly the form of the Sawada basic equation for the plasmon energy (cf. Ref. 58):

$$1 = \frac{8\pi\hbar^2 e^2}{\Omega |\vec{q}|^2} \sum_{|\vec{j}|\langle P_F} \sum_{|\vec{j}+\vec{q}|\rangle P_F} \frac{(\vec{j}\cdot\vec{q})/m + |\vec{q}|^2/2m}{(\hbar\omega)^2 - [(\vec{j}\cdot\vec{q})/m + |\vec{q}|^2/2m]^2}, \tag{36}$$

where P_F is the Fermi momentum corresponding to the radius of the Fermi sphere.

(ii) In the MO-LCAO-ZDO model of a periodic system (where LCAO is 'linear combination of AOs') the one-electron MO-kets have (which are labeled by irreducible representation $\vec{j}$) the form:

$$|\vec{j}\rangle = \frac{1}{\sqrt{N}} \sum_{\vec{\mu}} \exp[2\pi i(\vec{j}\cdot\vec{\mu})]|\mu\rangle. \tag{37}$$

For small values of $\vec{q}$ differences between one-electron integrals are given by:

$$h_{\vec{j}+\vec{q}} - h_{\vec{j}} = -\sum_{\vec{\mu},\vec{\nu}} h_{\vec{\mu}-\vec{\nu}} \left\{ \pi(\vec{q}\cdot(\vec{\mu}-\vec{\nu})) \sin\pi(\vec{j}(\vec{\mu}-\vec{\nu}))] \right.$$

$$\left. + \left[\frac{\pi}{2}(\vec{q}\cdot(\vec{\mu}-\vec{\nu}))\right]^2 \cos[\pi(\vec{j}\cdot(\vec{\mu}-\vec{\nu}))] \right\}, \tag{38}$$

where $\vec{\mu}$ and $\vec{\nu}$ are position vectors of the localized s-type orbitals and

$$\vec{j} = \frac{\vec{m}(\vec{j})}{L}, \quad \vec{q} = \frac{\vec{m}(\vec{q})}{L}, \tag{39}$$

with L being the number of atoms in the block for which Born–Kármán boundary conditions hold. $L = N^{1/d}$, where $d = 1, 2, 3$ is the dimension of the system with N atoms. The two-electron integrals in the MO-LCAO-ZDO approach are:

$$V(|\vec{q}|) = \frac{2}{N} \sum_{\vec{\mu},\vec{\nu}} \gamma(\vec{\mu}-\vec{\nu}) \cos[2\pi(\vec{q}\cdot(\vec{\mu}-\vec{\nu}))]. \tag{40}$$

$h_{\vec{\mu}-\vec{\nu}}$ in Eq. (38) and $\gamma(\vec{\mu}-\vec{\nu})$ in Eq. (40) are the one-electron and two-electron integrals for the localized orbitals. The expressions (34) and (40) for the two-electron integrals, as well as Eqs (35) and (38) for the relevant difference quantities, have very similar forms in the free-electron model and in the MO-LCAO-ZDO approximation. Consequently, the energy equation (36) for the free-electron model and the energy equation obtained by inserting relations (38) and (39) in the general equation (33) have analogical properties.

2.6.3 Oscillatory Behavior of Plasmons

In order to facilitate the understanding of the resonance phenomena given by Eq. (33) it is convenient to analyze the oscillatory character of the plasmon that is obtained if the $\vec{z}$ component of the full density operator $\hat{\varrho}_f$ is explicitly considered:

$$\hat{\varrho}_f(\vec{z}) = \sum_{\vec{u}} \hat{E}_{\vec{u}}^{(\vec{u}+\vec{z})} \tag{41}$$

(cf. definition (32)). In this case the operator equations are very useful, particularly if the large systems are treated.

If the assumptions (23), (28) and (30) are introduced in the expression for the commutator (cf. Eq. (16)) one obtains:

$$[\hat{H}, \hat{E}_{\vec{j}}^{(\vec{j}+\vec{q})}] = \Delta_{\vec{q}} h(\vec{j}) \hat{E}_{\vec{j}}^{(\vec{j}+\vec{q})} + \sum_{\vec{z}} V(|\vec{z}|)[\hat{E}_{\vec{j}}^{(\vec{j}+\vec{q}-\vec{z})} \hat{\varrho}_{\mathrm{f}}(\vec{z}) - \hat{\varrho}_{\mathrm{f}}(\vec{z}) \hat{E}_{(\vec{j}+\vec{z})}^{(\vec{j}+\vec{q})}], \qquad (42)$$

where $\Delta_{\vec{q}} h(\vec{j})$ is defined by Eq. (35).

Notice that the operators in Eq. (42) do not commute (cf. Ref. [61]). If the summation over $\vec{j}$ is carried out the second and third terms on the right-hand side of Eq. (42) compensate:

$$[\hat{H}, \hat{\varrho}_{\mathrm{f}}(q)] = \sum_j \Delta h_{\vec{q}}(\vec{j}) \hat{E}_{\vec{j}}^{(\vec{j}+\vec{q})}. \qquad (43)$$

The RPA has not been applied in the derivation of the general relation (43). Equation (43) shows that the extended Brillouin theorem with the full density operator (cf. Eq. (41)) does not describe the electron–electron interaction at all, since it does not depend on the integrals $V(|\vec{q}|)$. On the contrary, this interaction can be seen only if the double commutator $[\hat{H}, [\hat{H}, \hat{\varrho}_f, (\vec{q})]]$ is evaluated with the RPA-like assumption

$$\vec{z} = \pm\vec{q}, \qquad (44)$$

with the approximation

$$[\hat{n}_{\vec{q}}, \hat{\varrho}_f(\vec{q}] = \hat{E}_{(\vec{j}+\vec{q})}^{\vec{j}} - \hat{E}_{\vec{j}}^{(\vec{j}+\vec{q})} \doteq 0, \qquad (45)$$

which is acceptable for very large systems. Then the double commutator takes the form

$$[\hat{H}, [\hat{H}, \hat{\varrho}_{\mathrm{f}}(\vec{q})]] \doteq \sum_{\vec{j}} \{(\Delta_{\vec{q}} h(\vec{j}))^2 \hat{E}_{\vec{j}}^{(\vec{j}+\vec{q})} + V(|\vec{q}|) \hat{\varrho}_{\mathrm{f}}(\vec{q}) \hat{n}_j \Delta_{\vec{q}}^{(2)} h(\vec{j})\} \qquad (46)$$

with

$$\Delta_{\vec{q}}^{(2)} h(\vec{j}) = \Delta_{\vec{q}} h(\vec{j}) - \Delta_{\vec{q}} h(\vec{j} - \vec{q}). \qquad (47)$$

In the free-electron model the differences between two excitations $\Delta_{\vec{q}}^{(2)} h(\vec{j})$ do not depend on $\vec{j}$:

$$\Delta_{\vec{q}}^{(2)} h(\vec{j}) = \frac{|\vec{q}|^2}{m}. \qquad (48)$$

If we further assume that the term in Eq. (46) that depends on the differences of one-electron integrals is small in comparison with the term containing $\sum_j \hat{n}_{\vec{j}} = N$, then we have

$$[\hat{H}, [\hat{H}, \hat{\varrho}_{\mathrm{f}}(\vec{q})]] = \hbar^2 \omega_p^2 \hat{\varrho}_{\mathrm{f}}(\vec{q}), \qquad (49)$$

where

$$\omega_P = \sqrt{\frac{4\pi e^2 N}{m\Omega}} \qquad (50)$$

is the classical plasmon frequency.

In order to obtain the expression for time-dependent oscillatory behavior it is useful to insert the operator Eq. (49) into the generally valid expression for the second derivative of the transition matrix elements between two states.

$$\frac{\partial^2}{\partial t^2} \langle P, t|\hat{\varrho}_{\mathrm{f}}(\vec{q})|0, t\rangle = -\frac{1}{\hbar^2} \langle P, t|[\hat{H}, [\hat{H}, \hat{\varrho}_{\mathrm{f}}(\vec{q})]]|0, t\rangle = -\omega_P^2 \langle P, t|\hat{\varrho}_{\mathrm{f}}(\vec{q})|0, t\rangle, \qquad (51)$$

where the time-dependent many-electron kets $|P, t\rangle$ and $|0, t\rangle$ are solutions of the time-dependent Schrödinger equation. Equation (51) shows that a transition matrix element of electron density oscillates with frequency ω_P describing oscillation of electron charge.

From Eqs (49) and (51) it follows that if the energy $\hbar\omega_P$ is very large in comparison with all energies of one-electron transition matrix elements $\langle P|\hat{\varrho}_{\mathrm{f}}|0\rangle$, the oscillatory behavior is present. Notice that it was necessary to consider the double commutator of $\hat{\varrho}_{\mathrm{f}}$ with the Hamiltonian since the single commutator does not account for electron–electron interaction (cf. Eq. (43)) and therefore does not yield plasmon frequency.

2.6.4 Classical Limit for Plasmon Energy and Conditions for Existence of Plasmons

For understanding the basic nature of plasmons it is instructive to consider the limit $\vec{q} \to 0$ and the maximal $\vec{q}_{\mathrm{m}}$ for which the plasmon can appear. The limit $\vec{q} \to 0$ leads to a well-known classical frequency ω_P which has been found as a characteristic quantity of the double commutator of full electron density and the Hamiltonian (ef. Eq. (49)). We find it particularly useful to investigate if for $\vec{q} \to 0$ the Brillouin theorem yields the frequency ω_P as well. If this is true then three approaches, classical, Brillouin theorem and the quantum-mechanical oscillations of electron density, lead to the same result.

The analysis of the expression based on the Brillouin theorem provides a very illustrative description of how the plasmon phenomenon arises. For this purpose we introduce the following assumptions:

$$2(\vec{j} \cdot \vec{q}) \gg |\vec{q}|^2 \quad \text{and} \quad (\vec{j} \cdot \vec{q})/m + |\vec{q}|^2/2m \ll \hbar\omega, \qquad (52)$$

which are equivalent to those formulated in order to obtain Eq. (49).

Introducing assumptions (52) into the fundamental equation (36) based on the Brillouin single commutator Eq. (16), one obtains:

$$\begin{aligned}
\omega^2 &= \frac{8\pi e^2}{\Omega |\vec{q}|^2 m} \sum_{|\vec{j}| < (P_{\mathrm{F}}} \sum_{|\vec{j} + \vec{q}|) P_{\mathrm{F}}} |\vec{q}||\vec{j}| \cos \vartheta \\
&= \frac{8\pi e^2}{m\Omega |\vec{q}|^2} \int_0^{\pi/2} (|\vec{q}| P_{\mathrm{F}} \cos \vartheta)(|\vec{q}| \cos \vartheta)(\pi P_{\mathrm{F}}^2 \sin \vartheta \mathrm{d}v) \\
&= \frac{4\pi e^2}{m\Omega} \frac{2}{3} P_{\mathrm{F}}^3 = \frac{4\pi e^2 N}{m\Omega} = \omega_P^2,
\end{aligned} \qquad (53)$$

where ϑ is the angle between vectors $\vec{j}$ and $\vec{q}$ and P_{F} is the radius of the Fermi sphere. The classical plasmon frequency ω_P in Eq. (53) arises due to the different magnitudes of the three contributions in the integrand separated by brackets: the first bracket describes the individual one-electron transitions, the second one represents the narrow section of allowed transitions across the Fermi surface and the third one represents the surface of the Fermi sphere sheet with polar coordinate ϑ. For systems with high electron density the first two quantities are very small and the third one is very large. Equation (53) shows that for a limited density operator the plasmon arises with the classical plasmon energy only if there are contributions from an extremely large number of individual, one-electron almost-degenerate excitations, with low values of excitation energies, belonging to the same IREP.

Therefore, it is legitimate to raise the question whether the plasmon exists for larger excitation energies $\Delta\varepsilon$. In the case of finite systems, evidently $\Delta\varepsilon$ cannot be arbitrarily small. Consequently, even if all assumptions and approximations used in the derivations of the equations for plasmons are acceptable, the existence conditions can prohibit the appearance of a plasmon in small systems. Therefore, let us consider the existence condition for a plasmon using Eq. (33) for a periodic system:

$$1 < 2V(|\vec{q}|) \sum_{\substack{\pm j \in \mathcal{K} \\ \pm (\vec{j}+\vec{q}) \in \overline{\mathcal{K}}}} \frac{(\hat{n}_{\vec{j}} - \hat{n}_{(\vec{j}+\vec{q})}) \Delta\varepsilon_{\vec{j},(\vec{j}+\vec{q})}}{[\Delta\varepsilon_{\vec{j}m,(\vec{j}m+\vec{q})} + \delta]^2 - (\Delta\varepsilon_{\vec{j},(\vec{j}+\vec{q})})^2}, \tag{54}$$

where

$$\Delta\varepsilon_{\vec{j}m,(\vec{j}m+\vec{q})} = \max(\Delta\varepsilon_{\vec{j},(\vec{j}+\vec{q})}) \tag{55}$$

and δ is a small quantity of the order of the average of $\Delta\varepsilon_{\vec{j},(\vec{j}+\vec{q})}$. For a very large system the summation goes over to integration and $\delta \to 0$. The appearance of a plasmon in a periodic system requires that the right-hand side of Eq. (54) be larger than 1 for the given $\vec{q}$. If for a $\vec{q} = \vec{q}_{max}$ the right-hand side is equal to 1, then for $\vec{q} > \vec{q}_{max}$ a plasmon cannot exist in that system for this $\vec{q}$. Notice, that Sawada, Brueckner and Fukeda [59] have derived explicit conditions for $\vec{q}_{max}$ for the special case of the free-electron model for arbitrary P_F and $\vec{q}$ by integrating Eq. (36).

The second condition necessary to obtain a plasmon is that all electron–electron interactions have a mainly constructive character which can be fulfilled if the conditions (30) for integrals hold, which means that 'direct' excitations dominate over 'exchange' ones in the RPA (cf. Eqs (27)). This is possible to show in the high-density free-electron model for electron–hole transitions.

We showed the existence conditions for volume plasmons in the framework of the free-electron model but a generalization for surface plasmons can also be made in the framework of other models. Since the plasmon theory can be derived for the RPA, the existence of a plasmon can easily be checked by the RPA calculations. In agreement with the general theory presented, the *ab initio* RPA calculations carried out on alkali metal clusters of the size presented in this work and of considerably larger sizes do not exhibit plasmon-like excitations.

In summary, the plasmon can be taken as a quantum-mechanical analog of the classical resonance of sufficiently high electron density. Such a description of excitations can be appropriate for energy-loss spectroscopy and related methods. For dipole interaction with the electromagnetic field it seems that the existence conditions for plasmons cannot be fulfilled at all because the Brillouin theorem does not account for electron–electron interaction when the full density operator is considered. Existence conditions for plasmons are not satisfied for the small and moderate-size clusters, even in the framework of RPA, due to the not sufficiently large density of one-electron states near Fermi level not allowing for a large number of low-energy excitations. Moreover the exchange effects in small metallic systems are not negligible and therefore the constructive interference is not guaranteed.

2.7 CONCLUDING REMARKS

We have shown that an interplay between quantum-chemical predictions and experimental findings on optical response may offer complete information about structural and electronic properties of small metallic and mixed nonstoichiometric clusters as a function of their size and shape. Moreover the quantum-chemical investigations stimulated further refinements in the experiments as well as in theoretical solid-state-related approaches. The relatively small number of intense transitions in a large energy range obtained originally in depletion experiments of pure neutral and charged alkali metal clusters at high temperature [22, 23] was attractive for applications of relatively crude models [2, 52–55]. For example, in the framework of a simplified jellium picture, it was assumed that structural effects are of no importance. The situation has been changed since two-photon femtosecond spectroscopy measurements [15] applied to clusters larger than trimers became available, in which temperature effects are less critical and the lifetimes of different resonances were obtained. In addition, the control of temperature in depletion spectra of Na_n^+ clusters became feasible [16], showing that by lowering the temperature new spectroscopic features arose (cf. Figure 2.3) which shed light on the interpretations, since they could be connected directly with the structural properties. Consequently, in the jellium model [56, 57] also, the positions of nuclei have been introduced in order to obtain qualitative agreement with experimental results.

Theoretical interrogations of existence conditions for the appearance of plasmon phenomena presented in Section 2.6 have shown that they might be fulfilled only for much larger cluster sizes than originally expected.

In summary, quantum-chemical investigations have shown that characteristic features of the studied systems are not smooth functions of their size. In other words, the properties of a small cluster with a given number of atoms can be completely changed by the addition of a single atom. Cluster research is a field of surprising findings and as a consequence it stimulates the development of experimental techniques and computational quantum-chemical methods. In this respect the MCLR method and AIMD procedures represent important achievements. These complementary approaches are essential for studies of cluster properties since they have provided results that allow one to draw the following conclusions: the studied clusters assume defined molecular-like structures at low or moderate temperature ($T \leq 300$ K); their optical spectra are very sensitive to nuclear arrangements and therefore they can serve as fingerprints of structural properties; and a comparison of theoretical and experimental findings allows for the precise interpretation of the measured features.

At the same time we have illustrated with examples of clusters that the theoretical approaches involving the coupling between electronic structure and the motion of nuclei are useful tools removing the border between quantum chemistry and molecular dynamics. The development of new methods allowing the study of time-dependent phenomena involving electronic excitation and motion of nuclei [62] might become an attractive future direction of quantum chemistry.

2.8   APPENDIX

The *response matrices* **S** and **E** from Eq. (4) have a specific paired structure, as given below:

$$
\mathbf{S} = \begin{pmatrix}
\boldsymbol{\Sigma}^{oo} & \boldsymbol{\Sigma}^{oc} & \boldsymbol{\Delta}^{oo} & \boldsymbol{\Delta}^{oc} \\
\boldsymbol{\Sigma}^{co} & \boldsymbol{\Sigma}^{cc} & \boldsymbol{\Delta}^{co} & \boldsymbol{\Delta}^{cc} \\
-\boldsymbol{\Delta}^{oo} & -\boldsymbol{\Delta}^{oc} & -\boldsymbol{\Sigma}^{oo} & -\boldsymbol{\Sigma}^{oc} \\
-\boldsymbol{\Delta}^{co} & -\boldsymbol{\Delta}^{cc} & -\boldsymbol{\Sigma}^{co} & -\boldsymbol{\Sigma}^{cc}
\end{pmatrix}
\qquad
\mathbf{E} = \begin{pmatrix}
\mathbf{A}^{oo} & \mathbf{A}^{oc} & \mathbf{B}^{oo} & \mathbf{B}^{oc} \\
\mathbf{A}^{co} & \mathbf{A}^{cc} & \mathbf{B}^{co} & \mathbf{B}^{cc} \\
\mathbf{B}^{oo} & \mathbf{B}^{oc} & \mathbf{A}^{oo} & \mathbf{A}^{oc} \\
\mathbf{B}^{co} & \mathbf{B}^{cc} & \mathbf{A}^{co} & \mathbf{A}^{cc}
\end{pmatrix}
$$

$$
\sum_{pq,rs}^{oo} = \langle 0|[E_{qp}, E_{rs}]|0\rangle \qquad A^{oo}_{pq,rs} = \langle 0|[E_{qp},[H,E_{rs}]]|0\rangle
$$

$$
\sum_{pq,n}^{oc} = \langle 0|E_{qp}|n\rangle \qquad A^{oc}_{pq,n} = \langle 0|[E_{qp},H]|n\rangle
$$

$$
\sum_{m,rs}^{co} = \sum_{rs,m}^{oc} \qquad A^{co}_{m,rs} = A^{oc}_{rs,m}
$$

$$
\sum_{mn}^{cc} = \delta_{mn} \qquad A^{cc}_{mn} = \langle m|H|n\rangle - \delta_{mn}\langle 0|H|0\rangle
$$

$$
\Delta^{oo}_{pq,rs} = \langle 0|[E_{qp}, E_{sr}]|0\rangle \qquad B^{oo}_{pq,rs} = \langle 0|[E_{qp},[H,E_{sr}]]|0\rangle
$$

$$
\Delta^{oc}_{pq,n} = -\langle n|E_{qp}|0\rangle \qquad B^{oc}_{pq,n} = \langle 0|[E_{pq},H]|n\rangle
$$

$$
\Delta^{co}_{m,rs} = \Delta^{oc}_{rs,m} \qquad B^{co}_{m,rs} = B^{oc}_{rs,m}
$$

$$
\Delta^{cc}_{mn} = 0 \qquad B^{cc}_{mn} = 0
$$

where (oo) label the orbital–orbital parts of the matrices, (cc) the configuration–configuration parts, and (oc) the orbital–configuration mixing parts. $E_{ij} = e_{i\alpha,j\alpha} + e_{i\beta,j\beta}$ are spin-averaged excitation operators: $e_{\mu\nu}$ replaces spin-orbital ν by μ when acting on a configuration ('excites one electron from ν to μ').

2.9   ACKNOWLEDGMENTS

This work has been supported by the Deutsche Forschungsgemeinschaft (SFB 337, Energy transfer in molecular aggregates) and the Consiglio Nazionale delle Ricerche (CNR, Rome).

2.10   REFERENCES

[1]   *Science* **271**, 920–922 (1996).
[2]   W. de Heer, K. Selby, V. Kresin, J. Masui, M. Vollmer, A. Chatelain and W. D. Knight, *Phys. Rev. Lett.* **59**, 1805 (1987).

[3] A. P. Alivisatos, *Science* **271**, 933–936 (1996).

[4] J. I. Pascual, J. Mendez, J. Gomez-Herrero, A. M. Baro, N. Garcia, U. Landmann, W. D. Luedtke, E. N. Bogachek and H.-P. Cheng, *J. Vac. Sci. Technol. B* **13**(3), 1280–1284 (1995).

[5] R. N. Barnett and U. Landmann, *Nature* **387**, 788–790 (1997).

[6] V. Bonačić-Koutecký, P. Fantucci and J. Koutecký, *Chemical Review* **91**, 1035–1108 (1991) and references therein.

[7] V. Bonačić-Koutecký, L. Cespiva, P. Fantucci, C. Fuchs, J. Koutecký and J. Pittner, *Comments on Atomic and Molecular Physics* **31**(3–6), 233–290 (1995) and references therein.

[8] V. Bonačić-Koutecký, P. Fantucci and J. Koutecký, *Encyclopedia of Computational Chemistry*, John Wiley & Sons, Chichester, 1998.

[9] I. Boustani, W. Pewestorf, V. Bonačić-Koutecký and J. Koutecký, *Phys. Rev. B* **35**, 9437–9450 (1987).

[10] V. Bonačić-Koutecký, J. Gaus, M. F. Guest, L. Cespiva and J. Koutecký, *Chem. Phys. Lett.* **206**, 528–539 (1993).

[11] V. Bonačić-Koutecký, P. Fantucci and J. Koutecký, *Phys. Rev.* **B37**, 4369–4374 (1988).

[12] V. Bonačić-Koutecký, I. Boustani, M. Guest and J. Koutecký, *J. Chem. Phys.* **89**, 4861–4866 (1988).

[13] C. Wang, S. Pollack, J. Hunter, G. Alameddin, T. Hoover, D. Cameron, S. Liu and M. M. Kappes, *Z. Phys. D* **19**, 13–17 (1991); C. Wang, S. Pollack, T. Dahlseid, G. M. Koretsky and M. M. Kappes, *J. Chem. Phys.* **96**, 7931–7937 (1992).

[14] M. Broyer and Ph. Dugourd, *Comments on Atomic and Molecular Physics* **31**, 183–214 (1995) and references therein.

[15] T. Baumert, R. Thalweiser, V. Weiss and G. Gerber, in *Femtosecond Chemistry* (VCH Verlagsgesellschaft, Weinheim, 1995), pp. 397–432.

[16] C. Ellert, M. Schmitt, C. Schmidt, T. Reinert and H. Haberland, *Phys. Rev. Lett.* **75**, 1731–1734 (1995).

[17] V. Bonačić-Koutecký, M. M. Kappes, P. Fantucci and J. Koutecký, *Chem. Phys. Lett.* **170**, 26–34 (1990); V. Bonačić-Koutecký, J. Pittner, C. Scheuch, M. F. Guest and J. Koutecký *J. Chem. Phys.* **96**, 7938–7958 (1992); V. Bonačić-Koutecký, P. Fantucci, C. Fuchs, C. Gatti, J. Pittner and S. Polezzo, *Chem. Phys. Lett.* **213**, 522–526 (1993).

[18] Ph. Dugourd, J. Blanc, V. Bonačić-Koutecký, M. Broyer, J. Chevaleyre, J. Koutecký, J. Pittner, J.-P. Wolf and L. Wöste, *Phys. Rev. Lett.* **67**, 2638–2641 (1991); J. Blanc, V. Bonačić-Koutecký, M. Broyer, J. Chevaleyre, Ph. Dugourd, J. Koutecký, C. Scheuch, J.-P. Wolf and L. Wöste, *J. Chem. Phys.* **96**, 1793–1809 (1992).

[19] V. Bonačić-Koutecký, P. Fantucci and J. Koutecký, *Springer Series in Chemical Physics* Springer-Verlag: Heidelberg, **52**, 1994, 15–49.

[20] V. Bonačić-Koutecký, J. Pittner, C. Fuchs, P. Fantucci, M. F. Guest and J. Koutecký, *J. Chem. Phys.* **104**, 1427–1440 (1996).

[21] V. Bonačić-Koutecký, J. Pittner and J. Koutecký, *Chem. Phys.* **210**, 313–341 (1996) and references therein; V. Bonačić-Koutecký and J. Pittner, *Chem. Phys.* **225**, 173–187 (1997).

[22] C. Bréchignac, Ph. Cahuzac, F. Carlier and J. Leygnier, *Chem. Phys Lett.* **164**, 433 (1989); C. Bréchignac, Ph. Cahuzac, F. Carlier, M. de Frutos and J. Leygnier, *Z. Phys. D*, **19**, 1 (1991); C. Bréchignac, Ph. Cahuzac, F. Carlier, M. de Frutos and J. Leygnier, *Chem. Phys. Lett.* **189**, 28 (1992).

[23] R. Poteau, D. Maynau and F. Spiegelmann *Chem. Phys.* **175**, 289–297 (1993).

[24] P. Fantucci and V. Bonačić-Koutecký, to be published.

[25] J. Jellinek, V. Bonačić-Koutecký, P. Fantucci and M. Wiechert, *J. Chem. Phys.* **101**, 10 092–10 100 (1994).

[26] P. Fantucci, V. Bonačić-Koutecký, J. Jellinek, M. Wiechert, J. R. Harrison and M. F. Guest, *Chem. Phys. Lett.* **250**, 47–58 (1996).

[27] V. Bonačić-Koutecký, J. Jellinek, M. Wiechert and P. Fantucci, *J. Chem. Phys.* **107**, 6321, (1997).

[28] D. Reichardt, V. Bonačić-Koutecký, P. Fantucci and J. Jellinek, *Z. Physik D.* **40**, 486 (1997).

[29] D. Reichardt, V. Bonačić-Koutecký, P. Fantucci and J. Jellinek, *Chem. Phys. Lett.* **279**, 129–139 (1997).

[30] K. P. Lawley, 'Ab initio methods in quantum chemistry, Part I and II', in *Advances in Chemical Physics LXVII and LXIX*, ed. I. Prigogine and S. A. Rice (John Wiley & Sons, 1987).

[31] Y. Yamaguchi, J. D. Goddard, Y. Osamura and H. F. Schaefer III, *A New Dimension to Quantum Chemistry: Analytic Derivative Methods in Ab Initio Molecular Electronic Structure Theory*, Oxford University Press, New York, Oxford, 1994.

[32] A. D. Becke, *Phys. Rev. A* **38**, 3098–3100 (1988).

[33] C. Lee, W. Yang and R. G. Parr, *Phys. Rev. B* **37**, 785–789 (1988).

[34] K. K. Baeck, J. D. Watts and R. J. Bartlett, *J. Chem. Phys.* **107**, 3853 (1997).

[35] R. Car and M. Parrinello, *Phys. Rev. Lett.* **55**, 2471–2474 (1985); in *Simple Molecular Systems at Very High Density*, NATO SAI Series B: Physics, Plenum, New York, **186**, 455–476 (1989).

[36] R. N. Barnett, U. Landmann, A. Nitzan and G. Rajagopal, *J. Chem. Phys.* **94**, 608–616 (1991).

[37] B. Hartke and E. A. Carter, *J. Chem. Phys.* **97**, 6569–6578 (1992).

[38] C. Fuchs, V. Bonačić-Koutecký and J. Koutecký, *J. Chem. Phys.* **98**, 3121–3140 (1993).

[39] V. Bonačić-Koutecký, D. Reichardt, and J. Pittner, P. Fantucci and J. Koutecký, *Collect. Chem. Commun.* **63**, 1431–1446 (1998).

[40] U. Röthlisberger and W. Andreoni, *J. Chem. Phys.* **94**, 8129 (1991).

[41] G. Rajagopal, R. N. Barnett, A. Nitzan, U. Landmann, E. Honea, P. Labastie, M. L. Homer and R. L. Whetten, *Phys. Rev. Letters* **64**, 2933–2936 (1990); E. C. Honea, M. L. Homer and R. L. Whetten, *Phys. Rev.* **B47**, 7480–7493 (1993), and references therein.

[42] P. Labastie, J. M. L'Hermite, Ph. Poncharal and M. Sence, *J. Chem. Phys.* **103**, 6362–6367 (1995).

[43] Ph. Poncharal, J.-M. L'Hermite and P. Labastie, *Chem. Phys. Lett.* **253**, 463–468 (1996).

[44] H. Häkkinen, R. N. Barnett and U. Landmann, *Chem. Phys. Lett.* **232**, 79–89 (1995).

[45] R. N. Barnett, H. P. Cheng, H. Häkkinen and U. Landmann, *J. Phys. Chem.* **99**, 7731–7753 (1995).

[46] C. Ochsenfeld and R. Ahlrichs, *J. Chem. Phys.* **101**, 5977–5986 (1994); *Ber. Bunsenges. Chem.* **99**, 1191–1196 (1995); C. Ochsenfeld, J. Gauss and R. Ahlrichs, *J. Chem. Phys.* **103**, 7401–7407 (1995).

[47] B. Vezin, Ph. Dugourd, D. Rayane, P. Labastie, J. Chevaleyre and M. Broyer, *Chem. Phys. Lett.* **206**, 521 (1993); B. Vezin-Pintar, PhD Thesis, University of Lyon, 1994.

[48] B. Vezin, Ph. Dugourd, C. Bordas, D. Rayane, M. Broyer, V. Bonačić-Koutecký, J. Pittner, C. Fuchs, J. Gaus and J. Koutecký, *J. Chem. Phys.* **102**, 2727–2736 (1995).

[49] V. Bonačić-Koutecký, C. Fuchs, J. Gaus, J. Pittner and J. Koutecký, *Z. Physik D* **26**, 192 (1993).

[50] P. Labastie, private communication.

[51] P. Labastie and I. Reiche, private communication.

[52] W. Ekardt, *Phys. Rev. B* **31**, 6360 (1985).

[53] W. Ekardt, Z. Penzar and M. Sunjić, *Phys. Rev B* **33**, 3702 (1986).

[54] C. Yannouleas, R. A. Broglia, M. Brack and P. F. Bortignon, *Phys. Rev. Lett.* **63**, 255 (1989); C. Yannouleas, J. M. Pacheco and R. A. Broglia, *Phys. Rev. B* **41**, 6088 (1990); P.-G. Reinhardt, M. Brack and O. Genzken, *Phys. Rev. A* **41**, 5568 (1990); C. Yannouleas, E. Vigezzi and R. A. Broglia, *Phys. Rev. B* **47**, 9849 (1993).

[55] P. Borggreen, P. Chowdhury, N. Kebaili, L. Lundsberg-Nielsen, K. Lützenkirchen, M. B. Nielsen, J. Pedersen and H. D. Rasmussen, *Phys. Rev. B* **48**, 17507 (1993).

[56] W.-D. Schöne, W. Ekardt and J. M. Pacheco, *Phys. Rev. B* **50**, 11079 (1994).

[57] J. M. Pacheco and J. L. Martins, *J. Chem. Phys.* **106**, 6039 (1997).

[58] K. Sawada, *Phys. Rev.* **106**, 372–383 (1957).

[59] K. Sawada, K. A. Brueckner, N. Fukuda and R. Bout, *Phys. Rev.* **108**, 507–514 (1957).

[60] D. J. Thouless, *The Quantum Mechanics of Many-Body Systems*, Academic Press, New York, London, 1961.

[61] J. Koutecký and C. Scheuch, *Int. J. Quantum Chem.* **37**, 373 (1990).

[62] M. Hartmann, J. Pittner, V. Bonačić-Koutecký, A. Heidenreich and J. L. Jortner, *Chem. Phys.* **108**, 3096–3113 (1998).

3   Density Functional Theory and Car–Parrinello Molecular Dynamics for Metal Clusters

PIETRO BALLONE[†]

Institut für Festkörperforschung Forschungszentrum Jülich,

and

WANDA ANDREONI

IBM Research Division, Zurich Research Laboratory

3.1   INTRODUCTION

Clusters physics and computational density-functional theory (DFT) happily met at the beginning of the 1980s, and together enjoyed along season of growth and excitement, cross-fertilized by a large number of common issues and goals. Fifteen years later, their union is as strong as ever. This is due to a series of separate successes in experiment and theory, which are gradually merging.

Although the breakthroughs in the field of clusters have mainly been experimental, some models and computations based on DFT have played an important role in building the conceptual framework for these results. The first and most remarkable ones were DFT-based jellium and pseudopotential calculations, which provided the (now universally accepted) basis for explaining the shell structure and the collective excitations observed

---

† Present address: Department of Physics, University of Messina, Messina, Italy.

Metal Clusters. Edited by W. Ekardt
© 1999 John Wiley & Sons Ltd

in alkali-metal clusters [1–5]. The DFT approach took a long time to impose itself as a valid method for the determination of the geometries and electronic structures of small clusters [6] — in spite of the clear advantages it offers: a greater simplicity and lower computational cost than standard quantum-chemistry methods, and the inclusion of correlation effects (crucial for metallic systems) at a level far more suitable than that of post-Hartree–Fock (post-HF) or configuration-interaction (CI) approaches. Progress was hampered mainly by the formal analogy of the local density approximation (LDA) used for DFT with the jellium model, which is in principle more appropriate for extended systems and builds on a notion that was foreign to the tradition of theoretical chemistry. Moreover, distrust in LDA was heightened by the concurrent realization that the values obtained for the cohesive energies of dimers and simple organic molecules contained a significant error.

In the meantime, the development of the laser vaporization source for clusters [7] quickly generated a wealth of mass spectra for diverse elemental clusters (from carbon to silicon to iron), exhibiting e.g. specific magic numbers (MNs) and behavior to chemical reagents that called for an explanation. This shifted the focus of the theoretical investigation from the spectroscopy of the very small species to the determination of the relative stability of small- to medium-size aggregates. DFT–LDA started to become recognized as a useful method to resolve questions of this type. In fact, DFT–LDA was surprisingly successful in predicting structural and vibrational properties, namely quantities that depend on the derivatives of the binding energy curve rather than on the absolute value of the binding energy, as especially shown by the experience in solid-state physics.

The advent of DFT-based Car–Parrinello (CP) molecular dynamics (MD) [8] marked an important step forward because it provided the possibility to explore the complex structural patterns of atomic clusters with the least biased procedures and to simulate their dynamics [9]. This eventually led to the discovery of new structures and thus to the understanding of some special features of the mass spectra. In particular, it became possible to investigate how the shape and the electronic structure of a cluster change with temperature, which is especially relevant for metal aggregates [10]. It also became clear that DFT pseudopotential–plane-wave methods allowed one to treat clusters of diverse elements in the periodic table as well as of much larger sizes than traditional calculations had done. Thus, they were particularly suitable for the study of size-evolution patterns. Soon a CP-like technique was also applied to the simulation of interesting dynamical processes, such as cluster fragmentation, with results that could be, at least in part, compared with experiment [11].

Only recently, however, have advances in the DFT approach and extensions to the study of electronic excitations been made that will be beneficial to the linking of theory with the rapidly growing wealth of experimental data directly [12, 13].

This chapter is divided into two main parts. First, in Sections 3.2 to 3.5, we provide an outline of DFT with an emphasis on implementations, such as CP molecular dynamics, that are particularly important for the application to clusters. We discuss the computational schemes more commonly adopted as well as the extensions of the basic theory that were devised to calculate properties that can be measured experimentally. The discussion focuses on the main ideas, while most details are deferred to the specific publications. In Sections 3.6 to 3.11, we give an overview of the applications to metal clusters, namely 'clusters of those elements that form metals in the bulk' [14]. We only consider calculations based on modeling at the atomistic level. This necessarily limits the subject to small clusters, namely of up to ~100 atoms, and excludes simplified DFT models based on the jellium approximation (for which we refer to Refs [3, 5, 15]).

3.2 DENSITY FUNCTIONAL THEORY

Density functional theory is rapidly becoming the standard theoretical framework for parameter-free studies of the structure and adiabatic dynamics of atomic clusters. We shall provide a brief outline of this method below — by necessity nonrigorous and schematic.

The foundation of DFT is given by the well-known theorems of Kohn, Hohenberg and Sham [16], which state that the ground-state energy U of N electrons in an 'external' potential $V^{\text{ext}}(\boldsymbol{r})$ is the minimum of a unique and universal functional $E[\rho]$ of the electron density $\rho(\boldsymbol{r})$:

$$U = \min_{\rho} E[\rho]. \tag{1}$$

The variation over ρ spans every non-negative density whose integral is equal to the number N of electrons in the system:

$$\int \rho(\boldsymbol{r}) \, \mathrm{d}\boldsymbol{r} = N. \tag{2}$$

The proof, implications and subtleties of these theorems are discussed in several review papers and books [17]. Here it suffices to say that, by focusing on the ground state $\rho(\boldsymbol{r})$, DFT represents an enormous simplification as compared with traditional quantum-chemistry techniques (such as HF, CI or coupled-cluster) which are based on the many-electron wave function. On the one hand, this simplification vastly increases the range of sizes and complexity of the systems amenable to computational investigation. On the other hand, this advantage is partly offset by a severe limitation in the scope of the theory, which, in its basic formulation, is restricted to the electron ground state, leaving out the possibility of investigating electronic excitations. Extensions of DFT to excited states have been proposed [17, 18], and will to some extent be discussed in Section 3.5.2.

For the study of clusters, we are usually interested in the case in which the external potential is the Coulomb field of the atomic nuclei of charge $\{Z_i, i = 1, M\}$ and position $\{\boldsymbol{R}_i, i = 1, M\}$:

$$V^{\text{ext}}(\boldsymbol{r}) = \sum_{i=1}^{M} \frac{-Z_i}{|\boldsymbol{R}_i - \boldsymbol{r}|}, \tag{3}$$

which yields the contribution

$$U^{\text{ext}} = \int \rho(\boldsymbol{r}) V^{\text{ext}}(\boldsymbol{r}) \, \mathrm{d}\boldsymbol{r} \tag{4}$$

to the total energy $E[\rho]$.

In addition to U^{ext}, several other contributions are easily identified in $E[\rho]$, which is often written as

$$E[\rho] = U^{\text{ext}} + \frac{1}{2} \int \int \frac{\rho(\boldsymbol{r})\rho(\boldsymbol{r}')}{|\boldsymbol{r} - \boldsymbol{r}'|} \, \mathrm{d}\boldsymbol{r} \, \mathrm{d}\boldsymbol{r}' + \sum_{i<j}^{M} \frac{Z_i Z_j}{|\boldsymbol{R}_i - \boldsymbol{R}_j|} + T_{\text{s}}[\rho] + U_{\text{XC}}[\rho]. \tag{5}$$

The second and third terms in Eq. (5) are the energies associated with the classical electron−electron and ion−ion Coulomb interactions. The contributions that account for the quantum nature of the electrons are isolated in $T_s[\rho]$ and $U_{XC}[\rho]$. To define these terms, we have to introduce some of the details of the practical DFT implementations. In the standard Kohn−Sham (KS) formulation, the ground-state electron density is written in terms of independent electron orbitals $\{\Psi_i(\boldsymbol{r}), i = 1, N\}$:

$$\rho(\boldsymbol{r}) = \sum_{i=1}^{N} f_i |\Psi_i(\boldsymbol{r})|^2, \tag{6}$$

where f_i are the occupation numbers. Then, $T_s[\rho]$ is defined as the kinetic energy associated with the noninteracting electron orbitals

$$T_s[\rho] = -\frac{1}{2} \sum_{i=1}^{N} f_i \langle \Psi_i | \Delta | \Psi_i \rangle, \tag{7}$$

whereas the remainder, i.e. $U_{XC}[\rho]$, includes exchange and correlation contributions to the total energy. The exact $U_{XC}[\rho]$ is unknown for almost all systems of interest (the exceptions being the electron gas and a few other idealized models), and hence approximations to this contribution are required for applications. The simplest choice is the so-called local density approximation (LDA), which assumes that U_{XC} is given by

$$U_{XC}^{LD} = \int \rho(\boldsymbol{r}) \varepsilon_{XC}[\rho(\boldsymbol{r})] \, d\boldsymbol{r}, \tag{8}$$

where $\varepsilon_{XC}[\rho]$ is the exchange−correlation energy per particle of the homogeneous electron gas at density ρ, for which reliable expressions are available [19].

The extension of DFT to spin-polarized systems (of interest especially for transition-metal clusters) has been developed in Refs [20] and [21]. In particular, the generalization of Eq. (8) to the spin-polarized case [called local spin density approximation (LSDA)] can be straightforward [20, 22]. First, we define the density of spin-up (ρ_+) and spin-down (ρ_-) electrons separately; hence the exchange−correlation term becomes

$$U_{XC}^{LSD} = \int \rho(\boldsymbol{r}) \varepsilon_{XC}[\rho(\boldsymbol{r}); \zeta(\boldsymbol{r})] \, d\boldsymbol{r}, \tag{9}$$

where ρ is the total electron density

$$\rho(\boldsymbol{r}) = \rho_+(\boldsymbol{r}) + \rho_-(\boldsymbol{r}) \tag{10}$$

and ζ measures the local spin polarization

$$\zeta(\boldsymbol{r}) = \frac{\rho_+(\boldsymbol{r}) - \rho_-(\boldsymbol{r})}{\rho(\boldsymbol{r})} \tag{11}$$

In doing so, we implicitly assume that the ground-state wave function of the corresponding noninteracting system can be written as a single product of a spin-up and a spin-down Slater determinant [23, 24]. For most spin-polarized systems this assumption is an approximation, giving rise to the so-called spin-contamination problem familiar from HF-based methods. Several detailed studies, see e.g. [25], have shown that in LSDA, although certainly present, the spin-contamination error is less important than in wave-function-based quantum-chemistry approximations such as HF.

A further approximation is apparent in this simple extension to spin-polarized systems, namely that the spin-quantization direction is independent of the spatial coordinate r. This assumption is often satisfied in crystalline solids (see, however, Ref. [26]), but is much less justified for spin-polarized clusters because their relatively low symmetry allows exotic solutions for the spin structures, in which the magnetic moments centered on different atoms point in different directions. To include this effect in the DFT model, it is necessary to introduce a two-component spinor description for the wave function [20, 27]:

$$\Psi(r) = (\psi_1(r); \psi_2(r)), \tag{12}$$

so that the basic variable becomes the density matrix

$$\rho_{\alpha,\beta}(r) = \sum_i f_i \psi_{\alpha i}(r) \psi_{\beta i}^*(r). \tag{13}$$

In Eq. (13), α and β are spin indices. The position-dependent spin configuration is obtained from $\rho_{\alpha,\beta}$ by projection with spin-1/2 Pauli matrices. This generalized scheme has been applied to compute the ground-state properties of Fe_n ($n \leq 5$) in Ref. [27]. The results will be discussed in Section 3.9.

Computationally, the LDA and LSDA methods are very convenient, as the calculations are only as expensive as those based on the Hartree approximation. Despite their simplicity, these approximations are remarkably successful in describing ground-state properties of systems bound by metallic, covalent or ionic interactions [28]: for many of these systems the equilibrium geometry is predicted correctly, with interatomic distances being only a few percent shorter than the corresponding experimental values [17]. It is important to point out, however, that DFT–LDA is far less successful for systems bound by weak interactions, for which, for example, interatomic distances are often greatly underestimated. This problem is apparent in the results for dimers of IIA and IIB atoms [29]. The consequences for the larger aggregates are discussed in Section 3.7. It is also well known that LDA and LSDA overestimate atomization energies significantly. Lastly, dynamical properties such as harmonic frequencies are described by either scheme at a reasonable but not excellent level of accuracy, with an error in the vibrational frequencies of the order of a few percent for systems bound by metallic or covalent forces. Note that experimental accuracy is typically one order of magnitude better than any DFT determination of vibrational frequencies [30].

To overcome these limitations, it is necessary to include information on the global shape of the electron distribution ρ. The simplest step beyond LDA is provided by the so-called generalized gradient approximation (GGA) [31], which has the general form

$$U_{XC}^{GGA} = U_{XC}^{LDA} + \int \rho(r) f_{XC}[\rho(r), \nabla\rho] \, dr. \tag{14}$$

Unfortunately, the choice of f_{XC} is not unique, and several different approximations have been proposed in recent years [31]. In addition, the f_{XC} correction has often been split into its exchange and correlation contributions, each of which has been approximated independently. Finally, exchange and correlation gradient corrections (GC) of different origin have been combined in practical computations. Most widely used in applications are the approximations for f_{XC} proposed by Becke [32] and by Perdew and Wang [33] for the exchange, and by Perdew [34] and by Lee, Yang and Parr [35] for the correlation. Recently, the gradient-corrected exchange–correlation functional of Perdew, Burke and Ernzerhof [36] has received considerable attention.

In general, all these approximations substantially improve the description of atomization energies. The effect of GGA on other properties is not systematic, and may depend significantly on the specific choice of the GC. In most cases, GGA results are slightly better than LDA ones: interatomic distances are improved substantially (but remain too short) for weakly bonded systems, whereas for systems with stronger bonds, GGA predicts distances that tend to be greater than those found experimentally (see, for instance, Ref. [37]). Because the bond lengths are overestimated, the vibrational frequencies tend to be slightly underestimated by GC. A few computations for small molecules and radicals seem to suggest that in LSDA energy differences between different spin multiplicities are sometimes better. In summary, GGA constitutes an improvement over LSDA, although some quantitative inaccuracies remain for a number of properties.

Together with this relatively positive picture regarding the ground-state properties, it is important to point out some gray areas in the current approximations, in DFT itself, and, last but not least, in our knowledge of the DFT results beyond a range of well-explored systems. We focus our discussion on those problems that can affect the computational modeling of clusters. First of all, for lack of experimental data, a detailed comparison of computed and measured structural properties and vibrational frequencies is possible only for very small clusters, i.e. dimers, trimers and tetramers at most. Of course, the good results for these systems and for the extended solids strongly suggest that current local and semilocal DFT approximations are also reliable in between, but it is not yet clear whether the currently achievable accuracy is sufficient to describe the delicate balance of energies determining the boundaries between microclusters, mesoscopic systems and, finally, macroscopic systems. In this respect, the systematic underestimation of surface energies by LSDA and GGA is certainly a reason for concern [38].

Moreover, most of the comparisons have been carried out for simple metals. The computational analysis of transition-metal clusters is far less systematic, partly because of computational problems (discussed in Section 3.3) and partly because of the vastness of this subject, which covers a third of the periodic table and encompasses a large variety of electronic configurations. Computations for rare-earth-metal clusters have been sporadic because, in addition to the difficulties listed above for transition metals, LDA and GGA also fail to predict the electronic configuration for several f-electron systems [39] (including the isolated atoms) correctly. The validity of structural and dynamical LDA predictions is thus undermined. Finally, alloy clusters have been studied only marginally (Section 3.11), although it can be expected that these systems will display an intriguing variety of structures and chemical behaviors, and thus provide a strong test of DFT methods.

For finite systems and clusters in particular an important source of problems strictly related to the local and semilocal approximations is the asymptotic behavior of the potential felt by an electron that is well outside the cluster. The exact exchange–correlation potential converges asymptotically (and probably rather quickly [40]) to $-(Q+1)/r$, where Q is the net charge of the clusters in atomic units, and r is the distance from a suitable origin, which we assume to be the center of mass of the electron density. It is easy to verify that in the same limit the KS potential given by LSDA tends to $-Q/r$.

The asymptotic behavior of $V_{XC}(r)$ given by GGA is more binding, but still decays much faster with increasing r than the exact V_{XC} [41]. This problem, which especially affects the least-localized, highest-energy KS orbitals, has drastic effects for the description of the excess electron in anions: the highest occupied orbital in negatively charged clusters is sometimes predicted to be unbound, even for species that are known to be stable. As a consequence, several anionic clusters that have been extensively studied by photoelectron spectroscopy cannot be modeled in this way unless the highest-energy electron is localized by a fictitious external potential. In actual computations, electron states are usually localized by the use of a basis set or of boundary conditions (see Section 3.3). Properties of anionic clusters (such as the electron affinity, EA) computed under these conditions often compare well with experiments.

Although logically distinct, the error in the asymptotic part of V_{XC} is also related to the well-known problem DFT has in dealing with excitation energies [17], which can be stated as follows: the eigenvalues entering the KS equations cannot be interpreted as electron-orbital energies despite the fact that the HF eigenvalues, formally so similar, are related to physical energies by Koopman's theorem.

The limitation of DFT is twofold. The first aspect is a limitation of principle: DFT is restricted to ground-state properties, and there is no reason to expect that the KS eigenstates (occupied and unoccupied) built from the ground-state exchange–correlation potential contain information on the electron orbitals for the excited states. In fact, it has been proven that there is no simple relation between the exact exchange–correlation potentials for the ground state and for an excited state, thus ruling out a simple description of excitations on the basis of ground-state eigenvalues only [42]. The only exception to this negative result is the eigenvalue of the highest occupied state, which is generally believed to be equal to the negative of the ionization potential (IP) in exact DFT [43]. See, however, [44] for a different point of view.

The second problem is strictly related to the local and semilocal approximations for U_{XC}: they fail to reproduce the exact KS potential or the exact KS orbitals and eigenvalues well. As a consequence, even the eigenvalue for the highest occupied orbital is far from correct, i.e. minus the IP: the error for small aggregates is always significant (a discrepancy of $\sim 30\%$ is found in Nb [45] and even in the jellium model of alkali-metal clusters [46]), and it is almost the same in LSDA and GGA. This state of affairs has been the source of much frustration and a general distrust in the KS orbital energies as well as in the ability of DFT to predict excitation energies.

Although this negative point of view is largely justified, it is important to realize that at least some excitation energies are well within the reach of DFT. IPs, for instance, are defined as the difference of the ground-state energies of the systems having charge Q and $Q + 1$. These latter two energies are accessible to DFT, and IPs computed as an energy difference (vertical and adiabatic) compare fairly well with experiment, at both the LSDA and GGA levels. The same considerations apply for excitation energies that are total energy differences between electron configurations, each of which is the lowest-energy state of a given symmetry. Here again are both states amenable to a DFT computation, and therefore their energy difference can be compared to experimental excitation energies. Nevertheless, these special cases do not cover all the cases of interest (see Sections 3.5.2 and 3.5.4). Moreover, the separate computation of initial and final states is certainly less convenient than what is offered by schemes such as HF, for which a single ground-state computation is sufficient to describe the entire spectrum of available states via quantities (the eigenvalues) that are byproducts of the ground-state computation. For this reason, the development of methods that offer a combined description of ground and excited states is an active subject of research, although it is not obvious whether such a method will be available within the strict limits of DFT, or whether eigenvalues alone will ever be sufficient to describe optical properties with a sufficient level of accuracy. Simple schemes to estimate excitation energies by adding *ad hoc* corrections to the KS eigenvalues have been proposed [47], but have not been extensively tested in comparison with experiments.

At the very least, better approximations to the exact V_{XC} are required to compute more accurately what is accessible to DFT. Substantial progress is likely to require truly nonlocal approximations to U_{XC}. Several of them have been proposed, but only a few have been tested by explicit computations, and even fewer of these calculations have been devoted to clusters. Nowadays schemes that add local or semilocal correlation contributions to the HF (the so-called HF–KS schemes [48]) or to a mixture of HF and DFT exchange [49] are popular in the chemistry community. The justification of these methods is that the greatest error in U_{XC} can be attributed to the exchange, which can be computed exactly, although at a substantially higher computational cost than required by simple DFT approximations. Unfortunately, HF itself is a very poor approximation for metals [50], and the addition of correlation at the local or semilocal level is not sufficient to correct its drawbacks. For metal-cluster computations, an additional disadvantage may be that the quality of the description of the electronic structure depends on the size: hybrid schemes perform well for atoms [48], but their accuracy degrades with increasing size of the aggregates. As these problems are due to the long range of the unscreened HF exchange, corrections intended to mimic the electron screening of exchange have been proposed [51], but have not yet been tested for metal clusters. Closely related to the above is the optimized potential model [52], which also tries to solve the problem of approximating the HF exchange as well as possible. Computations for extended systems show, however, that — again, as is true for mixed schemes — this model (combined with simple approximations for the correlation) is also not a true and global solution to the LSDA or GGA problems, although some of its results (such as the KS eigenvalues) are significantly better [53], especially for finite systems. Hence, the optimized potential model could provide an interesting approximation for clusters.

Another approach that has been extensively discussed in the literature is the so-called self-interaction correction (SIC) first proposed by Lindgren [54] and then developed by Perdew and Zunger [55]. The starting point is the observation that in the expression (5) for the DFT total energy, the self-interaction of each electron is contained. Formally, this is true also in the standard HF equations, but there the self-contributions from the Hartree and the exchange terms cancel exactly. This exact cancellation is lost with approximate expressions for the exchange and, in general, does not hold for simple approximations of the correlation. This self-interaction term can be related to several of the basic problems of local and semilocal DFT schemes, including the incorrect asymptotic decay of V_{XC} far from a finite system. To correct this error, Perdew and Zunger proposed a scheme that enforces a cancellation of the self-interaction [55]. The SIC results for the total energy of atoms offer a major improvement over LDA. KS–SIC eigenvalues approximate electron excitation energies well, and, in particular, the highest eigenvalue is close to the IP. Moreover, the EAs of atoms also compare fairly well with experiments. Unfortunately, computations for larger (but still finite) systems are much less conclusive: total and atomization energies do improve, eigenvalues are closer to excitation energies, but the structural properties and vibrational frequencies computed within SIC sometimes deviate drastically from experimental data (see Ref. [56] for an application to Na clusters, the results of which are discussed in Section 3.6).

A stronger foundation underlies all methods related to the weighted-density approximation (WDA) [57] that attempt to provide an approximate description of the exchange–correlation hole surrounding each electron in the system. These schemes are bona fide density functional approximations, whose relation with the exchange–correlation energy and potential are established by the adiabatic connection developed in Refs [22, 58]. An important simplification of these methods is that only the spherical average of the exchange–correlation hole enters the determination of total energies; no higher angular components are required for this purpose. The WDA has yielded encouraging results for alkali-metal clusters within the spherical jellium model [59]. To our knowledge, however, it has never been used for realistic models of clusters. Because of its obvious connection with the homogeneous electron gas picture, from which almost all models of the exchange–correlation hole are derived, schemes based on WDA are likely to be a significant improvement over local approximations in the case of simple metals. However, that any of the simple prescriptions for the exchange and correlation hole will provide a good approximation for highly correlated systems, such as transition metals or rare-earth clusters, is unlikely. At present there is no formal and satisfactory solution for this problem.

In addition to the search for new and more reliable approximations to the exact DFT functional, an active subject of research is the development of simple schemes to model the potential energy surface of large ($\sim 10^3$–10^4 atoms) and low-symmetry systems. A short account of simplified (also often semiempirical) methods, which can be related to DFT, is given in Appendix A.

3.3 COMPUTATIONAL SCHEMES FOR DFT APPLICATIONS: THE ELECTRONIC PROBLEM

Having outlined the basic points of DFT, we now discuss the computational features that characterize the methods most frequently used for the study of clusters, with special emphasis on the techniques underlying the *ab initio* MD introduced by Car and Parrinello [8].

3.3.1 Kohn–Sham Equations

Starting from the energy functional $E[\rho]$, see Eq. (5), the first task is to compute the electronic ground-state energy U, which depends parametrically on the ionic coordinates $\{R_I\}$, and to define one point on the Born–Oppenheimer surface of the system

$$U[R_I] = \min_{\rho} E[\rho], \tag{15}$$

where $\rho(r)$ satisfies Eq. (2).

The extremum condition at a fixed number of electrons can be expressed by the Euler–Lagrange equation

$$\frac{\delta}{\delta \rho}[E - \mu \int dr\, \rho(r)] = 0, \tag{16}$$

where μ is a Lagrange multiplier introduced to impose the constraint in Eq. (2), and can be identified with the chemical potentials of the electrons.

If we assume the KS orbitals (with the constraint of orthonormality among them) as our basic variables instead of the density, we obtain a set of Schrödinger-like equations

$$\frac{\delta}{\delta\psi_i^*}\left[E - \mu_i \int d\mathbf{r}'\rho(\mathbf{r}')\right]$$

$$= \left[-\frac{1}{2}\Delta + V^{\text{ext}}(\mathbf{r}) + \int \frac{\rho(\mathbf{r}')\,d\mathbf{r}'}{|\mathbf{r} - \mathbf{r}'|} + \frac{\partial}{\partial\rho}(\rho\varepsilon) - \mu_i\right]\psi_i(\mathbf{r}) = 0 \qquad (17)$$

with a Hamiltonian H_{KS} that depends self-consistently on the solutions $\{\psi_i\}$ via the electron density ρ:

$$H_{\text{KS}} = -\frac{1}{2}\Delta + V^{\text{ext}}(\mathbf{r}) + \int \frac{\rho(\mathbf{r}')\,d\mathbf{r}'}{|\mathbf{r} - \mathbf{r}'|} + \frac{\partial}{\partial\rho}(\rho\varepsilon). \qquad (18)$$

Needless to say, this type of self-consistent problem is very well known in quantum-chemistry, as the set of equations (17) is closely related (but by no means equivalent) to the HF equations.

3.3.2 Basis Sets for Electronic Wave Functions

The standard strategy to tackle this problem is to select a set of basis functions $\{\phi_j(\mathbf{r})\}$ for the KS orbitals ψ_i,

$$\psi_i(\mathbf{r}) = \sum_j a_j^{(i)}\phi_j(\mathbf{r}), \qquad (19)$$

and a suitable starting guess ρ^{in} for the electron density. Then, it is possible to compute and diagonalize the Hamiltonian H_{KS}, and obtain a set of output orbitals $\{\psi_i\}$, which yields an output density ρ^{out} and thus a new H_{KS}. Under conditions often met in practice (and using techniques discussed, for instance, in [60]), it is possible to iterate this basic step until the input and output densities (or, equivalently, the input and output H_{KS}) differ by less than a given accuracy threshold δ.

Of course, the basis set $\{\phi_j(\mathbf{r})\}$ cannot be complete in practical computations (as the Hilbert space $\mathcal{H}$ is of infinite dimension), and there are two possibilities to optimize the overlap of the subspace spanned by $\{\phi_j(\mathbf{r})\}$ with the low-energy portion of $\mathcal{H}$.

3.3.2.1 Localized Basis Sets

One strategy is to select a set of basis functions that are reminiscent of atomic orbitals because chemical experience shows that most chemical bonds can be rationalized as simple combinations of atomic-like orbitals. This choice gives rise to the family of localized bases approaches familiar to the chemistry community. Popular basis sets consist of Gaussians, Slater orbitals, and numerical atomic-like functions. None of them is a complete set, which renders controlling the convergence of the calculations quite difficult. For light elements, or for valence electrons in pseudopotential schemes (see below), it is possible to limit the size of the orbital expansion to $\sim 20n$ (where n is the number of atoms), which leads to an inexpensive diagonalization of the Hamiltonian even for fairly large systems. On the other hand, the computation of matrix elements can be rather time-consuming, and the effort required to evaluate the Hartree contribution grows rapidly with system size (although important advances have recently been made, see Ref. [61]). Moreover, the basis functions generally are not linearly independent, and the H_{KS} matrix may become nearly singular if the overlap of functions centered on different atoms is large. As a result, computations based on localized basis sets may become unreliable unless the basis set is selected with great care and is extensively tested.

Another important problem is that, because $\{\phi_j\}$ is atom-centered and optimized for the dominant chemical configuration of the atomic species, the overlap of the subspace spanned by $\{\phi_j(r)\}$ with the ground-state orbitals ψ_i may depend significantly on the atomic positions and the chemical environment. As we shall see below, this problem becomes crucial when addressing the time evolution of the atomic positions. It can also constitute a major difficulty in systems whose electronic structure involves many different configurations with similar energies, as in transition-metal clusters.

3.3.2.2 *Plane-Waves*

The second route to solving Eq. (17), developed mainly by the solid-state community, relies on simple functions to express the KS orbitals. By far the most common choice is plane-waves. In this scheme, the orbitals are expressed by a finite and discrete Fourier series

$$\psi_i(r) = \sum_G c_G^{(i)} e^{iGr}. \tag{20}$$

The sum includes all the G vectors belonging to some simple and regular lattice in reciprocal space, up to a maximum modulus $|G|_{max}$ determined by the spatial resolution required by $\psi_i(r)$ [62].

Because plane-waves are simple and not specifically gauged to the functions to be developed, the expansion (20) requires a large number of them to reproduce the KS orbitals to an acceptable degree of convergence: typical applications to small transition-metal clusters may require $\sim 10^5$ plane-waves. Despite the amazing growth of computer power in recent years, applications of this size are still somewhat at the limit of our computational capabilities in terms of both CPU time and memory. However, the computation of matrix elements is straightforward and inexpensive, and energy contributions that presented bottlenecks in localized basis sets have also become easy. In fact, the Hartree potential can be calculated in reciprocal space, and the exchange–correlation energy and potential on an equispaced mesh. The evaluation of gradients and the Laplacian of the density required to apply GC is also straightforward. An additional advantage is that the quality of the basis set is determined by a single parameter, $|G|_{max}$, so that the convergence of the computation can be tested by systematically varying it. Moreover, the bias towards specific atomic positions and chemical bonds is very small. This is connected to a crucial property (discussed below) of plane-waves which renders this basis set the most suitable for MD applications of DFT (Section 3.4.1).

Besides these positive points, which led to the remarkable success of plane-wave implementations of DFT, there are also important drawbacks, in addition to the sheer dimension of the basis set required for many applications. First of all, the discretization of the Fourier expansion in reciprocal space implicitly defines a periodicity in real space: by using Eq. (20) we assume that the system under study is periodically repeated in space.

Although this reflects reality in applications to crystalline solids, it is an artificial boundary condition for isolated clusters and molecules, which can introduce spurious interactions between the periodic replicas and affect the result of the plane-wave computation. To some extent, the problem can be reduced at the cost of increased computational time by increasing the scale of the real-space periodicity, i.e. by adopting a finer grid in reciprocal space (the density of G vectors in reciprocal space is inversely proportional to the volume of the periodically repeated cell in real space). However, given the slow decay of the Coulomb potential, this cannot be an entirely satisfactory solution, especially for charged systems. Methods to solve this problem have been developed [63] that provide a representation of the Hartree potential that is free of spurious interactions between replicas, at a computational cost that is somewhat higher than the one required by standard periodic boundary conditions. Current applications to clusters often adopt this type of boundary conditions, following [64].

Another technical difficulty is that of describing electron states that are highly localized, such as the 3d states in transition metals, by means of plane-waves. This problem and currently adopted solutions are discussed in the following sections.

To conclude, we point out that in the past ten years, plane-wave algorithms implemented for vector computers have provided the most efficient and accurate tool to study clusters, and metal clusters in particular, within standard DFT schemes. However, with increasing size of the clusters of interest, and given the widespread use of massively parallel computers, the plane-wave technique might in the future be surpassed in efficiency by other methods. Several avenues are currently being investigated as promising alternatives, such as wavelets [65], which are particularly suitable for highly inhomogeneous systems (such as clusters), finite elements [66], and also Gaussian basis functions [67].

3.3.3 Pseudopotentials

The choice of the basis set is strictly related to another ingredient of several implementations of the DFT scheme, i.e. pseudopotentials. As is well known, the chemistry of the elements depends predominantly on the valence electrons, i.e. on those in the highest-energy incomplete atomic shell. It is natural, therefore, to include only the 'chemically active' electrons explicitly in the computation. Two different but closely related approaches have been introduced to exploit this basic simplification: (1) the 'frozen-core' approximation, which assumes that the core is not modified by the formation of chemical bonds, and (2) the pseudopotential formalism, which replaces the interaction between valence and core electrons by an 'external' potential acting on the former and does not explicitly include the latter.

In the plane-wave scheme, the introduction of pseudopotentials is strictly required because the number of plane-waves needed to describe core states is exceedingly high for all but the lightest elements.

Here we will not go into the details of the origin and the theoretical foundation of pseudopotential methods, as they can be found in [68]. Suffice it to say that, in terms of plane-wave applications, the use of pseudopotentials has a twofold aim: (1) to exclude the highly localized core electrons and (2) to cancel the rapid oscillations of wave functions close to the nucleus that are present even in valence states owing to an orthogonalization to the core states. A long evolution of the pseudopotential theory and practice converged in the 1980s to a widely accepted family of pseudopotentials [69] which are (i) norm-conserving, (ii) nonlocal, (iii) devised to reproduce exactly the energy eigenvalues of the *atomic* valence states and the corresponding wave functions outside a 'core' radius R_c, and (iv) able to approximate as well as possible those of the low-energy excited ones.

The norm-conservation property ensures that the norm of the true and pseudo *atomic* wave functions is the same outside the radius R_c, which has to be shorter than the distances relevant for bonding. This property is required to reproduce the electrostatics of the all-electron approach in the pseudopotential computation. R_c is an important parameter that determines the hardness of the pseudopotential and also its quality. Nonlocality means that the effect of the pseudopotential cannot be expressed by a simple multiplication in real space, and is imposed by the need to include the effect of orthogonalization to the core states. The latter depends on the symmetry of the valence state because s, p, d, ... orbitals must be orthogonalized explicitly only to core states of the same symmetry. As a consequence, the effective potential is different (and sometimes drastically so) for different angular momenta in the valence configuration. As the nonlocality is associated with orthogonalization to core states, the effective potentials for the different angular symmetries differ only close to the nucleus, and all converge quickly to the Coulomb part with increasing r ($-Z_v/r$, Z_v being the valence charge). With very few exceptions, present-day applications are performed with pseudopotentials that satisfy requirements (i)–(iv) above, although the detailed forms can differ significantly in practical implementations [70]. To make the computation parameter-free, most pseudopotentials are generated starting from *ab initio* (DFT) atomic calculations. As an example, we compare all-electron and pseudo wave functions for a heavy metallic element, namely Au, in Figure 3.1, and in Figure 3.2 we report angular-momentum-dependent pseudopotentials for Na and Fe.

The quality of an atomic pseudopotential is measured by its transferability, that is, its performance in describing a given atom in different valence–electron configurations (for example, isolated, but in different electronic configurations, in a molecule, a cluster or a solid). This property strongly depends on the way the core–valence interaction is

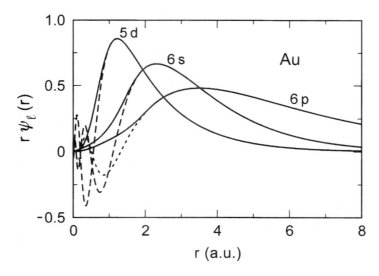

Figure 3.1 Pseudowave functions vs. all-electron wave functions for the gold atom

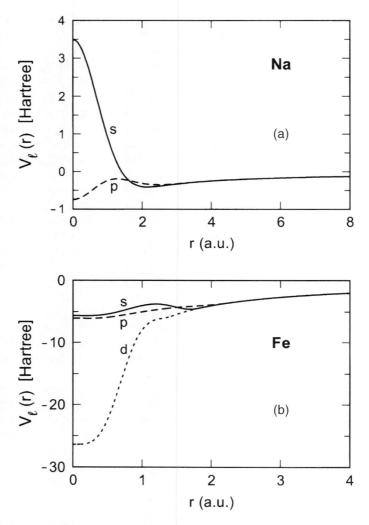

Figure 3.2 Angular-momentum LDA pseudopotentials for Na, as used in Ref. [123] (see also [267]), and for Fe, as used in Ref. [201]

constructed, and especially on the assumptions its derivation from the all-electron calculations relies on, such as the extent to which the core and valence electron densities can be considered decoupled and the extent to which the core states can be considered insensitive to the variation of the valence shell. The former may create severe problems if only few electrons are in the valence shells (most notably in alkali metals), but these can be largely remedied by simply reintroducing an albeit smeared-out core density into the pseudopotential formulation (in the so-called nonlinear core-correction (NLCC) approximation [71]). The latter can be further improved by including some of the core shells in the valence shell (e.g. by constructing a pseudopotential for Ag^{11+} rather than one for Ag^{+}).

Comparisons with high-quality all-electron results have shown that for most applications the loss of accuracy due to pseudopotentials is acceptable. Note, however, that comparing the relative quality of all-electron and pseudopotential results is not always straightforward. In fact, because of the very different energy scale for core and valence electrons, it is difficult to achieve the same level of computational accuracy and convergence with all-electron schemes as that of standard pseudopotential methods.

Finally, we point out a problem of a practical nature. It is apparent from this discussion that pseudopotentials are introduced to smooth the spatial variations in the KS orbitals in order to reduce the Fourier representation (20) to a manageable size. However, the degree to which this goal can be achieved is not completely arbitrary: some valence states such as the 3d's in the first transition-metal series and, to a lesser extent, the 2p's of second-row elements (O and F in particular) are intrinsically highly localized close to the nucleus. This spatial localization is a crucial property and cannot be reduced drastically by any simple pseudopotential technique without sacrificing the description of the chemical properties associated with these elements. This problem is alleviated by 'augmentation' techniques, such as the one introduced in Ref. [72], which involve supplementing the pseudopotential part with a localized charge component that evolves under simplifying assumptions when the localized states are engaged in a bond. This technique has been successfully applied to many transition-metal systems, including clusters [73]. However, part of the simplicity and computational appeal of pseudopotential methods is certainly lost in these 'augmented' schemes.

3.3.4 Diagonalization Techniques

The present discussion of the computation of the KS orbitals is focused on plane-wave implementations. Even after reducing the number of electrons to those in the valence shell and/or the required cutoff by a suitable set of pseudopotentials, the size of the basis for cluster applications is of the order of $10^4 - 10^5$ plane-waves, and this can easily exceed the capability of present-day computers for a straightforward diagonalization of the KS Hamiltonian. Actually, full diagonalization of H_{KS} (Eq. (18)) is not required because we are interested only in the $\sim N$ lowest-energy eigenvalues and eigenvectors that correspond to the occupied states. The computation of selected eigenvalues and eigenvectors of a large matrix is a well-developed technique, and is important for a variety of applications, not only for electronic-structure computations. We arrived at the self-consistent diagonalization problem of Eq. (17) from the minimum principle expressed in Eq. (15). We now take a step backward, and consider the minimization formulation again. Let us assume that we have developed the *occupied* KS orbitals in a finite basis, as expressed in Eq. (19). By choosing a starting set of expansion coefficients $\{a\}$, we can compute the density ρ and, after substitution into the energy functional, express $E[\rho]$ as a function of the (many) variables $\{a\}$. Then, the minimization required to compute the electronic ground-state energy can be performed with any of the techniques devised for optimizing functions having many variables. We emphasize that only the *occupied* states enter the computation (the unoccupied ones do not contribute to the energy). Once the minimum has been found, the optimal coefficients $\{a\}$ satisfy the necessary condition (17). This identifies the lowest-energy eigenvectors and eigenvalues of H_{KS}.

The simplest optimization algorithm is the so-called method of steepest descent, which, starting from a suitable initial electronic configuration, progressively reduces the energy by moving in the multidimensional parameter space pointing in the direction opposite to $\partial E/\partial a$. It turns out that the derivative $\partial E/\partial a^*$ (from which $\partial E/\partial a$ is easily obtained) can be expressed in terms of known functions:

$$\frac{\partial E}{\partial a_j^{*(i)}} = f_i \int dr\, \phi_j^*(r)[H_{KS}\psi_i(r)], \tag{21}$$

where $\phi_j(r)$ is the basis function associated with the coefficient $a_j^{(i)}$. For plane-waves this reduces to

$$\frac{\partial E}{\partial a_j^{*(i)}} = f_i \int dr\, e^{-iGr}[H_{KS}\psi_i(r)], \tag{22}$$

and the multidimensional gradient $\partial E/\partial a^*$ can be computed by Fourier transforming $H_{KS}\psi_i(r)$.

Equation (22) is the basis for all the iterative diagonalization techniques used for large-scale plane-wave computations. Efficient implementations do not rely on steepest descent because of the well-known limitations of this algorithm [74]. More efficient schemes accumulate information on the second derivatives $\partial^2 E/\partial a\partial a$ by a feedback process that monitors the progress of the minimization. The first of these algorithms to be used in plane-wave DFT computations was the conjugate gradient method [75]. Other methods implemented in standard plane-wave codes include the Lanczos algorithm, the closely related Davidson method [76], and the direct inversion of the iterative subspace (DIIS) [77]. Standard theorems of numerical analysis show that these methods approach a quadratic rate of convergence close to the minimum, and thus provide a significant improvement over steepest descent, which is known to converge only linearly [74]. On the other hand, the same theorems state that these refined algorithms are expected to converge in a number of steps comparable to the number of independent variables. In the case of plane-wave electronic-structure computations, this number can easily reach 10^6 (as given by the number of expansion coefficients times the number of occupied states), and the corresponding minimization effort would be overwhelming. Fortunately, practical computations have shown that convergence within a satisfactory degree of accuracy is often already achieved after a number of minimization steps of the order of $\sim 10^2$, making the direct minimization approaches the most powerful algorithms for large-scale electronic-structure computations. Details and subtleties of these methods, with particular emphasis on electronic-structure applications, are discussed in depth in Ref. [78].

3.4 COMPUTATIONAL SCHEMES FOR DFT APPLICATIONS: STRUCTURE AND DYNAMICS

3.4.1 Computation of Forces on the Ions

Equation (15) for the ground-state electron density can be differentiated with respect to the atomic coordinates to compute the forces acting on the atoms

$$F_i = -\frac{\partial U}{\partial R_i} = -\int \frac{\delta E[\rho]}{\delta\rho(r)} \frac{\partial\rho(r)}{\partial R_i}\, dr - \int \frac{\partial V_{ext}(r)}{\partial R_i}\rho(r)\, dr. \tag{23}$$

The first term can be rewritten as

$$\int \frac{\delta E[\rho]}{\delta\rho(r)} \frac{\partial\rho(r)}{\partial R_i}\, dr = 2\mathrm{Re}\sum_j f_j \int \frac{\partial\psi_j^*(r)}{\partial R_i} \frac{\delta E[\rho]}{\delta\psi_j^*(r)}\psi_j(r)\, dr. \tag{24}$$

This term vanishes under either of the following conditions: (1) the functional $E[\rho]$ is stationary with respect to variations of the KS orbitals, i.e. the $\{\psi_i\}$ are indeed eigenstates of the KS Hamiltonian H_{KS}; (2) the derivative $\partial\psi_j^*(r)/\partial R_i$ is identically zero.

Owing to the incompleteness of the basis and to numerical errors, the first condition is never met with a sufficient degree of accuracy in practical computations. The second condition is less demanding: the $\partial\psi_j^*(r)/\partial R_i$ partial derivative has to vanish unless the basis functions depend explicitly on the atomic positions. This is the case, for instance, for localized basis sets centered on the atoms [79]. Plane-waves, on the other hand, do not depend on the atomic coordinates, and therefore the force contribution given in Eq. (24) vanishes. In this case the computation of the forces is particularly simple, as it is reduced to the second contribution of the right-hand side of Eq. (23):

$$F_I = -\sum_i f_i \left\langle \Psi_i \left| \frac{\partial V_{\text{ext}}[R_K]}{\partial R_I} \right| \Psi_i \right\rangle, \tag{25}$$

which is reminiscent of the Hellman–Feynman relation of standard quantum mechanics. This feature has provided, until now, a decisive advantage for plane-waves in MD applications of DFT.

3.4.2 Car–Parrinello Molecular Dynamics

At this point, all the ingredients are in place for introducing *ab initio* MD. If we assumes we have minimized the functional $E[\rho]$ for a given set $\{R_I\}$ of nuclear coordinates, we can write the Hamiltonian governing the time evolution of the atomic positions as

$$H = \sum_I \frac{P_I^2}{2M_I} + U[\{R_I\}|\{a_G^{(j)}\}]. \tag{26}$$

The corresponding equations of motion are

$$\dot{P}_I = -\nabla_{R_I} U[\{R_I\}|\{a_G^{(j)}\}], \tag{27}$$

$$\dot{R}_I = \frac{P_I}{M_I}, \tag{28}$$

with the *ab initio* forces given by Eq. (25). In principle, it is possible to use the standard methods of classical simulation to integrate the equations of motion [80], starting from a set of initial positions and momenta, and updating them by a sequence of short time increments. At each step the new energy and forces are required, therefore the functional $E[\rho]$ has to be minimized and the corresponding forces have to be recomputed. This straightforward method of joining DFT and MD has recently been used for cluster simulations [81], but has a number of drawbacks: in order to ensure the conservation of energy and momenta, the minimization required at each step has to be very accurate, implying a large computational effort.

The wide acceptance and successful application of DFT–MD relies on a different point of view, described in the seminal paper by Car and Parrinello [8], see also [82]. In Eq. (23), we made the dependence of U on the expansion coefficients $\{a\}$ explicit for the KS orbitals. Strictly speaking, this notation is redundant, because these coefficients are those minimizing the functional $E[\rho]$ and, therefore, the only independent variables are the atomic positions. The notation itself, however, suggests that one considers U the potential-energy surface for the dynamical system as having both $\{R\}$ and $\{a\}$ as independent variables. Thus one can write the Hamiltonian for this extended system as

$$\hat{H}' = \sum_I \frac{P^2}{2M_I} + U[\{R_I\}|\{a_G^{(j)}\}] + \frac{\mu}{2} \sum_{i,G} \dot{a}_G^{(i)2} + \sum_{i,j} \Lambda_{i,j}(\langle \Psi_i | \Psi_j \rangle - \delta_{ij}), \qquad (29)$$

which yields the equations of motion

$$\mu \ddot{a}_G^{(i)} = -\frac{\partial U[\{R_I\}|\{a_G^{(j)}\}]}{\partial a_G^{*(i)}} + \sum_j \Lambda_{i,j} \frac{\partial \langle \Psi_i | \Psi_j \rangle}{\partial a_G^{*(i)}}, \qquad (30)$$

$$M_I \ddot{R}_I = -\nabla_{R_I} U[\{R_I\}|\{a_G^{(j)}\}]. \qquad (31)$$

The matrix $\Lambda_{i,j}$ is a generalization of the Lagrange multipliers entering Eq. (17).

The evolution of $\{R\}$ described by these equations is close to the one given by the original equations (27) and (28) if and only if the $\{\dot{a}\}$ are small and $\{a\}$ are close to those minimizing the $E[\rho]$ functional. One way to achieve this is to choose a low value for the mass parameter μ such that the time scale for the motion of the $\{a\}$ variables is much faster than that of the atomic coordinates. Then, it is possible to achieve a substantial adiabatic separation in the motion of the atomic and electronic degrees of freedom. In other words, by starting from well-converged DFT configurations and evolving the system in time according to Eqs (30) and (31), the transfer of kinetic energy from the ionic to the electronic degrees of freedom will be negligible over a time encompassing many vibrational periods of the atoms.

A few remarks are in order. First of all, although the kinetic energy associated with the electronic degrees of freedom can be kept low, it cannot be reduced to zero because the coefficients $\{a\}$ have to change in order to follow the time evolution of the ionic positions. The rate of kinetic energy transfer is determined by the electronic structure (the energy separation of occupied and unoccupied states is an important parameter) and by the μ/M ratio [83]: lower values of μ/M correspond to a better separation of the ionic and electronic time scales, and therefore to a better adiabatic evolution of the atomic positions. On the other hand, the time scale for the motion of the $\{a\}$ coefficient cannot be made too short, as otherwise the integration of the equations of motion would become prohibitively expensive. It turns out that in many cases of practical interest it is possible to achieve a good compromise between these two contradictory requirements, and to perform adiabatic dynamics for the ionic coordinates for times that are sufficient to sample the phase space of the system. In these favorable cases, the advantage of the adiabatic dynamics with respect to the step-by-step minimization is apparent: because of the second-order time derivative in Newton's equations, the coefficients $\{a\}$ are attracted by the low-energy regions of the phase space, with deviations from the minimum that have an oscillatory character rather than the exponential drift typical of all direct minimization methods [84]. Note also that the adiabatic dynamics is bound to fail when, during the time evolution, the energies of the highest occupied (HOMO) and lowest unoccupied (LUMO) molecular orbitals do cross. In this case the energy transfer between the ionic and electronic degrees of freedom cannot be neglected, both for reasons of principle (the time scales for the ionic and electronic evolution become comparable) and in practice: the electronic structure tends to be trapped in the metastable minimum having the wrong

HOMO and LUMO occupation, and thus leaves the Born–Oppenheimer surface. Needless to say, this problem is especially relevant for metal clusters, because these systems have a comparatively smaller HOMO–LUMO gap than other clusters or molecules. Nevertheless, for finite systems a vanishing HOMO–LUMO gap is the exception and not the rule, even for metallic elements, and experience based on many computations has shown that extensive adiabatic simulations can be performed for metal clusters.

A few details on the basic algorithms used to integrate the equations of motion are collected in Appendix B.

3.4.3 Optimization Strategies

The first and highly nontrivial problem in the computational analysis of clusters is to determine the ground-state structure as a function of size. Although it is difficult to make precise and rigorous statements that are valid for all but the simplest cases [85], this problem is believed to be in the NP (non-polynomial) class, i.e. the effort required for an exhaustive search of the ground-state geometry of a cluster of n atoms grows faster than any polynomial in n. On a more intuitive level, it is possible to grasp the difficulty of identifying the ground-state geometry by looking at the enumeration of the low-energy isomers of small clusters bound by pair potentials, reported in Ref. [86]. We anticipate that there is no general solution to this problem, although intriguing approaches have been proposed and used successfully for specific cases.

3.4.3.1 Simulated Annealing

The first nontrivial and still widely used algorithm is known as simulated annealing (SA) [87], and requires MD or Monte Carlo as its basic tool. The origin of the method is the similarity between combinatorial analysis and statistical mechanics. From a heuristic point of view, it is based on the idea of bringing the system into the relevant portion of the phase space by high-temperature simulation, and then cooling the cluster until it is trapped in a minimum of the potential energy. The MD annealing is often performed by resorting to Langevin dynamics [88], with a friction term and a stochastic external force that progressively dissipate the kinetic energy of the atoms. There is no proof that, at any finite annealing rate, the process will end up in the absolute minimum and thereby avoid the multitude of local metastable minima. However, by mimicking a thermal process, the SA algorithm allows a wide and efficient exploration of the phase space, with statistical weight biased towards the low-energy structures. The effectiveness of the method is limited by practical considerations: even for simple metal clusters, the temperature required to cross the energy barriers among minima is of the order of a few hundred degrees K. On the other hand, SA typically covers a few nanoseconds for classical simulations or, at most, a few picoseconds for *ab initio* MD. As a result, the annealing rates achieved in typical computations range between 10^{11} and 10^{14} K/s, which is several orders of magnitude faster than any quench achievable in experiments, even for small clusters produced in a supersonic expansion. Thus, it is not surprising to find amorphous structures for medium-sized clusters, whose existence is questionable [89]. Nevertheless, practical computations show that, at least for small clusters ($n \leq 50$) composed of a single metallic element, SA is able to produce low-energy structures efficiently, often with unexpected geometries. The situation might be different for alloy clusters because the annealing time (especially for *ab initio* simulations) is probably too short to allow a significant interdiffusion of the various atomic species.

3.4.3.2 Genetic Algorithms

In recent years, the optimization strategy known as 'genetic algorithm' (GA) has received increasing attention from the cluster physics community, following a pioneering application to the groundstate determination of medium-sized carbon clusters [90]. Like SA, GA attempts to mimic an optimization strategy successfully exploited by nature, i.e. the process of random mutations and selection in a self-reproducing, stationary population. Starting from an initial population (assumed, for instance, to be in thermal equilibrium) at time τ_0, the method specifies a mapping rule to associate an offspring to each pair (or larger group) of elements in such a way that a random mixture of the parents' features is passed on to the next generation. The rule includes a mechanism for 'errors', which play the role of mutations. The survival and reproductive rates are determined on the basis of a 'fitness' factor, measured, for instance, by the total energy of the clusters: low-energy elements survive and reproduce, whereas high-energy ones are statistically eliminated from the population. The total energy of the clusters is evaluated after a short local optimization of the structure, usually performed by quenched MD. The significant advantage of this method as compared with SA is that progress from one generation to the next is not based on the simulation of the time evolution of the atomic coordinates and, therefore, the long time drift is decoupled from the short time scales of the atomic motion that severely limits SA. The major limitations to GA again come from computational constraints: unbiased progress towards the global minimum of the potential energy can be achieved with reasonable certainty if and only if a large number of generations can be observed for a sizable population. Moreover, the local optimization stage between generations is a crucial step towards guiding the population drift towards the low-energy portion of the configuration space, and might therefore require considerable effort. As a consequence GA is better suited at present for searching the ground-state geometry of classical or semiempirical (such as tight-binding) potential models than for *ab initio* applications, although this is likely to change in the future.

3.4.4 Additional Remarks

The last algorithm we shall mention here is the application of artificial-intelligence methods to structural optimization. In this approach, the transformation from a generic structure to an optimized one is defined in terms of operators (neural networks) that have a highly nonlinear dependence on a set of parametric variables. The transforming operators are shaped in a learning stage, in which examples of useful transformations are used to set the parametric variables. After a sufficient number of examples has been assimilated, the machinery is ready for applications. Successes and failures are still used to improve the effectiveness of the method progressively. Encouraging results have been obtained in optimizing the geometry of organic and biological molecules [91]. To the best of our knowledge, no application to metal clusters has so far been reported.

We emphasize that in addition to a global optimization strategy, the precise determination of the structure requires an efficient and reliable local optimization algorithm. MD does not provide such a tool because it cannot optimize efficiently the degrees of freedom associated with soft restoring forces (and therefore long time relaxations), which are often present in clusters. The standard methods developed by the quantum-chemistry community (conjugate gradient and other quasi-Newton algorithms) must be adopted for the final structural refinement after the minimum (or minima) of interest has been located.

3.5 COMPUTATION OF PROPERTIES

3.5.1 Vibrational Analysis

The simplest spectroscopic property that can be computed by DFT is the vibrational density of states, which measures the response of the system to a periodic external perturbation coupled to the atomic (nuclear) coordinates. At $T = 0$ K this property is fully described by eigenvalues and eigenvectors of the dynamical matrix

$$D_{\alpha\beta}^{IJ}(\boldsymbol{R}_1^0, \boldsymbol{R}_2^0, \ldots, \boldsymbol{R}_N^0) = \frac{\partial^2 E[\boldsymbol{R}_1^0, \boldsymbol{R}_2^0, \ldots, \boldsymbol{R}_N^0]}{\partial \boldsymbol{R}_I^0 \partial \boldsymbol{R}_J^0}, \tag{32}$$

where α, β are Cartesian indices, I and J label the atoms in the system, and $[\boldsymbol{R}_1^0, \boldsymbol{R}_2^0, \ldots, \boldsymbol{R}_N^0]$ correspond to the ground-state geometry (or to a metastable geometry).

The computation and diagonalization of the dynamical matrix is straightforward, and does not require any extension of the basic theory. Second derivatives of the total energy with respect to the nuclear coordinates can be computed by perturbation theory or by finite differences approximations.

Density functional perturbation theory, as described in Ref. [92], has been applied to calculate phonon spectra of solids. It can easily be applied to clusters as well, the only limitation being that in the pseudopotential–plane-wave formalism, the computational cost is high, whereas the implementations within a localized basis formalism can be very involved.

Another simple method is the finite difference approach (implemented in most *ab initio* packages). Its accuracy is limited by the fact that the energy differences required for the determination of $\partial^2 E / \partial \boldsymbol{R}_I^0 \partial \boldsymbol{R}_J^0$ are small quantities compared to the total energy itself, and errors from rounding off may become large. As a consequence of these errors and their propagation in the diagonalization step, the low-frequency modes are determined with lower accuracy than the high-frequency modes. The reliability range of this simple method depends on the characteristic frequency scale of each system, and has to be determined by careful tests with different values of the atomic displacements.

At $T \neq 0$ the sharp lines corresponding to the harmonic modes are broadened by anharmonic effects until, at high temperature, the simple relationship between vibrational density of states and dynamical matrix is lost. In this regime, and especially for large aggregates, MD is the most suitable tool to compute the vibrational spectrum. Standard linear response theory within classical statistical mechanics shows that the spectrum $f(\omega)$ is given by the Fourier transform of the velocity–velocity autocorrelation function

$$f(\omega) = \sum_I \int_0^\infty \frac{\langle \boldsymbol{R}_I(t)\boldsymbol{R}_I(0)\rangle}{\langle \boldsymbol{R}_I(0)\boldsymbol{R}_I(0)\rangle} \cos(\omega t)\, \mathrm{d}t, \tag{33}$$

where $\langle \ldots \rangle$ indicates the statistical average over many trajectories, or, as is usually done, over different choices of the time origin along a unique MD trajectory [80]. This relation is deceptively simple: in practice it is often difficult to compute the velocity–velocity correlation function with sufficient statistical accuracy, especially in the short runs afford-able with *ab initio* MD. Methods to enhance the signal-to-noise ratio have been proposed and tested in practical computations [93]. The advantage they provide as compared with the straightforward evaluation of Eq. (33) is significant but not decisive.

It is important to recall that in CP molecular dynamics the mass parameter μ associated with the electronic variables introduces a source of inaccuracy for the determination of the frequencies: depending on its value (which usually ranges between 100 and 1000 electron mass units) it may lead to a non-negligible effect (typically a softening). Hence, one should estimate and subtract it before a precise comparison with experimental data can be made. In practice, this is rarely done [93].

Unfortunately, the usefulness of this analysis of vibrational properties is limited by the scarcity of experimental data for metal clusters. Beyond dimers, experimental results are rare, and concern almost exclusively clusters in frozen matrices.

3.5.2 Optical Properties

The photoabsorption spectrum $\alpha(\omega)$ of a cluster measures the cross-section for elec-tronic excitations induced by an external electromagnetic field oscillating at frequency ω. Experimental measurements of $\alpha(\omega)$ of free clusters in a beam have been reported, most notably for size-selected alkali-metal clusters [4]. Data for size-selected silver aggregates are also available, both for free clusters and for clusters in a frozen argon matrix [94]. The experimental results for the very small species (dimers and trimers) display the variety of excitations that are characteristic of molecular spectra. Beyond these sizes, the spectra are dominated by collective modes, precursors of plasma excitations in the metal. This distinction provides a clear indication of which theoretical method is best suited to analyze the experimental data: for the very small systems, standard chemical approaches are required (CI, coupled clusters), whereas for larger aggregates the many-body pertur-bation methods developed by the solid-state community provide a computationally more appealing alternative. We briefly sketch two of these approaches, which can be adapted to a DFT framework: (1) the random phase approximation (RPA) of Bohm and Pines [95] and the closely related time-dependent density functional theory (TD–DFT) [96], and (2) the GW method of Hedin and Lundqvist [97].

The linear response of a system (initially in the ground state $|0\rangle$) to a periodic one-body perturbation $\hat{Q}(\omega) = \hat{Q}[\exp(i\omega t) + \text{c.c.}]$ is fully described by the complex polarizability $\alpha(\omega)$, given by the well-known time-dependent perturbation-theory expression

$$\alpha(\omega) = -\sum_k |\langle k|\boldsymbol{Q}|0\rangle|^2 \left[\frac{1}{\hbar\omega - E_k + i\delta} - \frac{1}{\hbar\omega + E_k + i\delta} \right], \qquad (34)$$

where $i\delta$ is a small imaginary component, and E_k and $|k\rangle$ are the energy and the eigenstate of the kth excited state. The same complex polarizability can be expressed in terms of the particle–hole Green's function $G(\omega)$ [98]:

$$\alpha(\omega) = \int \hat{Q}(\boldsymbol{r}) G(\boldsymbol{r}, \boldsymbol{r}'; \omega) \hat{Q}(\boldsymbol{r}') \, d\boldsymbol{r} \, d\boldsymbol{r}' \qquad (35)$$

The connection with the experimentally measurable spectrum $S(\omega)$ is established by the relation

$$S(\omega) = \sum_k |\langle 0|\hat{Q}|k\rangle|^2 \delta(\hbar\omega - E_{k0}) = \frac{1}{\pi}\text{Im } \alpha(\omega) \quad (\omega > 0), \tag{36}$$

that is, the excitation energy is given by the poles of $\alpha(\omega)$ on the real, positive axis, whereas the corresponding residues measure the oscillator strength.

Needless to say, the exact $G(\omega)$ for the interacting system is not known, except for formal relations that cannot be applied in practice. Time-dependent DFT and RPA define a simple approximation for this Green's function, thus providing a scheme that is suitable for computations. The starting point is the noninteracting particle–hole Green's function $G^{(0)}$, defined by

$$G^{(0)}(\boldsymbol{r}, \boldsymbol{r}'; \omega) = \sum_i \sum_m \phi_i^*(\boldsymbol{r})\phi_m(\boldsymbol{r}')$$

$$\times \left[\frac{1}{\varepsilon_m - \varepsilon_i - \omega - i\delta} + \frac{1}{\varepsilon_m - \varepsilon_i + \omega + i\delta}\right] \phi_m^*(\boldsymbol{r}')\phi_i(\boldsymbol{r}), \tag{37}$$

where the sum over i is over the occupied states and the sum over m is over the unoccupied states. The terms ϕ and ε are independent electron eigenstates and eigenvalues, respectively. Finally, Green's function $G(\omega)$ is defined by the integral equation

$$G(\boldsymbol{r}, \boldsymbol{r}'; \omega) = G^0(\boldsymbol{r}, \boldsymbol{r}'; \omega) + \int G^0(\boldsymbol{r}, \boldsymbol{r}_1; \omega)V_{\text{eh}}(\boldsymbol{r}_1 - \boldsymbol{r}_2)G(\boldsymbol{r}_2, \boldsymbol{r}'; \omega)\,d\boldsymbol{r}_1\,d\boldsymbol{r}_2, \tag{38}$$

where the effective interaction $V_{\text{eh}}(\boldsymbol{r}, \boldsymbol{r}')$ is given by

$$V_{\text{eh}}(\boldsymbol{r}, \boldsymbol{r}') = \frac{e^2}{|\boldsymbol{r} - \boldsymbol{r}'|} + \frac{\delta^2 U_{\text{XC}}}{\delta\rho^2}\delta(\boldsymbol{r}, \boldsymbol{r}'). \tag{39}$$

The scheme can be cast in a somewhat simpler, but less transparent, matrix form by using the noninteracting orbitals as a basis for the integral relations listed above. First, one has to write the (many-electron) state $|n\rangle$ as a linear superposition of states obtained by single excitations above the ground state $|0\rangle$:

$$|n\rangle = \sum_{ph} (x_n^{ph} a_p^\dagger a_h - y_n^{ph} a_h^\dagger a_p)|0\rangle, \tag{40}$$

where $a^\dagger$ and a are creation and annihilation operators, respectively, and x_n, y_n are the coefficients (to be determined) of the linear superposition. The indexes h and p refer to occupied and unoccupied states, respectively. Then, the RPA equations can be summarized as

$$\begin{pmatrix} \mathbf{A} & \mathbf{B} \\ \mathbf{B}^* & \mathbf{A}^* \end{pmatrix} \begin{pmatrix} \mathbf{X}_n \\ \mathbf{Y}_n \end{pmatrix} = E_n \begin{pmatrix} \mathbf{X}_n \\ -\mathbf{Y}_n \end{pmatrix}, \tag{41}$$

where $\mathbf{A}$ and $\mathbf{B}$ are defined by

$$\mathbf{A} = \delta_{pp'}\delta_{hh'}(\varepsilon_p - \varepsilon_h) + \langle h'p|V|p'h\rangle, \tag{42}$$

$$\mathbf{B} = \langle pp'|V|hh'\rangle. \tag{43}$$

In principle, these relations can be used within a DFT computational scheme in a straightforward way. One has to (i) compute occupied and unoccupied (up to a given cutoff energy) KS orbitals and eigenvalues [99]; (ii) build the effective potential V and noninteracting Green's function G^0; (iii) solve the integral Eqs (38) and (39) for G, or, equivalently, solve the matrix equation (41), and, finally, (iv) compute $S(\omega)$ via Eqs. (35) and (36). Clearly, this path is computationally cumbersome, and, as a result, these equations have been solved mainly for the jellium model, starting with the early studies of Refs [2] and [100]. The results of these and other jellium computations are reviewed in Refs [5] and [15]. It is apparent that for medium- and large-size alkali-metal clusters, RPA contains all the qualitative features displayed by the experimental spectra. A quantitative comparison, however, cannot be performed at the jellium model level. Recently, attempts to include the effect of the atomic structure via a realistic pseudopotential model have been made with encouraging results (see Section 3.6) [101]. However, even this computation has to resort to drastic approximations on the effective interaction V (assumed to be spherically symmetric), which are likely to be acceptable in the case of alkali metals but cannot be applied to the other systems of interest.

More importantly, the basic approximation (40) becomes inadequate to represent the many-body effects in the excitation spectrum as soon as correlation becomes important, as is the case in transition metals. Even for the case of alkalis, for which the RPA results turn out to be fairly good, the application of RPA (which was developed in the HF framework) on top of DFT computations presents some basic problems and subtleties, which are discussed, for instance, in Ref. [5].

To overcome these limitations, methods beyond RPA are required. One scheme that has attracted interest in recent years is the so-called GW method [97]. A detailed description of this formalism is beyond the scope of our review, and we refer the reader to [39] for a recent and exhaustive discussion. Here we summarize only the main points. First of all, the central quantity in GW is again Green's function $G(\boldsymbol{r}, \boldsymbol{r}', \omega)$, which is given by

$$G(\boldsymbol{r}, \boldsymbol{r}', \omega) = G_0(\boldsymbol{r}, \boldsymbol{r}', \omega) + \int G_0(\boldsymbol{r}, \boldsymbol{r}'', \omega) \sum (\boldsymbol{r}'', \boldsymbol{r}''', \omega) G(\boldsymbol{r}''', \boldsymbol{r}', \omega), \qquad (44)$$

where Σ is the self-energy operator and G_0 is again the noninteracting Green's function. In turn, the self-energy is expressed in terms of the screened Coulomb potential W and a vertex operator Λ:

$$\sum (\boldsymbol{r}, \boldsymbol{r}', \omega) = \mathrm{i} \int G(\boldsymbol{r}, \boldsymbol{r}'', \omega) W(\boldsymbol{r}_3, \boldsymbol{r}_4, \omega) \Lambda(\boldsymbol{r}_3, \boldsymbol{r}', \boldsymbol{r}_4, \omega) \, \mathrm{d}\boldsymbol{r}_3 \, \mathrm{d}\boldsymbol{r}_4. \qquad (45)$$

Finally, the screened interaction is computed from the bare one by introducing the dielectric function $\varepsilon(\boldsymbol{r}, \boldsymbol{r}', \omega)$, and the vertex operator is approximated by some simple expression, usually replaced by the unit operator. From Eqs (38) and (39), the relation with RPA is apparent, the latter representing, in fact, a simple approximation to the GW relations. It is also apparent that GW is computationally much more demanding than RPA, especially if the set of equations (44) and (45) is solved self-consistently. In practice, all computations within the GW formalism resort to additional approximations and drastic simplifications. Concerning metal clusters, applications have been presented in Ref. [102] for the spherical jellium, and in Ref. [103] for a pseudopotential model of small sodium clusters. In both cases, the computation is based on noninteracting orbitals and energies provided by KS eigenstates and eigenvalues computed within LDA.

3.5.3 Static Dipole Polarizability

As a byproduct of the computation of photoabsorption spectra, it is possible to compute the static dipole polarizability α, given by the well-known sum rule

$$\alpha = \int_0^\infty d\omega S(\omega)/\omega, \tag{46}$$

which has been used extensively in spherical jellium computations. Recently, it has also been evaluated in an RPA-pseudopotential model [101]. In Eq. (46), we assumed the system to be spherical. In the general case, the linear response of a system to an external static electric field ε is given by

$$P = \alpha\varepsilon, \tag{47}$$

where P is the induced dipole moment and α is a matrix whose element $\alpha_{i,j}$, $(i, j = x, y, z)$ measures the dipole induced in the Cartesian direction i by an applied field oriented along j. The average of the three eigenvalues of α is the quantity to be compared with the experimental polarizability parameter α^{exp}, which in most cases is the average polarizability of a population of clusters with random orientation.

A precise theoretical and experimental determination of polarizability would provide an important probe of the electronic structure of clusters, as α is very sensitive to the presence of low-energy optical excitations. Accurate experimental data for a wide range of size-selected clusters are available only for sodium, potassium [104] and aluminum [105, 106]. Theoretical predictions based on DFT and realistic models do not cover even this limited sample of experimental data. The reason for this scarcity is that the evaluation of polarizability by the sum rule (46) requires the preliminary computation of $S(\omega)$, which, with the exception of Ref. [101], is available only for idealized models. Two additional routes exist to the evaluation of α, in close analogy with the computation of vibrational properties: static second-order perturbation theory and finite differences [107]. Again, the first approach has been used exclusively for the spherical jellium model. In this case, the equations to be solved are very similar to those introduced in Ref. [108] for the computation of atomic polarizabilities. Applications of this formalism to simple metal clusters are reported, for instance, in Ref. [109].

Finite differences, finally, is the simplest approach: the total energy and dipole moment are computed as a function of the strength of the external electric field. Then, a quadratic fit to the energy or, equivalently, a linear fit to the dipole moment as a function of the external field amplitude provide the polarizability α. This direct approach has been used for sodium clusters in Refs [110] (LSDA) and [111] (LSDA and GGA), and for aluminum in Ref. [106] (LSDA).

3.5.4 Photoelectron Spectroscopy

In recent years, photoelectron spectroscopy has been extensively used to probe the electronic structure of clusters. From an intuitive point of view, the technique measures the binding energy of electrons in the cluster. More formally, it probes the energy dependence of the response function connecting states with N (initial state) and $(N - 1)$ (final state) electrons.

To bring the electron binding energy into the range of visible and ultraviolet light sources, most of the experimental measurements have been performed on negatively charged clusters. In principle, the methods described in Section 3.5.2 for the calculation of optical spectra also cover photoelectron spectroscopy. In practice, no application to realistic models of clusters exists, and the analysis of experimental spectra has mainly been based on simplified (often oversimplified) assumptions.

Many interpretations of experimental measurements have been limited to the comparison with KS eigenvalues, based on the fact that for extended solids the KS band structure sometimes provides the correct qualitative features of photoemission experiments. Unfortunately, the deviation of the KS eigenvalues from physical excitation energies becomes stronger in going from extended to finite systems, because the N-dependence of the exchange–correlation potential becomes crucial, and because some errors implicit in standard approximations (like the self-interaction in LDA) are amplified by the reduced extension of the KS orbitals.

It is also important to mention that in finite systems and clusters in particular, the effect of ionic relaxation upon the removal of one electron can be sizable. This contribution is obviously neglected in any computation of the electron removal energies based on the KS eigenvalues determined for the starting ionic geometry and has to be estimated *a posteriori* to obtain a meaningful comparison with experiments.

The second major route to interpret photoelectron spectra has been via the computation of total energy differences. As noted in Section 3.2, this is fully justified for selected states. The lowest-energy peak corresponds, for instance, to the IP (if the initial state is neutral) or to the EA (for anionic species). In both cases, the corresponding excitation energy can be computed as a difference of ground-state energies, and this is well within the reach of DFT methods.

In general, energies connecting two states that are *both* the lowest-energy configuration for a given symmetry can be computed within DFT as a difference of ground-state energies. The majority of the peaks observed in photoelectron spectroscopy, however, do not fall into this category. Despite this basic limitation, the analysis of the entire photoelectron spectrum on the basis of total energy differences has been attempted, most notably in Ref. [112]. The intuitive point of view underlying these computations is that removing (in turn) one electron from each of the KS states and evaluating the total energy after allowing for limited electron relaxation corresponds to the experimental extraction of electrons from different bonding states. Unfortunately, this simple picture cannot be based on a rigorous foundation, although it probably captures many qualitative features of the spectra and avoids the gross underestimation of removal energies as (incorrectly) estimated by the KS eigenvalues.

With minor changes, the same considerations apply for the so-called transition-state method first introduced by Slater to compute excitation energies and applied to the analysis of cluster photoelectron spectra, e.g. in Ref. [113]. This method is based on the exact relation connecting the KS eigenvalues to the derivative of the total energy with respect to the occupation numbers [114]:

$$\frac{\partial E_{\text{tot}}}{\partial f_i} = \mu_i[\{f_j\}, j = 1, N], \tag{48}$$

where we made explicit the dependence of the eigenvalues on the occupation numbers. This relation can be used to evaluate the total energy difference upon switching off the electron in the ith KS orbital:

$$E_{\text{tot}}[f_i = 0] - E_{\text{tot}}[f_i = 1] = -\int_0^1 \mathrm{d}f_i \frac{\partial E_{\text{tot}}}{\partial f_i} = -\int_0^1 \mathrm{d}f_i \mu_i[\{f_j\}, j = 1, N]. \tag{49}$$

We emphasize again that the dependence of μ_i on the occupation number is crucial. Neglecting this dependence, Eq. (49) is equivalent to evaluating excitation energies from the KS eigenvalues. A much better approximation is given by

$$E_{\text{tot}}[f_i = 0] - E_{\text{tot}}[f_i = 1] = -\mu_i[f_i = 0.5]. \tag{50}$$

The state with occupation $f_i = 0.5$ is the so-called transition state, which gives this method its name. Generalizations approximating the integral over df_i by a discrete sum of $\mu_i[\{f_j\}, j = 1, N]$ over several intermediate occupations between 1 and 0 have also been proposed [114, 115]. In most cases, however, the advantage with respect to the simple relation (50) is minor. The weak point of the method is, again, the identification of the removal energy with $E_{\text{tot}}[f_i = 0] - E_{\text{tot}}[f_i = 1]$, which in most cases is not rigorously justified.

3.5.5 Electron Spin Resonance

Electron spin resonance (ESR) has sometimes been used to characterize electronic and structural properties of transition-metal clusters embedded in frozen rare-gas matrices. Neglecting the spin−orbit coupling, the interaction between electrons and the nuclear magnetic moment of each atom in the cluster can be expressed by the simple Hamiltonian [116, 117]:

$$H_{\text{Ie}} = \mathbf{S}^{\mathbf{I}} \mathbf{A} \mathbf{S}^{\mathbf{e}} \tag{51}$$

where $\mathbf{S}^{\mathbf{I}}$ and $\mathbf{S}^{\mathbf{e}}$ are the nuclear and electron spin operators, respectively. The operator $\mathbf{A}$ is the so-called hyperfine tensor, whose matrix elements are given by

$$A_{ij} = a\delta_{ij} + b_{ij}, \tag{52}$$

with

$$a = \frac{8\pi}{3} g_e \mu_e g_I \mu_I \rho_s(\mathbf{R}), \tag{53}$$

$$b_{ij} = g_e \mu_e g_I \mu_I \int \rho_s(\mathbf{r}) \frac{3 r_i r_j - \delta_{ij} r^2}{r^5} \, d\mathbf{r}. \tag{54}$$

The first term results from the Fermi contact interaction, while the second represents the long-range dipole−dipole interaction. In the equations above, g_e is the free-electron g factor, μ_e the Bohr magneton, g_I the nuclear gyromagnetic ratio, and μ_I the nuclear moment. Moreover, the nucleus is located at position $\mathbf{R}$, and the vector $\mathbf{r}$ has the nuclear position as its origin. Finally, $\rho_s(\mathbf{r}) = \rho_+(\mathbf{r}) - \rho_-(\mathbf{r})$ is the electron spin density. The only nontrivial input into these equations is precisely this last quantity, i.e. $\rho_s(\mathbf{r})$, which can be computed in the LSDA or another DFT approximation. The resulting Hamiltonian can be used to interpret the hyperfine structure measured in experiments. A recent application to metal clusters is reported in Ref. [118].

The application of these relations within a pseudopotential approach is confronted with the problem of evaluating the true spin density from the valence pseudocharge, which approaches the former only far from the nucleus. Unfortunately, as is apparent from the equations above, a precise determination of the spin density in the core region is crucial for an accurate determination of hyperfine parameters. Methods to reconstruct the all-electron spin density from its pseudovalence counterpart have been proposed and applied

to calculate hyperfine parameters in solid-state problems [119]. No cluster applications
have so far been reported. However, recent detailed tests using Martins−Trouiller (MT)
pseudopotentials on aromatic molecules, small aluminum clusters, and fullerene derivatives are not encouraging, because they show a significant dependence of the results on
the pseudopotential itself [120].

3.6 ALKALI METALS

Clusters of alkali metals and especially of sodium are the most studied of all. From
the theoretical point of view, sodium is the one most amenable to treatments with
simple models. The free-electron behavior known for the bulk phase has suggested
that jellium-like models could also be suitable for small-size aggregates. By means of
these models, in fact, a large variety of measurable properties have been calculated. This
in turn has allowed the approximations used to be tested at several levels [121]. Two
comprehensive and very instructive reviews have been dedicated to both experimental
and theoretical approaches to simple metal clusters with an emphasis on phenomenological aspects and jellium or jellium-derived models [4, 5]. Here we shall report on
DFT calculations that go beyond the assumption of a homogeneous, positively charged
background.

Early LSDA static pseudopotential approaches to sodium microclusters date back approximately 20 years [122], see Appendix C. It would be misleading to consider LDA
calculations as the natural extension of jellium models. However, the global validity
of the latter cannot but anticipate the success of the former. Clearly, these should also
clarify the role of the atomic structure in determining the electronic behavior of the clusters and the extent to which the inhomogeneity of the electron distribution is reflected
in the measurable properties. Many structural determinations are by now available for
the smaller aggregates, made at different levels of approximation and of accuracy (e.g.
[110, 111], see Appendix C). The most extensive investigation of sodium clusters so
far is the LDA−CP study of Ref. [123] (see Appendix C), which makes use of all
the features of the CP method. Namely, it uses dynamical SA to explore the potential-
energy surface, MD to simulate clusters at different temperatures, and detailed analysis
of the one-electron properties, which can be compared to the predictions of jellium-based
models.

Na_n clusters were considered for $n = 6, 8, 9, 10, 13, 18$ and 20. Several new insights
have been gained from this study. The extensive investigation of the potential-energy
surface for each of these clusters has allowed the identification of the pentagonal motif
as the common structural characteristic of sodium aggregates (see Figure 3.3), which
starts from Na_6, where the 3D pentagonal pyramid is still quasidegenerate with the 2D
compact arrangement shown on the right-hand side of Figure 3.3(a), and progressively
becomes dominant. Interestingly, the low-energy isomers are often those expected for
Lennard-Jones aggregates and/or assemblies of spheres, in accordance with the negligible
directionality of the bonding. Two structural characteristics are reported in Figure 3.4 as
a function of size: the average interatomic distance (Figure 3.4(a)), and the eccentricity η
defined as $\eta = 1 - I_{min}/I_{av}$ (Figure 3.4(b)), with I_{min} and I_{av} the minimum and average
values of the moments of inertia. The average distance appears to converge rapidly to the
bulk. The values for η are compared to those predicted by the ellipsoidal model [124].
Although the individual values are different, the trend is clearly the same. In particular,

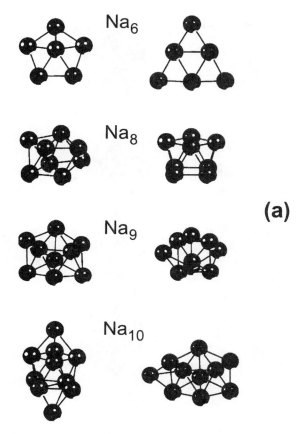

Figure 3.3a Low-energy isomers of Na clusters. Reprinted with permission from [123]. Copyright 1991 American Institute of Physics

the magic-number (MN) clusters of the spherical shell model ($n = 8, 18$ and 20) do not 'choose' the most spherical geometry compatible with their size.

We emphasize that, in spite of the often-claimed general agreement between the results of the SCF + CI approach (SCF: self-consistent field) [125] and these LDA calculations, there are remarkable differences, especially for the larger sizes, such as $n = 20$, for which SCF + CI computations have been performed starting from the geometries discovered in [123]. In fact, the compact geometries in Figure 3.3(b) became higher in energy than the open or more symmetric structures that LDA clearly disfavors. This discrepancy is probably related to the lack of convergence of the SCF + CI treatment of the electronic correlation, which is not sufficient to overcome the tendency of the HF approach to favor open atomic configurations.

Simulations at finite temperatures up to ~600 K reveal that the average eccentricity increases with increasing temperature (see Figures 3.6(a) and 3.7(a) for MN clusters). This disproves various conjectures according to which sodium clusters at high temperatures resemble spherical liquid droplets.

The discussion of the one-electron properties focused on the comparison of the Kohn–Sham orbital structure with the jellium-shell model. This was done by analyzing the

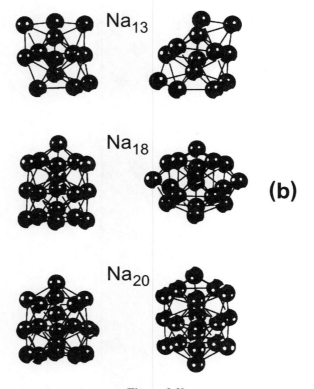

Figure 3.3b

cluster orbitals $\psi_i(r)$ in spherical harmonics centered at the cluster center,

$$\psi_i(\boldsymbol{r}) = \sum_{l,m} \phi^i_{lm}(r) Y_{lm}(\Omega), \tag{55}$$

which directly provides us with the weight ω^i_l of the l-component in the expansion of the ith wave function,

$$\omega^i_l = \sum_m \int [\phi^i_{lm}(r)]^2 r^2 \, dr, \tag{56}$$

and also the definition of the parameter E_L:

$$E_L \equiv 1 - \left(\sum_{i_{\text{occ}}} \sum_{l=0}^{l=L} \omega^i_l \right) N \tag{57}$$

that distinguishes between open and closed ($E_L = 0$) spherical shell configurations.

Figure 3.5(b) illustrates the radial part $\phi_{l0}(r)$ of the dominant components of the occupied wave functions for the MN Na_{20} in one of the low-energy isomers of Figure 3.3(b). The ordering is 1s, 1p, 1d, 2s, as expected, but the value of $E_{L=2}$, although relatively small, is not zero ($E_L = 0.056$), thus indicating the necessity of higher harmonics for the description of the electron states. Hybridization, on the other hand, is mainly of the s−d

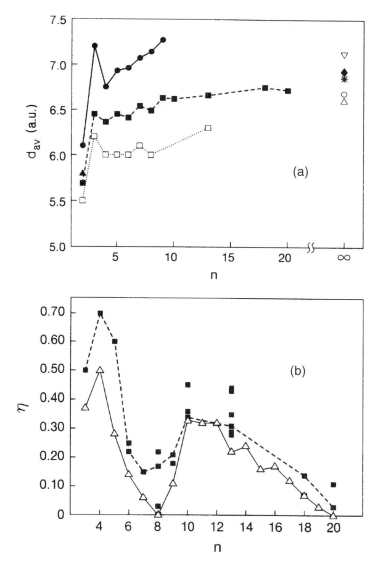

Figure 3.4 Na clusters: average interatomic distance (a) and eccentricity (b) as a function of size. Reprinted with permission from [123]. Copyright 1991 American Institute of Physics

type and pertains to the higher six states. Hybridization is more significant for non-MN clusters, such as Na_{10} and Na_{13}, but is also sizable for Na_{18}. Interestingly, the deviation from the spherical shell model increases with increasing temperature, as shown in Figures 3.6 and 3.7 for Na_8 and Na_{20}, respectively. This is in contrast with the widespread assumption that the validity of the jellium model is enhanced for a warm cluster [126].

Another useful analysis is the one shown in Figure 3.5(a), namely, the decomposition of the spherical component of the LDA electronic potential into its electrostatic, pseudopotential and exchange–correlation contributions. The degree of cancellation between the

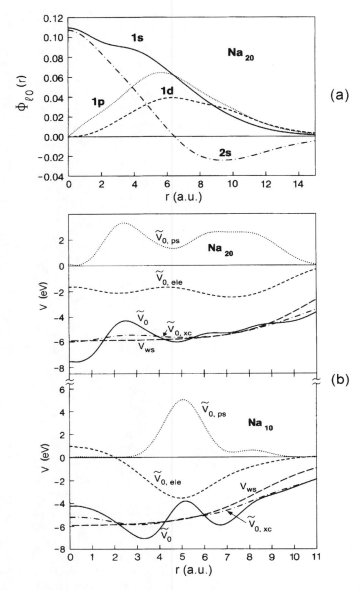

Figure 3.5 Na clusters: (a) Radial part of the dominant l-component of the one-electron eigenfunctions for Na$_{20}$, from the lower to higher occupied cluster orbitals: 1s, 1p, 1d and 2s. (b) Decomposition of the $l = 0$ component of the LDA potential ($\tilde{V}_0$) into the pseudopotential, electrostatic and exchange−correlation contributions, and comparison with the Wood−Saxon potential used in Ref. [1]. Reprinted with permission from [123]. Copyright 1991 American Institute of Physics

first two terms, which are the structure-dependent terms, gives an idea of the validity of the jellium picture as a first-order approximation. Figure 3.5(a) shows that in Na$_{20}$ such a cancellation is more effective than in Na$_{10}$. Comparison with the Wood-Saxon (WS) potential used in Ref. [4] shows to what extent this model can be considered a good

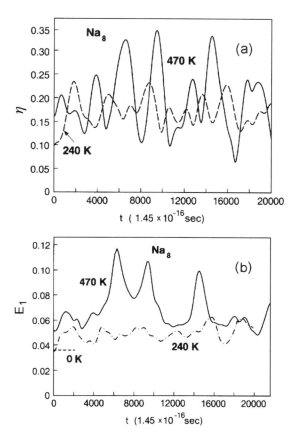

Figure 3.6 Na_8: time evolution of (a) the eccentricity parameter and of (b) the 'shell-model-error' parameter, calculated at two different temperatures. Reprinted with permission from [123]. Copyright 1991 American Institute of Physics

approximation to the actual potential. It generally has a shorter range, and closely approaches the exchange–correlation term.

Finally, the question of rigid or nonrigid behavior was considered. A first analysis was made from the direct examination of the atomic trajectories at several temperatures. Within the short time of the simulations, Na_6 at high temperatures was observed to undergo transformations between two distinct isomers (2D and 3D), and Na_8 was clearly visiting configurations all very similar, whereas for the larger clusters up to $n = 20$, owing to the multivalley character of the potential-energy surface, it was not possible to distinguish between fast isomerization and diffusive behavior. When, however, a criterion analogous to that of Lindeman for melting was assumed in order to distinguish between rigid and nonrigid behavior (see Figure 3.8), classification of the clusters at different temperatures was possible and consistent with the global character deduced from the direct observations.

There have been many questionable attempts to study 'phase transformations' in clusters of this size and in particular the improperly called 'melting'. Very recently, CP-like simulations (see Appendix C) of Na_{40} [127] have been reported, corresponding to ultrafast

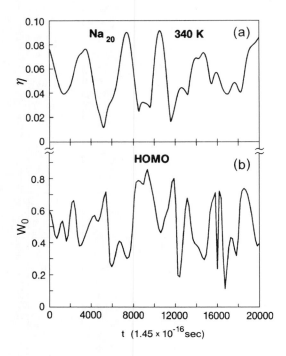

Figure 3.7 Na$_{20}$: time evolution of (a) the eccentricity parameter and of (b) the weight w_0 of the spherical component of the HOMO, calculated at 340 K. Reprinted with permission from [123]. Copyright 1991 American Institute of Physics

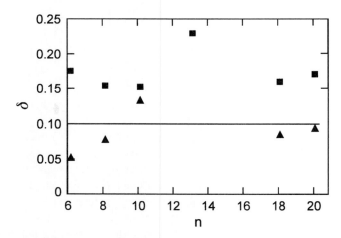

Figure 3.8 Mean fluctuation of the average nearest-neighbor distance calculated for Na$_n$ clusters at different temperatures (solid triangles = 'low temperature': $T = 200, 240, 260, 260$ and 340 K for $n = 6, 8, 10, 18$ and 20; solid squares = 'high temperature': $T = 600, 470, 440, 520, 570$ and 640 K for $n = 6, 8, 10, 13, 18$ and 20). $\delta = 0.10$ indicates the onset of enhanced diffusion. Reprinted with permission from [123]. Copyright 1991 American Institute of Physics

heating-up runs. Examination of the potential energy and of the root-mean-square displacement as a function of temperature revealed indications of a round-off transformation, which was characterized as 'melting'.

CP-like simulations have been applied to the study of the fission and evaporation process of sodium clusters [11, 128].

Owing to the lack of experiments, the determination of structure and electronic properties of lithium clusters is very limited compared to sodium [129]. They constitute, however, a source of other interesting issues. Establishing the effects of the zero-point motion on the structure and bonding of clusters of light elements such as lithium is important. These effects consist of a dominant contribution from the harmonic vibrations around the local potential minimum, and a smaller contribution due to the anharmonicity of the potential-energy surface. Needless to say, this second contribution is the most interesting one because it gives rise to quantum tunneling, and it is particularly important in Li because the light nuclear mass makes the elongation of the zero-point motion particularly wide. The simplest way to account for harmonic and anharmonic quantum effects in the ionic motion is via path integrals simulation [130]. We point out that, even in Li at very low temperature, the ions never get so close as to make exchange important, and quantum effects are limited to delocalization. The first exploratory investigation of this effect was carried out in Ref. [131], using path integrals together with a simple model potential (point ions in the Coulomb potential of a spherical negative background representing the electron density). The results show that quantum delocalization is important in the low-temperature regime (up to $\sim$200 K), blurring the distinction between different isomers, especially for clusters with incomplete ionic shells, that have (within the simplified potential model) a multitude of isomers almost degenerate in energy.

More recent calculations have used the path integral method combined with an LDA description of the electronic structure (Li_8 and Li_{20} in [132]; Li_4 and Li_5^+ in [133]). Again, quantum fluctuations are explicitly shown to be responsible for ionic delocalization and to render the distinction between short and long bonds meaningless. However, the number of nearly degenerate isomers displayed by the *ab initio* potential-energy surface is much smaller than that of the simple model used in Ref. [131], and, as a consequence, no qualitative effects are found in either the equilibrium geometries or the basic features of the electronic structure.

3.7 CLUSTERS OF THE IIA AND IIB ELEMENTS

Interest in clusters of the IIA (Mg, Ca, Sr, Ba and Ra) and IIB (Zn, Cd and Hg) elements has been motivated mainly by the fact that these systems may undergo a nonmetal–to–metal transition in their electronic structure as a function of size. The basis for this expectation is simple. On the one hand, bulk samples of these elements are clearly metallic, although, not surprisingly, key properties such as conductivity and plasma frequencies show that their electronic structure (and the Fermi surface in particular) deviates from the ideal free-electron picture much more than in the case of alkali metals [50]. On the other hand, experimental data show that the cohesive energy, equilibrium distances, vibrational frequencies and excitation energies for the homonuclear dimers of these elements are close to those expected for weakly bound van der Waals systems.

This state of affairs can be rationalized in simple terms. The atoms of these elements have a closed-shell ns^2np^0 electronic configuration, with a fairly large IP. The energy gap to the first excited state (the singlet ns^1np^1) is sizable (e.g. $\sim$2–4 eV [134]), although much smaller than the $ns^2np^6 \rightarrow ns^2np^5(n+1)s^1$ gap in rare-gas atoms [134]. When two atoms are brought together to form a dimer, the valence electrons occupy one bonding and one antibonding state, and therefore the 'band' contribution to the chemical bond vanishes [135]. Dispersion forces, always present, are in this case the only source of cohesion, in analogy to what is observed in rare gases. With increasing size of the aggregate, however, the atomic coordination increases and the occupied molecular states acquire some bonding character by hybridizing with the atomic empty p states, thus increasingly contributing to cohesion. Finally, electron states in the bulk are heavily hybridized, giving rise to a partially filled sp band and, therefore, to a metallic solid. It is therefore natural to conclude that a transition from 'van der Waals' to 'metallic' bonding has to take place in between [136], although there is no simple way to predict the critical size range or whether the transformation occurs gradually or abruptly like a phase transition. It is also clear from this discussion that the weak cohesion is not, by itself, a sufficient characterization for the bonding character of the dimer or the very small clusters, and that conclusive evidence has to be found in the electronic structure of these systems.

Two experimental papers on Hg_n clusters provide crucial information on the size evolution of the electronic properties in these systems [137, 138].

The first one [137] reports the ionization potential IP of Hg_n clusters from $n = 1$ to $n = 70$. For simple metals in the spherical jellium approximation, and neglecting electronic shell effects, IP converges to the bulk work function W for $N \rightarrow \infty$, with a size-dependent correction term proportional to $1/R$, where $R = (3N/4\pi\rho)^{1/3}$ is the radius of the jellium droplet, assuming a uniform electron density ρ estimated, for instance, from the bulk value. For mercury clusters the plot of the measured IP against $1/R$ displays a marked change of slope between the small ($n \leq 20$) and medium/large sizes, approaching the jellium-like limiting behavior only for $n \geq 50$. This change has been interpreted as evidence of a gradual van der Waals-to-metal transition taking place around $n \sim 40$. This interpretation is certainly reasonable, but it is not fully unambiguous because it is based on an oversimplified picture (the jellium model with a size-independent average electron density), which is unlikely to be fully appropriate even for metallic mercury.

The second, more direct, piece of experimental evidence is provided by the study of the auto-ionization process

$$Hg^*(5\,d^96s^26p^1) \longrightarrow Hg(5\,d^{10}6s^1) + e \tag{58}$$

reported in Ref. [138]. In this experiment, ultraviolet photons are used to promote one electron from the 5d shell to the 6p level. If the excitation energy is higher than the IP, Hg* decays by an electron-exchange process in which the 5d hole is filled and one electron is emitted. The efficiency of the process depends strongly on the localization of the valence s and p states, whereas the transition energies and line shape provide additional information on the electronic structure (see Figure 3.9). In agreement with the results of Ref. [137], these spectroscopic measurements reveal a significant change in the clusters' electronic properties starting at $n \sim 20$ and extending over a relatively wide size interval.

These two key experimental results underlie the interest in and excitement over mercury clusters and, by extension, clusters of all the IIA and IIB elements, which is apparent from the large number of publications devoted to these systems in the early 1990s. In particular, they motivated most, if not all, of the DFT-based computational studies. First, it

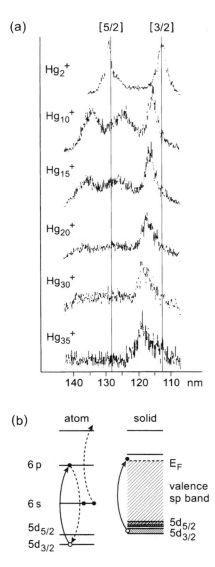

Figure 3.9 (a) Photoionization intensities for selected mercury clusters as a function of the incident photon wavelength in the energy range 8.5–11.5 eV. The vertical lines mark the corresponding atomic transitions. The dashed lines report the normalized intensities. (b) Diagram illustrating the photoionization process whose efficiency is reported in (a). Left: atomic energy levels; right: bulk mercury energy bands. Reprinted from [138]

should be emphasized that experimental evidence for a gradual but well-defined transition from van der Waals to metallic bonding is available only for mercury. Nevertheless, it is generally assumed that a similar transition takes place for all the other IIA and IIB elements. Given the similarity of the two extreme cases, i.e. the atom and dimer on the one hand and the metallic bulk on the other, it is reasonable to expect a similar behavior in between. However, because of the delicate balance of different energy contributions involved in the transition, it could very well be that the details of the transformation depend strongly on the specific element. In particular, among all these elements, the Hg atom is the one with the largest valence $s^2 \rightarrow s^1p^1$ gap (because of relativistic effects) and the one with the weakest cohesion (as suggested by the low melting point and relatively low critical temperature). For these reasons, it is possible that Hg is also the element with the widest range of stability for the pre-metallic state and, therefore, with the most apparent transition. The experimental indications, unfortunately, are ambiguous. Mass spectra and IPs for zinc and cadmium clusters [139, 140] reveal jellium-like features for $n = 10$ [141], suggesting that for this element the transition takes place even for the very small clusters. On the other hand, mass-spectra measurements for Mg, Ca and Ba [142] show that, up to very large clusters (n up to $\sim$10 000), the growth sequence is determined by atomic packing rules, which are common in rare-gas clusters. This result has been interpreted as an indication of an 'insulating' character for these clusters. The growth sequence alone, however, is not sufficient as an indication of the nature of the chemical bonding because the dominance of close packing of atoms could be due to other factors, including the experimental conditions.

The second mandatory remark concerns the applicability of DFT to these systems and to the related metal–nonmetal transition. At present, a systematic comparison between DFT approximations and experiments is possible only for the bonding parameters of the homonuclear dimers. Such studies are indeed available and have been motivated by the interest in the unusual optical properties of Hg vapors (and, to a lesser extent, of vapors of the other elements) relevant to laser technology [143] as well as by the opportunity to test approximate DFT methods on simple yet challenging systems that are fairly well characterized by experiments. We comment briefly only on the works related to the latter. The LDA prediction of a bound state for Be_2 [144], subsequently confirmed by experiments [145], has been one of the key results motivating the acceptance of DFT methods by the chemistry community. However, the same study made it clear that for Mg, Ca, Sr and Ba, for which experimental data were already available, LDA overestimates the binding energy significantly and underestimates the equilibrium distance. The same sequence of dimers has been used several times to test GGA schemes [146]. Although the quantitative results depend on the specific choice of the gradient-corrected functional, the results show that GGAs improve the agreement with experiment substantially. However, it is likely that this relatively good performance is due mainly to an accidental cancellation of errors, so that currently an unambiguous conclusion cannot be drawn. Comparison with accurate quantum-chemistry computations [147] is not conclusive either, because it is difficult to extract a simple bonding picture (i.e. to decide whether a dimer is 'metallic' or 'van der Waals') from the results of CI or quantum MC computations.

Regarding the larger sizes, we emphasize again that the most direct evidence of the van der Waals-to-metal transition should come from the analysis of the electronic structure, with the geometry, binding energy and IPs providing supplementary but not sufficient information. As the most obvious signature, i.e. sudden changes in the excitation spectrum, are not well described (or not described at all) by DFT in the current approximations, these do *not* provide a suitable theoretical approach. In addition, because of its nonlocal nature, the van der Waals attraction, believed to be the major bonding mechanism for the very small clusters, cannot be reproduced by either local or semilocal approximations of the exchange–correlation functionals [148]. Therefore, in the van der Waals regime, even structural and cohesive properties thus calculated should be considered with extreme caution. We shall now review the existing literature on the group-II clusters with the further warning that these studies are valid only for systems of size beyond the region dominated by van der Waals bonding, and that this hypothetical 'van der Waals'-to-metal boundary cannot even be located by present DFT schemes.

The growth sequence of Be clusters has been studied in Ref. [149] using LDA-MD for structural optimization, up to Be_{20} (see Appendix C). The optimal geometries turn out to be determined by close-packing rules. The relative stability of different cluster sizes, the (near) degeneracies of the KS eigenvalues, and the symmetry of the electron orbitals clearly correspond to the shell model. Surprisingly, the spin-polarized triplet states turn out to be almost degenerate with the singlet state. In one case (Be_9) it corresponds to the true ground state. The structures and stability of positively charged Be_n^+ and Be_n^{2+} clusters ($n \leq 5$) have been treated in Ref. [150], which focuses on the stability, detectability and, finally, the Coulomb explosion of the doubly charged clusters. As expected, the computation highlights the large increase in atomization energies upon removing one valence electron and the enhanced stability of linear geometries for the doubly charged clusters, which minimize the Coulomb self-energy by segregating the two unscreened charges at the opposite ends of the atomic chain.

LDA computations for Mg_n clusters (see Appendix C) are reported in Ref. [151] ($n \leq 7$) and extended to larger sizes (n up to 20) in Refs [152, 153]. The results provide a picture similar to that obtained for Be_n. The ground-state structures predicted by these various computations are in good agreement mutually. It is difficult to identify a dominant growth motif given the relatively narrow range of sizes for which extensive geometry optimization has been performed. The LDA result for Mg_4 is a regular tetrahedron, whose electronic structure and stability match the predictions of the spherical jellium model for a system with eight valence electrons [154]. As pointed out in [155, 156], the character of the preferred geometries for $n \simeq 10–20$ in LDA appears to be intermediate between sodium and silicon. In particular, the ground state of Mg_{10} [152, 153, 155] has the same structure (Figure 3.10(a)) as Si_{10} (tetracapped trigonal prism), however, with a smaller eccentricity [155] ($\eta(Mg_{10}) = 0.05$ vs. $\eta(Si_{10}) = 0.066$). As the number of valence electrons is 20, Mg_{10} corresponds to a magic number of the spherical jellium model and is isoelectronic with Na_{20}. It would be interesting to know from experiment whether Mg_{10} behaves as an MN cluster. Unfortunately, abundance spectra are unavailable for Mg aggregates in this size range. In Ref. [155] the same analysis of the cluster orbitals as reported for Na in Section 3.6 indicates sizable hybridization for the HOMO and a relatively small value for the spherical-shell-closing deviation parameter ($E_{L=2} = 0.049$), in close analogy to Na_{20}. However, regarding the electronic potential, the analysis of the spherical (dominant) component of the LDA potential in Figure 3.10(b) shows that the cancellation of the pseudopotential and electrostatic terms is much less effective than for Na_{20} (Figure 3.5(a)) and that modeling with a simple Wood–Saxon potential encounters serious difficulties.

Clusters of IIA elements heavier than Mg or IIB elements larger than the dimers remain virtually unexplored at the LDA and GGA level, although isolated computations have appeared. The most recent of these studies concerns the determination of ground-state geometries, cohesive energies and IPs for Ba_n ($n \leq 6$) within LSDA [157]. However, the use of a semiempirical atomic pseudopotential and that of a very small Gaussian basis set cast doubt on the reliability of the results. Moreover, the same pseudopotential is used in the MP2 (Møller–Plesset second-order) calculations performed for the sake of comparison, thus rendering the validity of such a comparison also questionable.

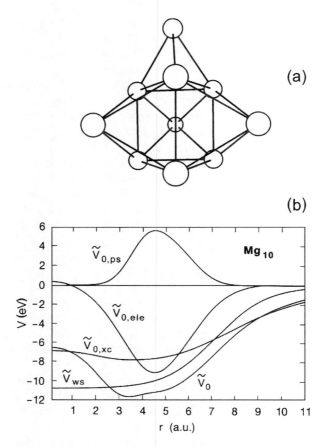

Figure 3.10 Mg_{10}: (a) ground-state structure and (b) same decomposition as in Figure 3.5(b). Reprinted from [155], © Springer-Verlag, Berlin, Heidelberg 1992

As explained above, both LDA and GGA descriptions of the structural and electronic properties of these clusters must be considered with caution. More work using more advanced approximations to the DFT functional is necessary to establish the extent to which such a picture corresponds to reality. It is important to note that many-body models going beyond the mean-field description offered by DFT–LDA give a very different picture of the size evolution in Hg_n clusters [158]: they predict a well-defined transition in the electronic structure and provide a fairly good fit to experimental observations.

3.8 IIIB METALS

Group-IIIB elements (i.e. B, Al, Ga, In and Tl) have an odd number of valence electrons in the fairly weakly bound s and p atomic shells. Unlike the group-II case, one could expect these elements to behave like simple metals, with properties determined by the average valence-electron density. Instead, the evolution of physical and chemical properties within the group is far from simple and monotonic.

First of all, as is often the case, the lightest element in the group (i.e. B) behaves quite differently from all the others, as it is a semiconductor with a fairly large band gap (1.5 eV, from Ref. [50]). For B, three complex crystal structures are known, all based on the packing of hollow B_{12} icosahedra [159].

As any solid-state textbook reports, bulk aluminum is a nearly free-electron fcc metal, characterized by a very high valence-electron density. Gallium, on the other hand, is liquid at temperatures close to room temperature, and displays two complex crystal phases at lower temperatures. Finally, indium and thallium recover most of the simple metal properties displayed by aluminum, with, however, a few peculiar properties that indicate an increased tendency towards covalent bonding in their chemistry.

This alternation of properties suggests that clusters of group-IIIB elements could display an intriguing variety of structures and properties. Unfortunately, extensive experimental data is available only for clusters of the simplest of these elements, i.e. aluminum. Less systematic but nevertheless abundant data has also been reported for indium. Boron, gallium and thallium clusters, by contrast, have been only marginally investigated in experiments.

For the sake of completeness, we first review DFT computations for boron clusters, although these systems do not belong to our definition of 'metal clusters', because bulk B is a semiconductor. They may be of interest for a comparison to see how the bonding evolves in the periodic table. Boron aggregates have indeed attracted much attention because of the remarkable stability of well-defined units in a variety of crystals: hollow B_{12} icosahedra in pure boron, B_6 octahedra and B_{12} cubo-octahedra in metal borides. For isolated clusters, experimental results for the cohesive energy, reactivity and IPs of small clusters (n up to $\sim$20) have been reported [160]. These data show that clusters with 5, 10 and 13 atoms are somewhat more stable than the others, although the differences in cohesive energies are not large on the scale of the high binding energy of boron. Moreover, the cohesive energy converges rapidly to the bulk value, suggesting that the clusters adopt a compact geometry even for the very small sizes.

Density functional computations in the LSDA and GGA have been reported for neutral and positively charged boron clusters for $n \leq 14$. Computations have been focused mainly on these two sizes because icosahedral B_{12} and B_{13} clusters are expected to exhibit a special stability. Unconstrained geometry optimization using LSDA CP molecular dynamics has been reported in Ref. [161] for B_{12} and in Ref. [162] for B_{13}. In both cases, the structures obtained by simulated annealing do not correspond to any simple and highly symmetric geometry. More recently, systematic computations for a wider size range [163] has revealed the high stability of planar and quasiplanar structures, which, almost without exception, are lower in energy than 3D isomers. The energy difference between 2D and 3D structures is beyond computational uncertainties, although the quantitative results seem to depend somewhat on the exchange–correlation functional adopted.

Aluminum clusters have been investigated using a variety of experimental techniques, providing abundance spectra [164, 165], spin multiplicities [166], IPs [164, 167], EAs, and static polarizability [105]. Reactivity studies have been reported for size-selected Al clusters in contact with a variety of small molecules (see, for instance, [167]), and the presence of different isomers in a population of clusters has been investigated by measuring the mobility of clusters in a buffer gas [168]. Finally, the electronic structure of these clusters has been probed by photoelectron spectroscopy on the anion species $Al_n{}^-$ [169, 170]. As mentioned above, bulk aluminum is remarkably close to a nearly free-electron metal. Experimental measurements suggest that the same picture remains valid for medium-large-size clusters: with only a few exceptions, steps and peaks in mass spectra, IPs and EAs correspond to shell-closing in the spherical jellium model. The free-electron behavior is underlined by the experimental observation of supershell effects in large aluminum clusters, showing that the ionic perturbation does not disrupt the coherent propagation of the electrons on a length scale comparable to the cluster radius (i.e. ~ 10 Å) [165].

The situation is less simple for small Al clusters: exceptions to the spherical jellium rules are more apparent both in terms of the mass spectrum and the size dependence of the IP. Moreover, for $15 \leq n \leq 40$, polarizability measurements, which provide an additional, although indirect, probe of the electronic structure, display a marked decrease of the measured α with respect to jellium model results, which interpolate the experimental data for larger sizes smoothly [105]. This observation has been interpreted as an indication of a change of the clusters' bonding as a function of size: for clusters of fewer than 40 atoms, important deviations from the jellium picture are observed, whereas for larger clusters the free-electron behavior is recovered. No simple explanation is available for this transition, although it is possible to guess its origin. In fact, a significant deviation from the jellium model is not surprising, in view of the strong interaction of valence electrons with cores of charge 3. The onset of a free-electron metal behavior for $n \geq 40$ might be due to the increased symmetry and regularity of the ionic positions with increasing size, thus reducing the electron–core scattering rate.

Unfortunately, no *ab initio* investigation has been performed in the size range covering the critical $n \sim 40$ region. Systematic density functional pseudopotential computations have been performed only for $n \leq 10$ [171, 172] (see Appendix C), although others have been reported for specific sizes up to $n = 55$ [89, 156] (see Appendix C). Once again, the comprehensive investigations of Refs [171, 172] reveal the importance of (quasi-)planar structures, although the size range of stable 2D geometries for aluminum is significantly less extended than that of boron. Figure 3.11 illustrates the results up to $n = 8$. The transition between nearly (buckled) 2D and 3D ground-state geometries occurs around $n = 5$, but the two families of structures remain very close in energy, at least up to $n = 10$.

Al_{13} has attracted special attention because of two unusual characteristics that occur simultaneously: its size corresponds to a magic number for the close-packing growth (either icosahedral or cubo-octohedral) and the number of valence electrons (39) is close to the shell-closing (40) of the spherical shell model. In fact, LDA calculations, both the dynamical SA in Ref. [89] and the extensive search starting from a number of different crystalline and noncrystalline geometries in Ref. [156], predict a quasiicosahedral structure for the ground state, with a weak distortion (D_{3d} [156]) driven by the Jahn–Teller effect. Al_{13} is thus special among the 13-atom clusters, which generally do not assume such highly symmetric atomic configurations (see, for example, [156]), and also anomalous with respect to the structural trend shown at least up to $n = 10$, because quasiplanar structures are clearly unstable. Dynamical SA has also been applied to Al_{55} [89], the size which in icosahedral growth comes next. Although in this case the number of valence electrons does not justify the assumption of a symmetric structure, the results of Ref. [89], namely, a highly disordered arrangement, is probably an artifact of the very short annealing (see Section 3.4.3.1).

The analysis of neutral and ionized Al_n clusters has been extended in Ref. [172] to Ga_n for the same size range ($n \leq 10$) (see Appendix C). The peculiarities of bulk Ga, i.e. the

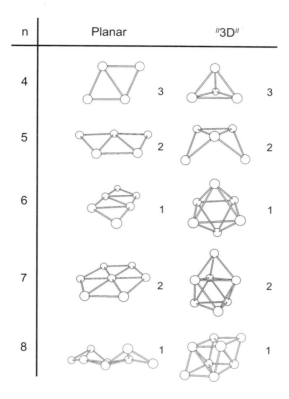

n	Planar		"3D"	
4		3		3
5		2		2
6		1		1
7		2		2
8		1		1

Figure 3.11 Al_n clusters: planar and 3D isomers from LDA calculations. The number next to each structure corresponds to the spin multiplicity. The '2D' structure of Al_8 is buckled. Reprinted with permission from [172]. Copyright 1993 American Institute of Physics

low melting point and the complex crystal structures, are usually explained in terms of the atomic configuration, IIIB, which has filled 3d shells. In clusters, this peculiarity is manifested mainly in the short Ga–Ga interatomic distances and in the presence of bonding angles close to 90°. Apart from these effects, the picture provided by these computations for Ga_n is remarkably similar to the one for Al_n. Earlier computations for smaller Ga clusters were reported in Ref. [173]. The results of this first LDA study are superseded and complemented by those of Ref. [172]. We remind the reader that experimental results for Ga clusters are limited to abundance spectra [174] and photoelectron spectroscopy [170]. A partial comparison of experimental data with the results of Ref. [172] is provided in Ref. [170].

To the best of our knowledge, no systematic investigation of cluster properties has been carried out by DFT methods for IIIB elements heavier than Ga or for sizes larger than the dimer and trimer.

3.9   MAGNETISM IN TRANSITION-METAL AND RARE-EARTH-METAL CLUSTERS

Among the recent developments of metal-clusters physics, one of the most exciting is certainly the experimental measurements of magnetic deflections of size-selected clusters

(n up to $\sim 10^3$) in a molecular beam, carried out mainly by de Heer and collaborators [175]. Not surprisingly, Fe, Co and Ni, which are ferromagnetic in the bulk, are the elements more extensively investigated. We refer to [176] for a recent and detailed account of the experimental situation for these elements and for an outline of the major open questions. Besides Fe, Co and Ni, large magnetic moments have been measured by magnetic deflection also for clusters of the ferromagnetic f-electron metals Gd [177], Tb and Dy [178] and, interestingly, for Rh [179], which is not ferromagnetic in the bulk. In addition to the results of magnetic deflection, photoelectron spectroscopy investigations of the electronic structure of Ni [180], and Fe [181] and Co [182] clusters have been reported, together with several reactivity studies [183].

Perhaps the simplest problem concerning these systems is the determination of the size dependence of the spin moment in clusters of elements that are ferromagnetic in the bulk. Starting from the known magnetic moments of atoms and solids, it is easy to realize that the average moment *per atom* μ has to decrease with increasing cluster size [184]. Indeed, the experimental results show that clusters of the ferromagnetic elements display magnetic moments per atom that are intermediate, and converge slowly and nonmonotonically to that of the bulk [185, 186]. These data for Fe, Co and Ni are shown in Figure 3.12.

As mentioned in Section 3.2, small clusters may be expected to display a wide variety of exotic spin structures, involving a change in the magnetization direction as a function of position within the aggregate. Noncollinear magnetic structures are the exception in solids, because the high symmetry of crystals together with the small coupling of spin and lattice (as compared to the spin–spin exchange interaction) render the parallel and antiparallel spin configurations the only relevant ones. In small clusters the situation could be different, partly because the anisotropy energy is likely to be larger than in solids [187] (where it is often suppressed by symmetry), and partly because in most small clusters a multitude of inequivalent atoms exist, allowing a variety of different spin structures to be similar in energy.

The low-energy dynamics of the clusters' spin moment are also of interest. In extended systems, the lowest-energy modes are the long-wavelength spin waves, or magnons. In small systems, for which long-wavelength magnons cannot exist, the low-energy modes are those related to the spin-lattice orientation, and are determined mainly by the so-called anisotropy energy (see Ref. [188]). At very low temperatures, important quantum effects in the spin dynamics (tunneling among almost degenerate directions) may be expected, analogous to what has been observed in supported magnetic particles [189].

The temperature dependence of the magnetic moment in small systems is also of importance in understanding finite-size effects on magnetic phase transitions. Experimental data reported in [185] show that for Co and Ni, as expected, the ferromagnetic transition is smeared over several hundred degrees in small particles of $10^2 - 10^3$ atoms and that the Curie temperature is lower (although only slightly) than the bulk value.

Finally, ferromagnetic solids usually display zero total moment (unless they are artificially magnetized) because macroscopic systems are split into domains of different magnetic orientation in order to minimize the long-range dipole–dipole magnetic interaction. Up to $n \sim 10^4 - 10^5$ atoms, the dipole–dipole magnetic energy is much smaller than the exchange energy giving rise to ferromagnetism (see, for instance, [188]), and these systems are expected to be monodomain (or *superparamagnetic* in the terminology of magnetic materials).

On the theoretical side, many of the issues raised by the experiments are rather well understood at the qualitative level. The increase of the magnetic moment with decreasing

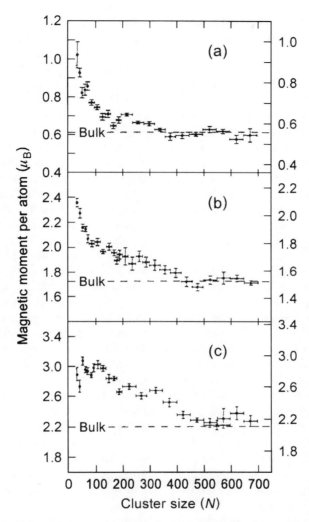

Figure 3.12 Average magnetic moment of size-selected clusters in a molecular beam ((a) nickel, (b) cobalt, and (c) iron). The estimated temperature of the clusters is $\sim$100 K. Reprinted with permission from [185]. Copyright 1994 by the American Association for the Advancement of Science

size was predicted well before the experiments [190] and analyzed in detail on the basis of Hubbard Hamiltonians (see, for instance, [191]). The low-energy spin dynamics in clusters has been discussed on the basis of both classical and quantum-spin Hamiltonians, taking into account the spin-lattice coupling due to the anisotropy energy [192]. Finally, finite-size effects on the thermal properties of magnetic aggregates, and on the ferro–paramagnetic transition in particular, have been analyzed in advance of the experiments using the Ising and Heisenberg models [193].

However, if we go beyond the qualitative level into the quantitative analysis of the experimental results (not to mention the prediction of new phenomena), the situation is far less satisfactory. Regarding DFT studies, we notice a small series of early computations, performed using small, localized basis sets and neglecting the relaxation of atomic positions [190]. All these computations agreed in predicting a quick convergence to the bulk of electronic and magnetic properties. After a few years, interest was suddenly revived in the 1990s by the publication of magnetic deflection results [175]. Nevertheless, despite the considerable efforts by several groups, only the simplest problem, i.e. the determination of the ground-state geometry and spin for Fe_n, Co_n and Ni_n as a function of size, has been addressed to some degree of confidence and completeness for clusters with up to $n \sim 7$. Few and far less systematic computations have been performed for n up to ~ 13. For larger sizes, computations can be carried out only by adopting symmetry constraints of questionable validity. For $n \leq 7$, a few LSDA results are available also for excited spin configurations, providing information on the geometry–spin relation and the spectrum of spin excitations. However, especially for these properties, the computational exploration of these systems is still at a preliminary stage. Comparison with the results reviewed in Ref. [176] immediately reveals that the size ranges explored by theory and experiments (most measurements are limited to $n > 15$ for technical reasons) only just overlap, and that no detailed comparison of magnetic properties is possible beyond the dimers.

One of the reasons of the wide gap separating experimental information and DFT studies is that the LSDA description of spin properties has often been regarded with suspicion, both because of fundamental limitations (i.e. spin contamination) and because of its inability to predict the ground state of the prototype ferromagnet, i.e. iron [194], correctly. In addition, several theoretical predictions of large moments in small clusters of nonmagnetic transition elements that failed to be confirmed by experiments, did not help DFT in gaining wide acceptance in this field. A second and more decisive reason is the extreme complexity of computations for 3d transition-metal clusters, including spin optimization and structure relaxation. Mere inclusion of the spin, as described in the LSDA, doubles the size of the computation and hence does not pose an overwhelming problem. However, to determine the global energy minimum with respect to geometry and spin requires several (expensive) structural optimizations, one for each spin multiplicity that could represent the ground state (usually three or four for n up to 7; more for larger sizes). In addition, 3d orbitals are highly localized, and their expansion in plane-waves is correspondingly demanding. Moreover, the error introduced by the pseudopotential approximation in the magnetic structure may be more important than for other properties, because the spin multiplicity depends sensitively on the valence–core exchange interactions [195]. Localized functions basis sets do not overcome these difficulties completely, because the crucial interplay of localized 3d and extended 4s valence orbitals represents a challenge for such sets.

Despite these undeniable problems, DFT still provides the most accessible path to a parameter-free description of the magnetic properties of small transition-metal clusters. As mentioned in Section 3.2, the spin contamination problem, although certainly present, does not appear to be quantitatively very significant. The failure of current DFT approximations to predict the geometry and magnetic moment of bulk iron (for which several structures are nearly degenerate) does not imply a similar weakness for small systems. For these systems the available electron eigenstates are more widely spaced in energy, and therefore energy differences among different spin multiplicities are larger than in the bulk. In fact, the spin-polarized GGA description of the iron, cobalt and nickel dimers is relatively good [196], despite a significant overestimation of the binding energy, and the spin state of these dimers agrees with the experimental results. For all these reasons, there is hope that current DFT schemes could provide a fair (although not without occasional failures) description of how the spin depends on the size and geometry in small ferromagnetic clusters, and that only the computational problems remain as the main obstacle for an extensive study of these properties.

With these considerations in mind, we review the recent papers, noting that none of these studies considers the spin–orbit interaction. Moreover, magnetism is attributed only to spin, the orbital part of the magnetic moment being neglected. Therefore, the resulting magnetic moment stems from the unequal filling of spin-up and spin-down KS states, as illustrated in Figure 3.13.

Clusters of the prototype magnetic metal, i.e. iron, have been investigated most extensively by density functional methods. In addition to the several studies of the iron dimer [197], the first extensive DFT optimization of geometry and spin state for Fe_n (n up to 4) has been reported in [198]. Further studies of the size–geometry–spin relation in iron clusters with n up to 7 have been reported in Refs [196, 199–201] (see Appendix C).

DFT results for the geometry, spin, cohesive energy, IP and vibrational frequencies for Co_n have been reported in Ref. [202] for $n \leq 4$ and in Ref. [196] for $n \leq 5$. Both studies employ an LSDA and spin-polarized GGA (see Appendix C). The same scheme has been applied to Ni_n clusters (up to $n = 5$) [196]. LSDA results for the structures of Ni_n ($n = 2$–6, 8 and 13) are also reported in [203] (see Appendix C).

It is very encouraging that all these studies agree about the general features and trends, although quantitative discrepancies remain between the various computational approaches. The ground-state structures of Fe_n, Co_n and Ni_n (see Figure 3.14) are three-dimensional as soon as $n \geq 4$, where there are important Jahn–Teller distortions to the high density of electronic states close to the highest occupied level. The magnetic moment per atom is significantly higher than the bulk value and decreases fairly quickly, but not monotonically, with increasing cluster size. For all these systems, there is an apparent correlation between spin multiplicity and interatomic distances: the higher the spin moment, the longer

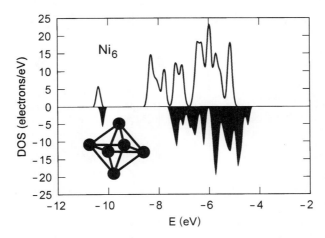

Figure 3.13 Spin-up (upper panel) and spin-down (lower panel) density of occupied (KS) electron states for the lowest-energy isomer of Ni_6. Computations have been performed within DFT–LSD–GGA. The computed Kohn–Sham eigenvalues have been convoluted with a 0.15-eV-wide Gaussian (after P. Ballone and R. O. Jones, unpublished work)

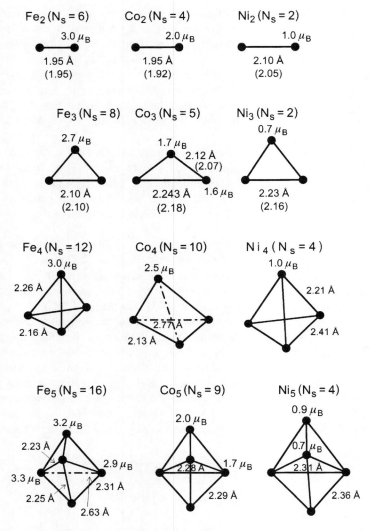

Figure 3.14 Lowest-energy isomers and atomic magnetic moments of transition-metal microclusters as predicted by LSD + GGA computations. Reprinted from [196], with kind permission from Elsevier Science—NL, Sara Burgerhartstraat 25, 1055 KV Amsterdam, The Netherlands

the interatomic distances, which corresponds to the decrease of the electron kinetic energy required to build a high-spin configuration. In several of these systems, the spin moment is not uniformly distributed between the atoms, reflecting the presence of inequivalent sites. Not surprisingly, the contribution of the d levels to bonding (identified in Ref. [196] by projecting molecular states onto atomic orbitals) is highest for Fe and lowest for Ni, as the d shell becomes progressively filled and, therefore, less reactive. Moreover, for each element the bonding due to the d states decreases with increasing cluster size, reflecting the general trend of increasing nearest-neighbor distances and decreasing d–d overlap with increasing atomic coordination.

Comparison with experimental results is possible only for the dimers. As anticipated above, the spin-polarized GGA results for spin state, atomization energies, IPs and EAs are reasonably good. Not surprisingly, the most problematic dimer is Fe_2, which has the highest d component in its molecular orbitals. Vibrational frequencies agree less well with experiments, although at present it is not possible to decide whether this problem is numerical in nature or due to more fundamental reasons.

In this short review of DFT studies on small ferromagnetic clusters, the computation recently reported in Ref. [27], which introduces noncollinear magnetism in the determination of the spin structure of small iron clusters, deserves special mention. The method has been briefly sketched in Section 3.2. The results for the geometry, cohesive energy and total spin moment are close to those of previous studies. The novel feature is that, as could be expected, noncollinear magnetism is present in several spin configurations of small transition-metal clusters (see Figure 3.15 for representative examples), although for the ground state it only occurs for $n = 5$. These authors observed that noncollinear magnetism tends to increase the magnetic moment localized on each atom, suggesting that by changing the direction of local magnetization the system retains an almost free-atom-like atomic moment up to higher sizes. Several points in the basic formulation still need to be clarified and improved by further studies. For instance, it is unclear why the spin moment of the ground state of Fe_5 is not an integer multiple of the electron moment, or how one can ensure that the total spin is an eigenvector of the total spin operator. Finally, it might be that at this level of detail the limitations of LDA or GGA exchange–correlation potentials become apparent, and the quantitative picture could be different for improved DFT approximations. Nevertheless, it is clear that this formulation opens the possibility of studying *ab initio* several of the basic issues mentioned in the introductory part of this section.

Of the f-electron metals that are ferromagnetic in the bulk (i.e. Gd, Tb and Dy), only Gd has been explored by DFT computations for cluster magnetism. In particular, computations

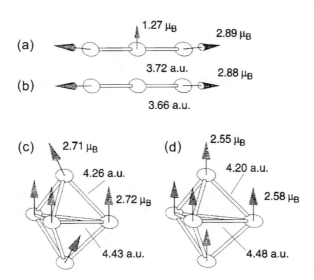

Figure 3.15 Atomic and magnetic structure of low-energy isomers of Fe_3 and Fe_5 computed within DFT–LSD–GGA. Configuration (c) is the ground-state structure for Fe_5. All the others are excited magnetic states. Reprinted from [27]

have been reported for Gd_{13}, including a model analysis of canted magnetism [204]. Unfortunately, the poor description of the f-electrons by current DFT approximations is a source of serious concern, especially because this problem is more pronounced in small systems. A careful analysis of this issue is in order before DFT methods can be used to predict properties of rare-earth clusters.

Of the many nonferromagnetic transition elements (especially 4d) for which high spin moments have been predicted in clusters, only Rh has been unambiguously found by experiments to be ferromagnetic [179, 205]. Here again, the measured values are significantly lower than the theoretical prediction, although it is not yet certain whether this discrepancy must be ascribed to a lack of accuracy in the theoretical or in the experimental approach. Besides rhodium and palladium, the element whose electronic structure is closest to the requirements for magnetic polarization is vanadium. However, despite several early predictions of ferromagnetism in small V_n clusters, this possibility has been ruled out both by experiments [206] and by recent LSD and GGA computations [207].

Whereas ferromagnetism in small clusters has been investigated intensively in the past few years, comparatively little effort has been devoted to the size evolution of antiferromagnetic structures. The prototype bulk antiferromagnet is chromium, which has a bcc structure and a Néel temperature of 311 K. This element also displays an interesting electronic structure for the dimer, characterized by the antiferromagnetic coupling of the two atoms, a strong d–d bond, and an exceptionally short equilibrium distance $r_e = 1.68$ Å [208]. Note that LSDA provides a good description of this unusual bond [209]. An LSDA investigation of the geometry and electronic structure of Cr_n clusters [210] suggests that the unusual and strong bonding of Cr_2 is reflected by a sudden transition of the growth sequence at $n \sim 11$. For $n \leq 11$, the structures appear to be dominated by the dimer motif: even-numbered clusters are collections of weakly interacting dimers, odd-numbered clusters present, in addition, a long bond to the lonely atom. For $n > 11$, the dimers lose their identity, and, somewhat surprisingly, the ground-state electronic structure and geometry are reminiscent of the bcc bulk even for $n \sim 15$. In view of these results, it is unfortunate that perfect antiferromagnetic ordering cannot be measured by magnetic deflection, and that even partial antiferromagnetic ordering, leaving a small, staggered magnetization, can be difficult to detect directly. Magnetic-deflection measurements of chromium clusters [206], setting an upper limit of $0.77\mu_B$ per atom, are consistent with the results of the computational study.

Bulk manganese has several crystal polymorphs characterized by complex unit cells and a variety of magnetic structures, including two distinct antiferromagnetic phases [159]. Owing to its low stability, the Mn_2 dimer is not well characterized by experiment, but is generally believed to be given by the antiferromagnetic coupling of the two atoms [211]. The description of its electronic structure has been a serious challenge for approximate DFT computations, which have produced a variety of different results for the equilibrium distance, the binding energy, the total and the atomic spin multiplicity. Up to now, none of the standard approximations (LSDA, GGA or hybrid schemes) seems to provide a satisfactory description of this dimer, if we assume that the accepted experimental bonding parameters and spin multiplicities are correct. Despite these important problems, the structure and magnetic properties of larger clusters (up to Mn_5) have been discussed in Ref. [212] on the basis of the hybrid B3LYP scheme [49] for the exchange–correlation functional. This study predicts a steep increase in cohesion with increasing size as well as a ferromagnetic coupling of the atomic spins, each retaining the free-atom moment.

Such a huge spin polarization is unusual in condensed systems, and would be extremely interesting if confirmed by experiments.

To conclude this section on the magnetism in transition-metal clusters, we mention that many studies have been performed at the TB–Hubbard Hamiltonian and many-body potentials levels. A few of the TB–Hubbard Hamiltonian computations have been very valuable as model studies [158, 213], elucidating the relation between the clusters' moment and the relevant physical parameters such as the number of d electrons, the Coulomb, exchange and hopping integrals (U, J and t, respectively, in the terminology of the Hubbard model), and the ground-state geometry. On the other hand, the various TB studies that attempted a quantitative description of specific systems did not provide much useful information, because the results appear to depend too strongly on the form of the Hamiltonian and on the choice of the parameters, and because correct and incorrect results cannot be identified without higher-level *ab initio* computations. The same comment applies to studies based on many-body potentials including magnetic contributions: current knowledge about the many-body origin of magnetism excludes the validity of any simple model of this kind for small particles [214].

3.10 NOBLE AND NONMAGNETIC TRANSITION-METAL CLUSTERS

Interest in a number of these clusters is widespread for several specific reasons. For example, evidence exists that catalytic centers of Pt and Ir particles on solid supports are microaggregates of 4–6 atoms [215]. Although this does not imply that such small bare clusters also exhibit catalytic activity, it has been a stimulus to investigate their properties [216]. In the case of silver, evidence that aggregates as small as four atoms can be catalytic centers in AgBr emulsions for the latent-image formation in the photographic process was found about 15 years ago [217]. Again, it is by no means proven that there is a direct link between the chemistry/physics of small silver clusters in the gas phase and the latent-image formation [218] in the real material. However, this triggered a number of studies at the time (see e.g. [219]). More recently, gold clusters have attracted much attention because of the synthesis of nanostructures formed by self-assembling such units coated with alkane- or alkyl-thiols [220]. Currently, the discovery that small gold aggregates (1 to 6 nm) supported on titania catalyze the oxidation of carbon monoxide seems to have revived the interest and enthusiasm in this field [221].

However, experimental and especially theoretical studies of all these clusters are fragmentary and incomplete. LDA or GGA to DFT in principle faces no major problems to treating the nonmagnetic transition-metal clusters and heavy noble metals. With local or semilocal approximations to DFT [222], on the other hand, the cohesive energies of these systems with mixed s–d chemical bonding are difficult to reproduce. Technical difficulties regarding both the choice of basis sets, when localized functions are chosen, and/or of the pseudopotentials should not be underestimated. For the heavier metals, they are related to the increase of the electronic shells, to the different localization of the d and s orbitals in the valence shell, and to the polarization of the core electrons following changes in the configuration of the valence electrons. For the details, the reader is referred to Appendix C. Here we recall only that for each specific case these problems can be overcome, at least in part. As for alkali and sp metals, LDA-DFT again predicts compact structures (at least in the cases treated so far), a strong contraction of interatomic distances with decreasing size that is consistent with the change in coordination and with 'scattered' experimental data, and IPs (and also EAs) that are in fair-to-good agreement with experiment. Not much is known about the effects of gradient corrections. Calculations using LSDA and B-LYP GC have recently been performed for Nb_n with $n = 4, 8, 9$ and 10 [223], which allows a direct comparison. As expected, B-LYP GC resulted in a lowering of the binding energies with respect to LDA and an increase of the interatomic distances. Interestingly, the energy differences between various geometrical isomers showed that B-LYP GC also weakened the LDA tendency to prefer compact structures. Variations in IPs and EAs turned out to be significant (a few tenths of an eV, which represents more than 10% for the IPs and up to 25% for the electron detachment energies). In the pseudopotential or effective-potential approaches, scalar relativistic effects are accounted for in the core–valence interaction. A detailed comparison for the dimers and small clusters (see e.g. Refs [118, 223]) shows that neglecting these effects results in an underestimate of the interatomic distances, which becomes large (about 6%) for cases such as gold [224]. All-electron DFT calculations beyond the dimers are rare. This is regrettable, especially when compared to efforts made using LDA for the solids (see Ref. [224] and references therein) and using post-HF methods for the gold trimer and tetramer [225]. One exception is the recent determination of magnetic hyperfine interactions for Ag_5 and for the heptamers of Cu, Ag, Au, and of mixed Cu–Ag composition [118]. These calculations were made with localized basis sets, using both LDA and Becke–Perdew (BP) GC exchange–correlation functionals and taking into account relativistic effects whose role is crucial (see Section 3.5.5). Only selected geometries were considered and optimized at the spin-restricted level of approximation. Especially for the heptamers, however, the structural issue is not a severe one. Alkali- and noble-metal atoms generally tend to prefer the compact arrangement of the pentagonal bipyramid (see [219]). The good agreement with experimental data confirmed the validity of the computations as well as the predictions for the geometries.

Within the pseudopotential–plane-wave framework, only a very limited investigation exists for Cu, Ag and Au clusters. It would be interesting to see how the geometries and electronic structure change on passing from alkali to noble metals, and, in particular, to what extent the jellium-based picture can be extended. However, this question has not been pursued within DFT schemes to the extent it deserves. The only exception is LDA calculations on small Cu_n clusters with $n = 2, 3, 4, 6, 8$ and 10 [73] (see Appendix C). Although the search for equilibrium structures was limited, no remarkable difference in the energy ordering and in the scale of isomer energy differences was found with respect to sodium. However, hybridization of s and d states was found to be sufficient to perturb the jellium-like picture more than in sodium. Figure 3.16 shows the decomposition of the density of the occupied states in the $l = 0, 1$ and 3 components for $n = 6, 8$ and 10.

DFT computations using localized basis functions have sporadically been applied to silver and gold microclusters. The early LSDA approach [226] (see Appendix C) to silver dimer and trimer (neutral and cationic) was a comparative study of results obtained with both 1e- and 11e-pseudopotentials. It clarified the importance of the 4d-electron states in determining structural and bonding properties, which is partly related to their polarization.

More recently, nonrelativistic all-electron GGA calculations were extended up to the silver hexamer [227] and used BP GC (see Appendix C). Some interesting discrepancies with the predictions of HF CI calculations of Ref. [219] were found: in particular the onset of 3D stable geometries at $n = 6$ was not confirmed. Although the full source

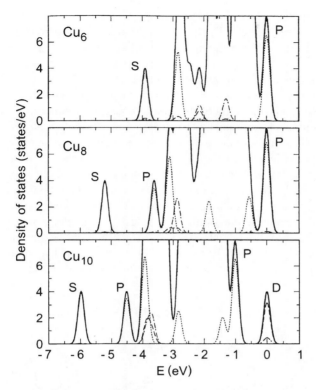

Figure 3.16 Cu clusters: decomposition of the valence electron density of states into s, p and d contributions. Reprinted from [73], with kind permission from Elsevier Science–NL, Sara Burger-hartstraat 25, 1055 KV Amsterdam, The Netherlands

of the disagreement may be more complex, the authors of [227] attributed it mainly to two limitations of [219]: the explicit neglect of the participation of the 4d electrons in the bonding owing to the use of a 1e- effective core potential and the restriction of the structural optimization made with high-symmetry constraints.

In the case of gold, an attempt has been made to determine the convergence of several cluster properties, such as the cluster radius, the cohesive energy, the IP, and core-level shifts, with increasing size (up to $n = 147$) by means of AE–GGA (AE: all-electron) calculations [228] (see Appendix C). The significance of the results is, however, marred by the fact that only selected, highly symmetric structures (icosahedra, cubo-octahedra and octahedra) were considered and that no structural relaxation beyond radial expansion was allowed.

Of the family of nonmagnetic transition metals, niobium clusters are those for which most theoretical investigations have been made, although these were mainly restricted to the structural properties: neutral clusters Nb_n, with n up to 7 (LDA [229]), with $n = 8-10$ (LDA [230, 231], LDA and B-LYP GC [223]), anionic clusters up to $n = 8$ (LDA [113]), anionic and cationic clusters with $n = 8-10$ (LDA and B-LYP GC [223]). Reference [223] (see Appendix C) presents an extensive structural study using CP molecular dynamics as well as local optimization procedures, and is corroborated by electronic and vibrational spectra as well as by an extension to cations and anions, thus allowing comparison with experimental data. The structural pattern found is different from that found for alkali- and noble-metal clusters. Starting from the tetramer, the geometries of the lowest-energy isomers are three-dimensional (Figure 3.17). For the octamer, capping octahedra is definitely more favorable energetically than arranging the atoms on the vertices of an octahedron. At the level of the 9- and 10-atom clusters, on the contrary, the lowest-energy structures of Na and Nb are similar. However, the energy ordering of other isomers is dissimilar and especially the scale of the isomer energy differences. In fact, the potential-energy surface found for sodium is very flat: this is not compatible with a system like niobium, where 'd' orbitals dominate the chemical bonding.

The dominant contribution of the 'd' atomic orbitals to the electronic structure of the occupied valence states and the dependence of the latter on the specific geometry of the cluster are clearly shown in Figure 3.18. An interesting finding regards Nb_9: two quasidegenerate isomers are obtained, in agreement with experimental evidence of the presence of two distinguishable isomers. Indeed, in their electronic properties and IPs significant differences are found, which are also consistent with experimental observations.

An important difference exists between the results of Ref. [113] and those of Ref. [223] for the Nb_8 anion. Although they agree in the prediction of the ground-state geometry,

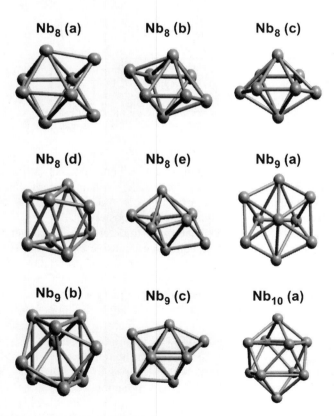

Figure 3.17   Nb clusters: low-energy isomers, reprinted from [223]. They have been described as two different cappings of the octahedron for Nb_8, a bicapped pentagonal bipyramid (left) and a tricapped octahedron (right) for Nb_9, and a bicapped antiprism for Nb_{10}

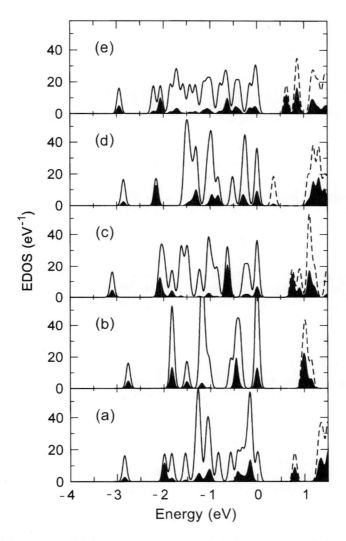

Figure 3.18 Nb$_8$ isomers: density of electronic states (solid line for the occupied; dashed for the unoccupied) and its projection onto s and p atomic orbitals (black). Reprinted from [223]

in Ref. [223] additional isomers at low energy were found that had not been obtained in Ref. [113], independent of whether LSDA or spin-polarized GGA was used. The reason for this is not clear. It may be the result of the different implementation of the DFT computations (see Appendix C) or the optimization strategy (both use local optimization from several initial configurations, but in [223] it was implemented by CP–MD search). Hence, based on the geometries in [113], photoelectron spectra [115] have been interpreted and the presence of high-energy isomers has been invoked. Such an interpretation appears to be questionable.

DFT-based investigations of the chemisorption on these clusters are not available — in spite of the strong need for them. The determination of preferred configurations for oxygen on Nb_4 [231] and the comparative investigation of the electronic and vibrational properties of Pt_3 and Pt_3-based carbonyls [232] can be regarded as attempts in this direction within the LDA–CP method.

3.11　HETEROATOM AND ALLOY CLUSTERS

Binary aggregates $A_n B_m$ are usually called 'heteroatom clusters' when $m = 1$ (or sometimes when $m \ll n$), and 'alloy clusters' otherwise. As mentioned in Section 3.4.3, the optimization of their structure is much more demanding than for the one-component clusters. This additional difficulty together with the practical inexistence of experiments explains the extreme limitation of theoretical work devoted to alloy clusters. Beyond the specific interest in their physical and chemical properties and in the effects of the small size compared to metal alloys, such systems would also be interesting as test cases of widely used model potentials. Some LDA calculations were performed with the CP method [233] on $Na_n K_n$ microclusters ($n = 5$ and 10), namely, both dynamical SA and MD at different temperatures. These simulations revealed a nontrivial behavior in the form of a clear tendency to segregate with the larger-size alkalis on the outermost shell, down to low temperatures. This phenomenon can be understood as resulting from an electronic mechanism that drives potassium to the lower-density regions and/or from an effect of atomic-size mismatch, in analogy with what happens in rare-gas mixtures, where the atoms having a larger size are located on the surface of the cluster. Segregation also seems to characterize the equilibrium geometries of $Na_n Cs_{2n}$ (n up to 30) [234], calculated within the SAPS (spherically averaged pseudopotential) model [121].

Early interest in heteroatom clusters having alkali metals as the 'host' was academic rather than dictated by precise observations. The main question regarded the extent to which the jellium-derived shell model retained its validity. However, this question was approached on the basis of oversimplified structural models in which the heteroatom (typically a closed-shell alkali-earth such as Mg) was located at the center of the cluster [235, 236]. In this hypothetical scheme, the perturbation of the electronic structure relative to that of the isoelectronic alkali cluster is somewhat trivial: for instance, in the $Na_n Mg$ system the presence of Mg would only alter the sequence of levels of the shell jellium model from 1s, 1p, 1s, 2s, ... (appropriate to sodium clusters) to 1s, 1p, 2s, 1d, ... (see also [236]). This would lead to the prediction that $Na_6 Mg$ and $Na_8 Mg$ are MNs.

Later, CP calculations for the $Na_n Mg$ system ($n = 6$–9 and 18) revealed quite a different scenario [237] (see Appendix C). In fact, in the case of the 8-electron system, for instance, only an isomer with high symmetry such as the octahedron and with Mg at the center has a shell-model-like electronic configuration. However, this geometry (as well as others of the kind for the larger clusters) is too high in energy to be a probable one. In the low-energy isomers, Mg was never at the cluster center, but rather a few atomic units away from the center of mass (with the notable exception of the C_{3v} isomer of $Na_7 Mg$, Figure 3.19(a)). More specifically, the low-energy isomers did not correspond to geometries of Na_n with Mg entering as an interstitial impurity atom but rather to substitutional isomers of Na_{n+1}. As shown in Figure 3.19(a), two growth routes could be identified, each corresponding to adding a 'new' sodium atom as a cap, but differing significantly, for example, in the symmetry: one ('pentagonal') has local fivefold symmetry, the other ('tetrahedral') has threefold symmetry and is based on a tetrahedron as the central seed. The former is analogous to the growth sequence found for pure Na clusters [123], whereas

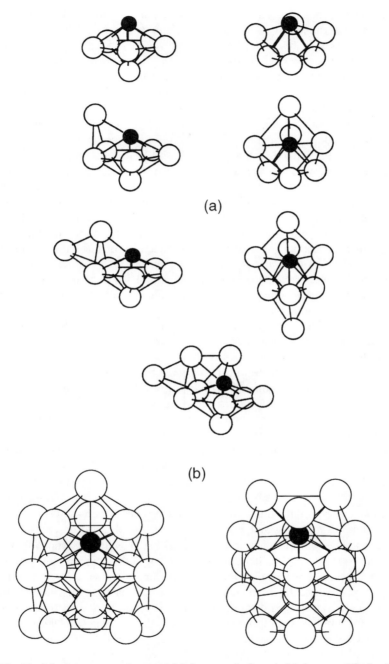

Figure 3.19 $Na_n Mg$ low-energy isomers (a) for $n = 6-9$ and (b) for $n = 18$. In (a) the two isomers correspond to the pentagonal (left) and the tetrahedral (right) growth paths. Reprinted from [237], with kind permission from Elsevier Science–NL, Sara Burgerhartstraat 25, 1055 KV Amsterdam, The Netherlands

the latter (slightly higher in energy) is energetically unfavorable in that case. For $n > 7$, Mg is no longer coordinated with all the sodium atoms. Figure 3.19(b) shows two degenerate low-energy isomers found for $Na_{18}Mg$, both with Mg fully embedded in the host cluster. The pentagonal growth continue at this size, as one of the lowest-energy isomers (on the left) can indeed be viewed as a substitutional isomer of Na_{19}. The other low-energy isomer (on the right) has a structure that consists of slightly bent layers in an hcp type of stacking. Whereas in the energetically unfavorable Mg-centered structures (and in the C_{3v} isomer of Na_7Mg, Figure 3.19(a)), the electronic configuration closely follows the predictions of the shell models, hybridization between different l-components was present in the lowest-energy isomers. Indeed, among $n = 6, 7$ and 8, no clear magic number has been observed experimentally [238], and at $n = 8$, moreover, a lowering of intensity occurs. In contrast to the Na_nMg system, a clear magic number at $n = 8$ ($N = 10$) is observed in the abundance spectra of Na_nZn and K_nMg, as predicted by jellium-shell models. By modeling the divalent atom with core sizes smaller than Mg [239], one could verify that, if the impurity is sufficiently smaller than the host atoms, it prefers to sit at the center of the cluster, thus reproducing the electronic level diagram of the jellium-shell model. To investigate the effects of the ionic charge of the impurity on the structure of the undoped cluster, CP calculations were extended to another $N = 10$ system, namely Na_7Al [239]. In particular, dynamical SA resulted in an impurity-centered low-energy structure similar to the C_{3v} isomer of Na_7Mg (Figure 3.19(a)), but with Al off-center by ~0.3 a.u. The electronic configuration turned out to be a closed-shell one. The same was found to hold for several isomeric and quasidegenerate forms, leading to the prediction of a magic number for $n = 7$ in the Na_7Al abundance spectrum.

An LDA investigation of the equilibrium geometries and of the electronic structure in Na_nPb systems was also performed [240] (see Appendix C). Structural optimization was accomplished with dynamical SA (which, at variance with the CP simulations, was made using Langevin MD) followed by conjugate-gradient minimizations. The motivation was the availability of mass spectra that in this case revealed a spectacular stability for the $n = 6$ cluster [241]. It was found that in this case it could be explained in terms of two related features of the calculated ground-state isomer: its high symmetry (O_h) and the closed-shell electronic configuration. This turned out to correspond to the shell-model picture, with, however, the 1s orbital largely split off from the rest and localized on Pb.

LSDA studies of equilibrium structures and electronic properties were also performed on $AlLi_n$ clusters ($1 < n < 8$, and $n = 17$) [242], using steepest-descent-like minimizations for the structural optimization and dynamical SA as well (see Appendix C). In these systems, at least up to $n = 5$, the heteroatom again does not occupy an internal position. In this size range, the evolution of the electronic structure is shown to be determined by the closing of the 3p shell of Al, which takes place at $n = 5$ and is responsible for the drop in the IP at $n = 6$. This drop would also have been predicted by the shell model, but this is only a coincidence. In fact, the electronic structure is definitely at odds with its predictions. In $AlLi_5$, the HOMOs are localized around Al, which acts as the electron-attracting center, and the chemical bonding of Al with the lithium atoms is found to be covalent. Further calculations led to the proposal of $AlLi_5$ as being the 'core' of larger metal clusters, such as $AlLi_{17}$, and also as the building block of 'molecular' clusters such as Al_2Li_{10}.

Finally, let us mention a recent calculation of the special heteroatom cluster $Al_{13}Na$ [243] (see Appendix C), which corresponds to a 40e system, namely, an MN of the shell model, and is isoelectronic to the Al_{13}^- anion discussed in Section 3.8. The result is not surprising: Al_{13} is electron-eager enough to close the electronic shell, corresponding to a relatively high value for the IP. This stabilizes a (distorted) icosahedron with Na as capping atom on a threefold site that closely corresponds to a shell-model-like sequence for the energy levels.

3.12   CONCLUSIONS

Our survey should have clarified how and to what extent DFT-based approaches can be applied to metal clusters. The multiple efforts of the past 10 years or so have certainly demonstrated that density functional theory with simple approximations for the exchange and correlation energy is a powerful method for an extensive exploration of the (Born–Oppenheimer) potential-energy surface of clusters, in particular of metal clusters. Moreover, coupled to the MD scheme of Car and Parrinello, it provides a computationally affordable tool for the (at least in principle) unbiased search for equilibrium structures via simulated annealing and for the determination of dynamical and thermal properties. From our overview of the applications, it soon becomes clear that the power of the CP method has not been exploited to the extent it could have been. The great majority of applications were structural determinations, and even in these, dynamical SA was only rarely used to satisfaction. Nevertheless, new insights into the physics and chemistry of some metal clusters have been obtained. In particular, these efforts have provided a reliable description and understanding of shell effects in alkali-metal clusters, led to the discovery of unforeseen geometries and the identification of paths in the structural growth of aggregates of simple metals such as Na_n, Al_n, Na_nMg and $AlLi_n$ as well as of some transition metals, and allowed the first (although still preliminary) exploration of the magnetic properties of Fe_n, Co_n and Ni_n clusters.

The extension of the theory to time-dependent phenomena is (slowly) expanding the reach of DFT to the domain of optical and spin spectroscopies. This advance is expected to greatly increase the value of DFT computations in general as a guide in the interpretation of experimental data. At the same time, it will enhance the chance of having benchmarks and obtaining useful feedback for the theoretical and computational schemes.

To close this chapter, it is important to discuss the impact DFT-based computational studies have had on the global development of metal-clusters physics in recent years and to estimate their potential for the future. As stated in almost every book or review on cluster physics, its primary aim is to describe the evolution of geometrical and electronic properties with increasing size from molecules to bulk solids. The scope and importance of this research field have been enhanced by the rapid development of nanotechnologies, based on the peculiar properties of micro- and mesoscopic structures. It is undeniable that most of the progress towards the long-term goals has been promoted by experiments, with computational studies trailing behind. The reason for this is that the *ab initio* characterization of even a single cluster in the $n \sim 100$ size range on the basis of atomistic models is still a computational 'tour de force', requiring resources that are available to only a handful of institutions and research groups. As a result, key questions have essentially been avoided, such as those concerning the growth to macroscopic (atomic and magnetic) structures, the thermodynamic phases of medium-sized clusters, their chemical reactivity, and their interaction with condensed phases such as surfaces or liquids.

An important impulse must come from experiments to justify the use of large computer resources to investigate metal clusters, such as, for example, either the discovery of intriguing effects that cannot be understood otherwise or a specific technological interest. Unfortunately, practical applications of size-selected clusters have so far failed to become a reality.

In fact, nowadays large-scale DFT-based computations are feasible on parallel computers and have been applied to the study of the structure and dynamics of few nanoscale systems using the CP method. These include metallofullerenes [244], peptide nanostructures [245], carbon nanotubes [246], sodium nanowires [247] and, more recently, large organic molecules on metal surfaces [248].

In the absence of either such a strong stimulus from experiment or a major breakthrough in computational techniques, this somewhat unsatisfactory situation will persist, and the global picture of clusters will still have to be assembled from a multitude of detailed studies. This collective undertaking can succeed if and only if its foundations are strong, and if the various contributions are mutually compatible and consistent. For this reason, throughout this chapter we have stressed the importance of having a clean theoretical framework and of ensuring computational rigor and thoroughness.

3.13   APPENDICES

3.13.1   Appendix A: DFT-related Methods

We briefly describe three methods that can be considered as related to DFT and that are sometimes used for cluster computations: the Thomas–Fermi (TF) approximation, the so-called DFT-based tight-binding (TB), and the many-body potentials of the embedded-atom type.

The TF approximation [249] was the precursor of modern DFT, as it expresses the total energy in terms of the electron density only. In contrast to the KS method, TF does not resort to independent electron orbitals for the computation of the electron kinetic energy, which is assumed to be a function of the electron density. The simplest version of TF corresponds to an LDA for the kinetic energy:

$$T[\rho] = \int \rho(\boldsymbol{r})t[\rho(\boldsymbol{r})]\,\mathrm{d}\boldsymbol{r},\tag{59}$$

where $t[\rho(\boldsymbol{r})]$ is the kinetic energy per electron of the (noninteracting) homogeneous electron gas at density $\rho(\boldsymbol{r})$:

$$t[\rho(\boldsymbol{r})] = \frac{3}{10m}[3\pi^2\rho(\boldsymbol{r})]^{2/3}.\tag{60}$$

Better approximations can be obtained by including a gradient correction, in analogy to what is done for the exchange–correlation energy. Unfortunately, simple approximations for $t(\boldsymbol{r})$ are far less successful that those for $\varepsilon_{\mathrm{XC}}$, and the limitations of TF are well documented in the literature (see, for instance, Ref. [17]). For the original TF method, the electron density diverges at the nuclear position and decays too slowly outside a finite system, no negatively charged ion can be bound, and, finally, no molecule or cluster can be stable with respect to dissociation into the constituent atoms. Several improvements proposed over time have overcome these limitations [17], but none provides a quantitative description of the electronic properties that approaches the accuracy of present-day KS schemes. Nevertheless, for large systems, TF sometimes provides the only computationally affordable approach, and for this reason it has been extensively used for clusters computations, especially in the spherical jellium model. These applications are discussed in Ref. [5].

In fact, the computational advantage of a 'density-only' scheme such as TF over the KS approach is so important that this type of approximation has been recently revived in connection with CP simulations [250]. Applications to simple metal clusters are reported in Ref. [251]. Unfortunately, the scheme developed in [250] seems to be limited to nearly free-electron systems such as alkali metals and aluminum, and for clusters the description is not quantitative, even for relatively simple properties such as equilibrium geometries.

Tight binding is a well-known solid-state method often used to interpolate electronic bands in terms of a few Hamiltonian matrix elements [50]. It can be derived from DFT by (i) expanding the KS orbitals on a small basis of localized, atomic-like orbitals, and (ii) introducing some (sometimes drastic) approximation for the multi-center integrals entering the energy functional (5). These two approximations result in an important reduction of the computational cost as compared with standard KS methods, which, unfortunately, is largely compensated by the corresponding loss of accuracy and predictive power. The so-called DFT–TB method [252] has also been developed in connection with the CP molecular dynamics in Ref. [253]. A similar method has been proposed independently in Ref. [254]. Both schemes have been used extensively for simulated carbon and other semiconductor clusters, while their application to metal clusters has been sporadic (see Ref. [255] for an example).

Far more extensive has been the application of semiempirical TB methods, in which the matrix elements are used as free parameters to fit experimental data, to metal clusters (and to transition-metal clusters in particular). However, a discussion of these applications is beyond the scope of our review.

Finally, interatomic potentials of the embedded atom [256] and effective medium type [257] have been used to simulate medium-to-large ($n = 10^2 - 10^5$) transition- and noble-metal clusters at nonzero temperatures. The relation of these schemes to DFT is discussed in Ref. [257]. Almost without exception, all applications rely on a semiempirical approach, starting from the functional form for the potential derived from DFT, and including a few free parameters that are used to fit experimental data. For small systems ($n \leq 10^2$), shell and other quantum effects (such as Jahn–Teller distortions and directionality in the bonds), not properly included in these models, are likely to be important, and the reliability of all these semiempirical potentials is therefore questionable. The situation could be more favorable for larger aggregates, as these models are known to be rather successful for extended systems, especially in the case of noble metals. Needless to say, these models are the only ones that allow at least a qualitative investigation of the structural and dynamical properties of large-size clusters at an affordable computational cost, and for this reason have been used extensively. However, even a short discussion of the large number of classical potential studies of metal clusters is beyond the scope of our review.

3.13.2 Appendix B: Details of CP Molecular Dynamics

We collect here a few technical details concerning the integration of the equations of motion derived from the CP Lagrangian.

For reasons of simplicity, low memory requirement and long time stability, the algorithm commonly used to integrate the CP equations of motion is the Verlet algorithm [80]:

$$x(t + \delta t) = 2x(t) - x(t - \delta t) + \delta t^2 \ddot{x}(t), \tag{61}$$

$$\dot{x}(t) = \tfrac{1}{2}[x(t + \delta t) - x(t - \delta t)], \tag{62}$$

where x is either an ionic or an electronic variable and δt the time step of the numerical integration. The error on x involved in this discretization of the time evolution is of order δt^4. A typical feature of the Verlet algorithm, apparent from the two equations above, is that the velocity at time t can be evaluated only after the position at time $t + \delta t$ has been computed. Recent implementations rely on the velocity Verlet algorithm [80], which is formally equivalent to the simple Verlet algorithm, but allows the simultaneous evaluation of positions and velocities. For both algorithms, typical values for δt range between 0.07 and 0.24 fs. This time step is somewhat shorter than those used in classical MD simulations, as it is constrained by the requirement of accurately integrating the fast electronic degrees of freedom.

The problem of integrating the equation of motion for degrees of freedom characterized by different time scales is alleviated by multiple-time-step algorithms [258]. In the case of the CP equations, two strategies can be envisaged. The simpler one is to use a different time step for the ionic (slow) and electronic (fast) degrees of freedom, as done in Ref. [259]. The second is to identify, among the electronic degrees of freedom, those limiting the stability of the time integration, and to evolve those last in time following a more accurate algorithm [260]. In the case of plane-wave implementations, only the second strategy seems to be able to provide a significant advantage over the simple Verlet algorithm.

Because of unavoidable numerical errors, the orthonormality condition for the KS orbitals is progressively violated with increasing time during a CP simulation, unless this condition is explicitly enforced at each time step. To conserve energy, the orthonormalization has to result from the application of an Hermitian operator. This excludes the simple Gram–Schmidt algorithm, whose effect on the KS states depends on their ordering. The orthogonalization method most commonly used in plane-wave CP implementations is the iterative algorithm described in Ref. [82]. Alternatively, it is possible to enforce the orthonormalization by writing the KS orbitals as the product of a unitary transformation on a set of orthonormal 'reference' orbitals, as detailed in Ref. [261].

For systems with a small but systematic violation of adiabaticity, the transfer of energy from the ions to the electrons results in the progressive decrease of the ionic temperature, while the electronic coefficients increasingly deviate from the Born–Oppenheimer surface. In such a case, the stability of the simulation can be improved by introducing additional degrees of freedom mimicking the presence of two distinct thermal 'baths' in contact with the ionic and electronic variables [262]. The 'bath' for the electronic degrees of freedom absorbs the excess energy flowing into the KS orbitals, bringing them back to the electron ground-state energy. At the same time, the ionic temperature is kept constant by thermal contact with a bath at fixed temperature T. This technique is an obvious extension to the CP dynamics of the well-known Nosé algorithm of classical MD [263]. More sophisticated algorithms based on the same idea are discussed in Ref. [264].

Most CP simulations of clusters aimed at investigating their behavior at finite temperature are made on the microcanonical ensemble. Only rarely have constant-temperature simulations using the Nosé algorithm [263] been reported.

3.13.3   Appendix C: Computational Details

In this appendix, we briefly recall details of the computational schemes adopted in the articles discussed in Sections 3.6 to 3.11. In a number of papers, unfortunately, important details are missing. This often renders the assessment of the reliability of the results difficult and their reproduction not possible. The methods/codes most commonly used are the following.

C1. Plane-waves and l-dependent DFT atomic pseudopotentials (CP and related). Distinct calculations differ in the choice of the exchange–correlation functional, the construction of the pseudopotentials and the energy cutoff of the plane-wave expansion, as well as in the size of the periodically repeated cell. Some calculations, especially those of charged clusters, use Hockney's boundary conditions [63, 64].

C2. Gaussian basis sets (obtained by fitting to numerical atomic pseudo wave functions) and l-dependent DFT atomic pseudopotentials (original code introduced in Ref. [6]). So far, all calculations are based on the LSDA of the exchange–correlation functional. Auxiliary Gaussians are employed to represent the electron density in the Hartree potential and also the exchange–correlation potential.

C3. Linear combination of Gaussian-type orbitals (LCGTO)—either all-electron or 'model' core potentials [229] (deMon–KS code, introduced in Ref. [265]). Auxiliary Gaussians are used as in C2. Calculations are based on either LDA or GGA.

Sometimes other schemes that are also based on LCGTOs are used. In the following, we give some examples from the various sections.

In Section 3.6:
Sodium clusters

- Reference [6]: C2 scheme; LSDA; 1e ($l = 0, 1$) pseudopotentials from Ref. [266].
- Reference [110]: C2 scheme; extension of the auxiliary Gaussians to improve the fitting of the polarization charge.
- Reference [111]: C3 scheme; all electron, LSDA and also GCs (from either Becke 88 [32] or Perdew–Wang (PW86) [33] for the exchange, and Perdew [34] for the correlation). LCGTO: double-zeta valence plus polarization (DZVP).
- Reference [123]: C1 scheme; LDA; 1e ($l = 0, 1$) pseudopotentials given in the paper (see errata in Ref. [267]) with nonlinear core corrections (see Figure 3.2(a)). For the sake of comparison, the results without nonlinear core corrections are also given. Fcc cell with $a = 54$ a.u.; plane-wave cutoff of 15 Ry.
- Reference [127]: C1 scheme as implemented in [64]; LDA; MT; s, p pseudopotentials; plane-wave cutoff of 5 Ry.

In Section 3.7:
Beryllium

- Reference [149]: C1 scheme; LSDA; pseudopotentials from Ref. [266]; simple cubic cell with $a = 22$ a.u.; plane-wave cutoff of 20.9 Ry.

Magnesium

- Reference [151]: C2 scheme; LSDA; pseudopotentials from Ref. [266].
- Reference [152]: C2 scheme; LDA and BP GCs; pseudopotentials from Ref. [266]; plane-wave cutoff of 15 Ry.

- Reference [153]: C1 scheme; LDA; pseudopotentials from Ref. [266]; simple cubic cell with $a = 46$ a.u.; plane-wave cutoff of 8 Ry.
- References [155, 156]: C1 scheme; LDA; s, p pseudopotentials given in Ref. [156]; fcc cell with $a = 54$ a.u.; plane-wave cutoff of 10 Ry.

In Section 3.8:
Aluminum

- References [171, 172]: C1 scheme; LDA; s, p pseudopotentials from Ref. [266]; simple cubic cell with $a = 30$ a.u. ($n < 6$) or 36 a.u. ($n \geq 6$); plane-wave cutoff of 10.6 Ry.
- Reference [156]: C1 scheme; LDA; s, p pseudopotentials (analytic given in the paper); fcc cell with $a = 54$ a.u.; plane-wave cutoff of 10 Ry.
- Reference [89]: C1 schemes; plane-wave cutoff of 4 or 6 Ry ('softened' pseudopotentials); no other details given.

Gallium

- Reference [172]: the same as for Al; C1 scheme; LDA; s, p pseudopotentials from Ref. [268].

In Section 3.9:
Iron, cobalt, and nickel

- References [196, 199, 200, 202]: C3 scheme; all-electron; in Ref. [196], basis sets of DZVP quality, optimized explicitly for DFT calculations; LSDA and GGA (PW86 for the exchange [33] and Perdew [34]).

Chromium

- Reference [210] LSDA; AE-frozen core; double numerical basis set augmented by polarization functions. The code uses the so-called D-mol method developed in Ref. [272].

Nickel

- Reference [203]: C2 scheme; pseudopotentials from Ref. [266].
- Reference [201]: C1 scheme; pseudopotentials constructed with the MT procedure [70] (see Figure 3.2(b)); simple cubic cell with $a = 20$ a.u.; plane-wave cutoff of 70 Ry.

In Section 3.10:
Copper

- Reference [73]: C1 scheme; LDA; soft-core Vanderbilt pseudopotentials [269]; simple cubic cell with $a = 25.9$ a.u.; cutoff of 15 Ry, and augmentation density with a cutoff of 130 Ry.

Niobium

- Reference [223]: C1 scheme; MT; scalar-relativistic 13e pseudopotentials with s, p, d radii given in the paper; spin-polarized LDA and B-LYP GC; plane-wave cutoff of 60 Ry; Hockney's boundary conditions. The effect of a 5e-pseudopotential was studied, with and without nonlinear core corrections.
- Reference [229]: C3 scheme with model core potential; 11e in the valence shell; LSDA and BP GC functionals.

- Reference [113]: C3 scheme with model core potential and core-projection operators for valence wave functions as explained in the paper; Gaussian basis set (given in the paper) 11e in the valence shell; LSDA and BP GC functionals.

Silver

- Reference [226]: C2 scheme with scalar-relativistic 11e s, p, d pseudopotentials (given in the paper); also given are results with 1e pseudopotentials from [266] for the sake of comparison; LDA.
- Reference [227]: DGaus code [270]: AE-LCGTO (DFT-optimized Gaussians); spin-polarized BP GC.

Gold

- Reference [228]: AE-LCGTO-based code as developed originally in Ref. [271]; scalar-relativistic; LSDA and GGA corrections to the energy; many details given in the paper.

In Section 3.11:
NaK

- Reference [233]: C1 scheme; LDA; s, p pseudopotentials; fcc cell with $a = 54$ a.u. and a plane-wave cutoff of 5 Ry.

NaMg

- Reference [237,239]: C1 scheme, details as in Section 3.6 for Na (see above).

NaPb

- Reference [240]: C1 scheme; LDA; MT; s, p pseudopotentials for both Na and Pb. In the latter the frozen core is [Xe], and NLCC taken into account. $R_c = 2.3$ and 2.5 for $l = 0, 1$ in Na, and 3.18 for both $l = 0$ and 1 in Pb; bcc with $a = 30$ a.u.; plane-wave cutoff of 9 Ry.

AlLi

- Reference [242]: C1 scheme; s, p pseudopotentials for both Li and Al constructed via the MT procedure; plane-wave cutoff of 15 Ry. $R_c(\mathrm{Al}) = 2.1$ and 2.5 a.u. for $l = 0, 1$ respectively; $R_c(Li) = 2.2$ and 2.8 a.u. for $l = 0, 1$ respectively. Hockney's boundary conditions.

AlNa

- Reference [243]: C1 scheme; LDA; s, p pseudopotentials from [266]; plane-wave cutoff of 11.5 Ry; fcc cell with $a = 38.5$ a.u.

3.14   ACKNOWLEDGMENTS

We wish to thank H. Grönbeck for useful discussions, and C. Bolliger for her patient and careful editing of the text.

3.15 REFERENCES AND NOTES

[1]   W. D. Knight, K. Clemenger, W. A. de Heer, W. A. Saunders, M. Y. Cho and M. L. Cohen, *Phys. Rev. Lett.* **52**, 2141 (1984).
[2]   W. Ekardt, *Phys. Rev. B* **31**, 6360 (1985).
[3]   W. Ekardt, W.-D. Schöne and J. M. Pacheco, see Chapter 1 of this book.
[4]   W. A. de Heer, *Rev. Mod. Phys.* **65**, 611 (1993).
[5]   M. Brack, *Rev. Mod. Phys.* **65**, 677 (1993).
[6]   For the early applications for DFT to clusters, see J. L. Martins, J. Buttet and R. Car, *Phys. Rev. B* **31**, 1804 (1985) and references therein.
[7]   T. G. Dietz, M. A. Duncan, D. E. Powers and R. E. Smalley, *J. Chem. Phys.* **74**, 6511 (1981).
[8]   R. Car and M. Parrinello, *Phys. Rev. Lett.* **55**, 2471 (1985).
[9]   R. Car, M. Parrinello and W. Andreoni, in *Microclusters*, ed. S. Sugano, Y. Nishina and S. Ohnishi (Springer, Berlin, 1987), pp. 134–141.
[10] W. Andreoni, in Proc. Enrico Fermi School on *The Chemical Physics of Atomic and Molecular Clusters*, Varenna 1988, ed. G. Scoles and S. Stringari (Società Italiana di Fisica Publ., Bologna, 1990), pp. 159–175.
[11] C. Yannouleas, R. N. Barnett and U. Landman, see Chapter 4 of this book.
[12] S. J. A. van Gisbergen, F. Koostra, P. R. T. Schipper, O. V. Gritsenko, J. G. Snijders and E. J. Baerends, *Phys. Rev. A* **57**, 2556 (1998) and references therein.
[13] R. Bauernschmitt, R. Ahlrichs, F. H. Hennrich and M. M. Kappes, *J. Am. Chem. Soc.* **120**, 5052 (1998) and references therein.
[14] This definition, although commonly adopted, can sometimes be ambiguous and misleading. In fact, the notion of metal versus nonmetal is not obvious in clusters, because the standard definitions in terms of conductivity, Fermi surface, etc. do not apply for finite systems.
[15] N. Van Giai, *Prog. Theor. Phys. Supplement* **124**, 1 (1996).
[16] P. Hohenberg and W. Kohn, *Phys. Rev.* **136**, B864 (1964); W. Kohn and L. J. Sham, *Phys. Rev.* **140**, A1133 (1965).
[17] (a) G. P. Srivastava and D. Weaire, *Adv. Phys.* **36**, 463 (1987); (b) R. O. Jones and O. Gunnarsson, *Rev. Mod. Phys.* **61**, 689 (1989); (c) R. G. Parr and W. Yang, *Density Functional Theory of Atoms and Molecules*, Oxford University Press, Oxford, 1989; (d) R. M. Dreizler and E. K. U. Gross, *Density Functional Theory*, Springer, Berlin, 1990.
[18] E. K. Gross, J. F. Dobson and M. Petersilka, in *Density Functional Theory II*, Topics in Current Chemistry, Vol. **181**, ed. R. F. Nalewajski (Springer, Berlin, Heidelberg, 1996), pp. 81–172.
[19] Quantum Monte Carlo (QMC) data for the electron gas have been published in D. M. Ceperley and B. J. Adler, *Phys. Rev. Lett.* **45**, 566 (1980); G. Ortiz and P. Ballone, *Phys. Rev. B* **50**, 1391 (1994). The applications of LDA rely on simple interpolations of the QMC data. The most commonly used of these interpolations have been proposed by S. H. Vosko, L. Wilk and M. Nusair, *Can. J. Phys.* **58**, 1200 (1980), and in Ref. [55]. More recently, a new interpolation has been introduced by S. Goedecker, M. Teter and J. Hutter, *Phys. Rev. B* **54**, 1703 (1996).
[20] U. von Barth and L. Hedin, *J. Phys. C* **5**, 1629 (1972).
[21] M. M. Pant and A. K. Rajagopal, *Solid State Commun.* **10**, 1157 (1972).
[22] O. Gunnarsson and B. I. Lundqvist, *Phys. Rev. B* **13**, 4274 (1976).
[23] T. Ziegler, A. Rauk and E. J. Baerends, *Theor. Chim. Acta* **43**, 261 (1977).
[24] U. von Barth, *Phys. Rev. A* **20**, 1693 (1979).
[25] J. A. Pople, P. M. W. Gill and N. C. Handy, *Int. J. Quantum Chem.* **56**, 303 (1995); D. Harris and G. H. Loew, *J. Am. Chem. Soc.* **101**, 3959 (1997).
[26] R. Nathans and S. J. Pickart, in *Magnetism*, ed. G. T. Rado and H. Suhl (Academic Press, New York, 1963), p. 211.
[27] T. Oda, A. Pasquarello and R. Car, *Phys. Rev. Lett.* **80**, 3622 (1998).
[28] A precise assessment of the accuracy of DFT schemes is rendered difficult by several additional approximations of numerical origin that are always involved in practical computations; see the discussion in Section 3.3.2 on basis sets, geometry optimization, etc.

[29]   The equilibrium distances predicted by LDA for Be_2, Mg_2, Cd_2 and Hg_2 are 2.36, 3.33, 3.05 and 2.99 Å, respectively [146]. The corresponding experimental values are 2.47, 3.89, 4.81 and 3.63 Å, from K. P. Huber and G. Herzberg, *Constants of Diatomic Molecules*, Van Nostrand Reinhold, New York, 1979.

[30]   Note that data from observed spectra may contain anharmonicity effects that are not accounted for in the calculation.

[31]   M. Ernzerhof, J. P. Perdew and K. Burke, in *Density Functional Theory I*, Topics in Current Chemistry, Vol. 180, ed. R. F. Newajski (Springer-Verlag, Berlin, Heidelberg, 1996).

[32]   A. D. Becke, *J. Chem. Phys.* **84**, 4524 (1986); *Phys. Rev. A* **38**, 3098 (1988) (also referred to as B86 and B88, respectively).

[33]   J. P. Perdew, in *Electronic Structure of Solids '91*, ed. P. Ziesche and H. Eschrig (Akademie Verlag, Berlin, 1991), p. 11 (referred to as PW91); see also J. P. Perdew and Y. Wang, *Phys. Rev. B* **33**, 8800 (1986) (referred to as PW86).

[34]   J. P. Perdew, *Phys. Rev. B* **33**, 8822 (1986); Erratum **34**, 7406 (1986).

[35]   C. Lee, W. Yang and R. G. Parr, *Phys. Rev. B* **37**, 785 (1988).

[36]   J. P. Perdew, K. Burke and M. Ernzerhof, *Phys. Rev. Lett.* **77**, 3865 (1996).

[37]   See, for instance, A. Dal Corso, A. Pasquarello, A. Baldereschi and R. Car, *Phys. Rev. B* **53**, 1180 (1996), for solids; F. W. Kutzler and G. S. Painter, *Phys. Rev. B* **45**, 3236 (1992), and J. P. Perdew, K. Burke and M. Ernzerhof, *Phys. Rev. Lett.* **80**, 891 (1998), for diatomic molecules.

[38]   See for instance, G. Boisvert and L. J. Lewis, *Phys. Rev. B* **56**, 7643 (1997), for a recent computation of surface energies and relaxations of the clean (100) surface of Cu.

[39]   F. Aryasetiawan and O. Gunnarsson, *Rep. Prog. Phys.* **61**, 237 (1998).

[40]   The exact exchange–correlation potential for light atoms has been computed from 'exact' electron densities provided by CIs and quantum MC computations or other high-quality correlated wave functions. See, for instance, Ref. [41](b), or C. J. Umrigar and X. Gonze, *Phys. Rev. A* **50**, 3827 (1994).

[41]   (a) The asymptotic behavior of the GC based on the Becke exchange is $-a/r^2$, where a is a constant much smaller than 1 (see Ref. [146]; E. Engel, J. A. Chevary, L. D. Macdonald and S. H. Vosko, *Z. Phys. D* **23**, 7 (1992)). (b) A gradient-corrected scheme with the correct asymptotic behavior is discussed in R. Van Leeuwen and E. J. Baerends, *Phys. Rev. A* **49**, 2421 (1994). Note that, while the exchange–correlation potential has a simple analytic form, the corresponding exchange–correlation energy cannot be expressed in closed form.

[42]   J. P. Perdew and M. Levy, *Phys. Rev. Lett.* **51**, 1884 (1983); L. J. Sham and M. Schlüter, *Phys. Rev. Lett.* **51**, 1888 (1983).

[43]   M. Levy, J. P. Perdew and V. Sahni, *Phys. Rev. A* **30**, 2747 (1984).

[44]   R. Neumann, R. H. Nobes and N. C. Handy, *Mol. Phys.* **87**, 1 (1996); L. Kleinman, *Phys. Rev. Lett.* **56**, 12042 (1997).

[45]   H. Grönbeck, A. Rosén and W. Andreoni (unpublished).

[46]   This is certainly the most favorable case for LDA. Still, we find that the highest eigenvalue for a $20e$ sphere at $r_s = 4$ is −2.9 eV, whereas the IP is 4.02 eV.

[47]   S. B. Trickey, *Phys. Rev. Lett.* **56**, 881 (1986).

[48]   E. Clementi, S. J. Chakravorty, G. Corougin and V. Carrevetta, in *MOTECC. Modern Techniques in Computational Chemistry*, ed. E. Clementi (ESCOM, Leiden, 1989), pp. 589–613, and references therein.

[49]   A. D. Becke, *J. Chem. Phys.* **98**, 5648 (1993).

[50]   See, for instance, N. W. Ashcroft and N. D. Mermin, *Solid State Physics*, Holt-Saunders International Editions, New York, 1976.

[51]   F. R. Vikajlovic, E. L. Shirley and R. M. Martin, *Phys. Rev. B* **43**, 3994 (1991); G. Engel, *Phys. Rev. Lett.* **78**, 3515 (1997).

[52]   R. T. Sharp and G. K. Horton, *Phys. Rev.* **90**, 317 (1953); often computations use the approximation to the OEP model proposed by J. B. Krieger, Y. Li and G. J. Iafrate, *Phys. Rev. A* **45**, 101 (1992).

[53]   D. M. Bylander and L. Kleinman, *Phys. Rev. Lett.* **74**, 3660 (1995) and *Phys. Rev. B* **52**, 14566 (1995).

[54]   I. Lindgren, *Int. J. Quantum Chem.* **5**, 411 (1971).

[55]   J. P. Perdew and A. Zunger, *Phys. Rev. B* **23**, 5048 (1981).

[56] E. S. Fois, J. I. Penman and P. A. Madden, *J. Chem. Phys.* **98**, 6352 (1993).

[57] J. A. Alonso and L. A. Girifalco, *Solid State Commun.* **24**, 135 (1977); O. Gunnarsson, M. Jonson and B. I. Lundqvist, *Phys. Rev. B* **20**, 3136 (1979).

[58] J. Harris and R. O. Jones, *J. Phys. F* **4**, 1170 (1974).

[59] L. C. Balbas, A. Rubio, J. A. Alonso and G. Borstel, *J. Chem. Phys.* **86**, 799 (1989); L. C. Balbas, J. A. Alonso and A. Rubio, *Europhys. Lett.* **14**, 323 (1991).

[60] W. E. Pickett, *Comput. Phys. Rep.* **9**, 115 (1989).

[61] M. C. Strain, G. E. Scuseria and M. J. Frisch, *Science* **271**, 51 (1996).

[62] A high value of $|G|_{max}$ corresponds to a large number N_{pw} of plane-waves in the expansion ($N_{pw} \sim |G|_{max}^3$) and to a high degree of flexibility and completeness of the basis set.

[63] R. W. Hockney, *Methods Comput. Phys.* **9**, 136 (1970).

[64] R. N. Barnett and U. Landman, *Phys. Rev. B* **48**, 2081 (1993).

[65] K. Cho, T. A. Arias, J. D. Joannopoulos and P. K. Lam, *Phys. Rev. Lett.* **71**, 1808 (1993); S. Wei and M. Y. Chou, *Phys. Rev. Lett.* **76**, 2650 (1996).

[66] J. Bernholc, E. L. Briggs, D. J. Sullivan, C. J. Brabec, M. Buongiorno Nardelli, K. Rapcewicz, C. Roland and M. Wensell, *Int. J. Quantum Chem.* **65**, 531 (1997).

[67] G. Lippert, J. Hutter and M. Parrinello, *Mol. Phys.* **92**, 477 (1997).

[68] H. Ehrenreich, F. Seitz and D. Turnbull, eds, *Solid State Physics*, Advances in Research and Applications, Vol. 24, Academic Press, New York, London, 1970.

[69] D. R. Hamann, M. Schlüter and C. Chiang, *Phys. Rev. Lett.* **43**, 1494 (1979).

[70] One of the prescriptions that is currently used most extensively is that introduced by N. Troullier and J. L. Martins, *Phys. Rev. B* **43**, 1993 (1991).

[71] S. G. Louie, S. Froyen and M. L. Cohen, *Phys. Rev. B* **26**, 1738 (1982).

[72] D. Vanderbilt, *Phys. Rev. B* **41**, 7892 (1990).

[73] C. Massobrio, A. Pasquarello and R. Car, *Chem. Phys. Lett.* **238**, 215 (1995).

[74] W. H. Press, B. P. Flannery, S. A. Teukolsky and W. T. Vetterling, *Numerical Recipes*, Cambridge University Press, Cambridge, 1986.

[75] I. Stich, R. Car, M. Parrinello and S. Baroni, *Phys. Rev. B* **39**, 4997 (1989).

[76] Y. Saad, *Numerical Methods for Large Eigenvalue Problems*, Manchester University Press, New York, 1992.

[77] P. Császár and P. Pulay, *J. Molec. Struct.* **114**, 31 (1984); J. Hutter, H. P. Lüthi and M. Parrinello, *Comput. Mater. Sci.* **2**, 244 (1994).

[78] M. C. Payne, M. P. Teter, D. C. Allan, T. A. Arias and J. D. Joannopoulos, *Rev. Mod. Phys.* **64**, 1045 (1992).

[79] R. Pulay, *Mol. Phys.* **17**, 197 (1969).

[80] M. P. Allen and D. J. Tildesley, *Computer Simulation of Liquids.*, Clarendon, Oxford, 1989.

[81] D. Reichardt, V. Bonačič-Koutecký, P. Fantucci and J. Jellinek, *Chem. Phys. Lett.* **279**, 129 (1997); B. Hartke and E. A. Carter, *Chem. Phys. Lett.* **189**, 358 (1992).

[82] For reviews see: G. Galli and M. Parrinello, in *Computer Simulations in Material Science*, NATO ASI Series E: Applied Sciences, Vol. 205 ed. M. Meyer and V. Pontikis (Kluwer, Dordrecht, 1991), pp. 283–304; and R. Car, in *Monte Carlo and Molecular Dynamics of Condensed Matter Systems*, ed. K. Binder and G. Ciccotti (Italian Physical Society Publ., Bologna, 1995), pp. 601–634.

[83] G. Pastore, E. Smargiassi, and F. Buda, *Phys. Rev. A* **44**, 6334 (1991).

[84] M. C. Payne, *J. Phys. Cond. Matter* **1**, 2199 (1989).

[85] The equivalence of the search for the ground-state structure of an n-atom cluster and the 'traveling salesman' problem is proved by L. T. Wille and J. Vennik, *J. Phys. A* **18**, L419 (1985), for the case of pair potentials.

[86] M. R. Hoare and J. A. McInnes, *Adv. Phys.* **32**, 791 (1983).

[87] S. Kirkpatrick, C. D. Gelatt Jr and M. P. Vecchi, *Science* **220**, 671 (1983).

[88] N. Binggeli, J. L. Martins and J. R. Chelikowsky, *Phys. Rev. Lett.* **68**, 1992 (1992).

[89] J-Y. Yi, D. J. Oh and J. Bernholc, *Phys. Rev. Lett.* **67**, 1594 (1991).

[90] D. M. Deaven and K. M. Ho, *Phys. Rev. Lett.* **75**, 288 (1995).

[91] K. J. Jalkanen, S. Suhai and H. Bohr, in *Theoretical and Computational Methods in Genome Research*, ed. S. Suhai (Plenum Press, New York, 1997).

[92] S. Baroni, P. Giannozzi and A. Testa, *Phys. Rev. Lett.* **58**, 1861 (1987) and **59**, 2662 (1987).

[93] J. Kohanoff, W. Andreoni and M. Parrinello, *Phys. Rev. B* **46**, 4371 (1992).

[94]   W. A. Harbich, S. Fedrigo and J. Buttet, *Chem. Phys. Lett.* **195**, 613 (1992).
[95]   D. Bohm and D. Pines, *Phys. Rev.* **92**, 609 (1953).
[96]   A. Zangwill and P. Soven, *Phys. Rev. A* **21**, 1561 (1980).
[97]   L. Hedin and S. Lundqvist, in *Solid State Physics. Advances in Research and Applications,* Vol. **23**, ed. F. Seitz, H. Ehrenreich and D. Turnbull (Academic Press, New York, London, 1969), pp. 2–181.
[98]   A. L. Fetter and J. D. Walecka, *Quantum Theory of Many Particle Systems*, McGraw-Hill, New York, 1971.
[99]   The computation of the unoccupied states is not strictly required, see S. Baroni and A. Quattropani, *Nuovo Cimento Soc. Ital. Fis.* **5D**, 89 (1985); A. A. Quong and A. G. Eguiluz, *Phys. Rev. Lett.* **70**, 3955 (1993).
[100] D. E. Beck, *Phys. Rev. B* **35**, 7325 (1987).
[101] J. M. Pacheco and J. L. Martins, *J. Chem. Phys.* **106**, 6039 (1997).
[102] S. Saito, S. B. Zang, S. G. Louie and M. Cohen, *Phys. Rev. B* **40**, 3643 (1989).
[103] G. Onida, L. Reining, R. W. Godby, R. Del Sole and W. Andreoni, *Phys. Rev. Lett.* **75**, 818 (1995).
[104] W. D. Knight, K. Clemenger, W. A. de Heer and W. A. Saunders, *Phys. Rev. B* **31**, 2539 (1985); W. D. Knight, W. A. de Heer, K. Clemenger and W. A. Saunders, *Solid State Commun.* **53**, 445 (1985).
[105] W. A. de Heer, P. Milani and A. Châtelain, *Phys. Rev. Lett.* **63**, 2834 (1989).
[106] P. Milani, I. Moullet and W. A. de Heer, *Phys. Rev. A* **42**, 5150 (1990).
[107] It is well known that a system in a uniform electric field does not have a well-defined ground state. Evaluating α by perturbation theory or finite differences therefore requires a limiting process with an external field whose intensity vanishes as its range grows to infinity. For finite systems, such as clusters, this limit does not present fundamental problems.
[108] G. D. Mahan, *Phys. Rev. A* **22**, 1780 (1980).
[109] D. E. Beck, *Phys. Rev. B* **30**, 6935 (1984).
[110] I. Moullet, J. L. Martins, F. Reuse and J. Buttet, *Phys. Rev. Lett.* **65**, 476 (1990) and *Phys. Rev. B* **42**, 11598 (1990).
[111] J. Guan, M. E. Casida, A. M. Köster and D. R. Salahub, *Phys. Rev. B*, **52**, 2184 (1995).
[112] C. Massobrio, A. Pasquarello and R. Car, *Phys. Rev. Lett.* **75**, 2104 (1995) and *Phys. Rev. B* **54**, 8913 (1996).
[113] R. Fournier, T. Pang and C. Chen, *Phys. Rev. A* **57**, 3683 (1998).
[114] D. P. Chong, *Chem. Phys. Lett.* **232**, 486 (1995).
[115] H. Kietzmann, J. Morenzin, P. S. Bechthold, G. Ganteför, W. Eberhardt, D. S. Yang, P. A. Hackett, R. Fournier, T. Pang and C. Chen, *Phys. Rev. Lett.* **77**, 4528 (1996).
[116] A. Abragam and B. Bleany, *Electron Paramagnetic Resonance of Transition Ions*, Clarendon, Oxford, 1970.
[117] Note that the computation of hyperfine interactions is much more complex when spin–orbit effects are included, see for instance, E. van Lenthe, P. E. S. Wormer and A. van der Avoird, *J. Chem. Phys.* **107**, 2488 (1997).
[118] E. van Lenthe, A. van der Avoird and P. E. S. Wormer, *J. Chem. Phys.* **108**, 4783 (1998).
[119] C. G. van de Walle and P. E. Blöchl, *Phys. Rev. B* **47**, 4244 (1993); M. Boero, A. Pasquarello, J. Sarnthein and R. Car, *Phys. Rev. Lett.* **78**, 887 (1997).
[120] M. Boero and W. Andreoni (unpublished).
[121] The simple electronic structure of sodium also renders the application of other types of models relatively easy and extendable to relatively large sizes. See e.g. Hückel calculations and MC structural search (R. Poteau and F. Spiegelmann, *Phys. Rev. B* **45**, 1878 (1992) and *J. Chem. Phys.* **98**, 6540 (1993); Erratum **99**, 10 089 (1993)); or the so-called spherically averaged pseudopotential (SAPS) model (M. D. Glossman, J. A. Alonso and M. P. Iñiguez, *Phys. Rev. B* **47**, 4747 (1993)). This is a simplified atomistic scheme, in which the 'external potential' (written as the sum of the atomic pseudopotentials) acting on the electrons is developed in spherical harmonics around the cluster center of mass, and only the spherical component is retained in the solution of the KS equations.
[122] J. Flad, H. Stoll and H. Preuss, *J. Chem. Phys.* **71**, 3042 (1979).
[123] U. Röthlisberger and W. Andreoni, *J. Chem. Phys.* **94**, 8129 (1991).

[124] K. Selby, M. Vollmer, J. Masui, V. Kresin, W. A. de Heer and W. D. Knight, *Phys. Rev. B* **40**, 5417 (1989).

[125] V. Bonačič-Koutecký, J. Pittner and J. Koutecký, *J. Chem. Phys.* **104**, 1427 (1996).

[126] In a recent paper by C. Yannouleas and U. Landman (*Phys. Rev. Lett.* **78**, 1424 (1997)), the smearing-out of the shell-closing markings in the IPs, observed especially in K and Ag clusters, has been interpreted as a temperature effect and explicitly ascribed to the electronic entropy contribution. This interpretation is based on a liquid droplet model modified for the inclusion of shell-corrections.

[127] A. Rytkönen, H. Häkkinen and M. Manninen, *Phys. Rev. Lett.* **80**, 3940 (1998).

[128] R. N. Barnett, U. Landman, A. Nitzan and G. Rajagopal, *J. Chem. Phys.* **94**, 608 (1991).

[129] For recent calculations see: R. Rousseau and D. Marx, *Phys. Rev. A* **56**, 617 (1997); R. O. Jones, A. I. Lichtenstein and J. Hutter, *J. Chem. Phys.* **106**, 4566 (1997).

[130] R. P. Feynman, *Statistical Mechanics.* Benjamin, Reading, MA, 1972.

[131] P. Ballone and P. Milani, *Phys. Rev. B* **45**, 11222 (1992).

[132] R. Rousseau and D. Marx, *Phys. Rev. Lett.* **80**, 2574 (1998).

[133] R. O. Weht, J. Kohanoff, D. A. Estrin and C. Chakravarty, *J. Chem. Phys.* **108**, 8848 (1998).

[134] The first excitation energies for Be, Mg, Ca, Sr, Ba, Zn, Cd and Hg are 2.72, 2.71, 1.88, 1.76, 1.12, 4.01, 3.73 and 4.67 eV, respectively (A. A. Radzig and B. M. Smirnov, *Reference Data on Atoms, Molecules, and Ions*, Springer, Heidelberg, 1985). For the sake of comparison, we also list the corresponding values for He, Ne, Ar, Kr, Xe, which are 19.82, 16.62, 11.55, 9.92 and 8.32 eV, respectively.

[135] It is worth noting that the positively charged dimers $M_2{}^+$ have a binding energy of the order of 1 eV/atom.

[136] The terms 'van der Waals' and 'metallic' bonding are used here in an intuitive and nonrigorous way. We emphasize that the onset of 'metallic' bonding does not exclude important contributions to cohesion from dispersion forces.

[137] K. Rademann, B. Kaiser, U. Even and F. Hensel, *Phys. Rev. Lett.* **59**, 2319 (1987); B. Cabaud, A. Hoareau and P. Melinon, *J. Phys. D* **13**, 1831 (1980).

[138] C. Bréchignac, M. Broyer, Ph. Cahuzac, G. Delacretaz, P. Labastie, J. P. Wolf and L. Wöste, *Phys. Rev. Lett.* **60**, 275 (1988).

[139] I. Katakuse, T. Ichihara, Y. Fujita, T. Matsuo, T. Sakurai and H. Matsuda, *Int. J. Mass. Spectrom. Ion Processes* **69**, 109 (1986).

[140] M. Ruppel and K. Rademann, *Chem. Phys. Lett.* **197**, 280 (1992); K. Rademann, M. Ruppel and B. Kaiser, *Ber. Bunsenges. Phys. Chem.* **96**, 1204 (1992).

[141] The jellium-like features coexist with several others that do not fit into the jellium picture at all, see the discussion in Ref. [4].

[142] Mg: T. P. Martin, T. Bergmann, H. Göhlich and T. Lange, *Chem. Phys. Lett.* **176**, 343 (1991). Ca: T. P. Martin, U. Naher, T. Bergmann, H. Göhlich and T. Lange, *Chem. Phys. Lett.* **183**, 119 (1991). Ba: D. Rayane, P. Melinon, B. Tribollet, B. Chabaud, A. Hoareau and M. Broyer, *J. Chem. Phys.* **91**, 3100 (1989).

[143] M. Krauss and F. H. Mies, in *Excimer Lasers,* Topics in Applied Physics, Vol. **30**, ed. C. K. Rhodes (Springer-Verlag, Heidelberg, 1979).

[144] R. O. Jones, *J. Chem. Phys.* **71**, 1300 (1979).

[145] V. E. Bondybey and J. H. English, *J. Chem. Phys.* **80**, 568 (1984).

[146] See, for instance, G. Ortiz and P. Ballone, *Phys. Rev. B* **43**, 6376 (1991).

[147] Highly accurate quantum-chemistry results are available for Be_2 (CI: B. H. Lengsfield III, A. D. McLean, M. Yoshimine and B. Liu, *J. Chem. Phys.* **79**, 1891 (1983); diffusion quantum MC: C. Filippi and C. Umrigar, *J. Chem. Phys.* **105**, 213 (1996)) and Mg_2 (CI: H. Partridge, C. W. Bauschlicher, Jr, L. G. M. Pettersson, A. D. McLean, B. Liu, M. Yoshimine and A. Komornicki, *J. Phys. Chem.* **92**, 5377 (1990)).

[148] Note that this limitation concerns approximate DFT schemes, while exact DFT, which by definition provides the true ground-state energy, also includes dispersion forces. For an indication of how to include van der Waals into approximate DFT, see, for instance, W. Kohn, Y. Meir and D. E. Makarov, *Phys. Rev. Lett.* **80**, 4153 (1998).

[149] R. Kawai and J. H. Weare, *Phys. Rev. Lett.* **65**, 80 (1990).

[150]   S. N. Khanna, F. Reuse and J. Buttet, *Phys. Rev. Lett.* **61**, 535 (1988). Note that positively charged species, all characterized by standard chemical bonds, are not affected by the problems mentioned above for the neutral species.

[151]   F. Reuse, S. N. Khanna, V. de Coulon and J. Buttet, *Phys. Rev. B* **41**, 11743 (1990).

[152]   V. de Coulon, P. Delaly, P. Ballone, J. Buttet and F. Reuse, *Z. Phys. D* **19**, 173 (1991); P. Delaly, P. Ballone and J. Buttet, *Phys. Rev. B* **45**, 3838 (1992).

[153]   V. Kumar and R. Car, *Phys. Rev. B* **44**, 8243 (1991).

[154]   On the basis of LDA calculations, it has been proposed that a new form of bulk Mg can be built by 'bringing together' Mg_4 clusters (S. N. Khanna and P. Jena, *Phys. Rev. Lett.* **69**, 1664 (1992)). Such a proposal is, however, highly questionable.

[155]   W. Andreoni and U. Röthlisberger, in *Nuclear Physics Concepts in the Study of Atomic Cluster Physics*, Lecture Notes in Physics, Vol. **404**, ed. R. Schmidt, H. O. Lutz and R. Dreizler (Springer-Verlag, Berlin Heidelberg, 1992), pp. 353–357.

[156]   U. Röthlisberger, W. Andreoni and P. Giannozzi, *J. Chem. Phys.* **96**, 1248 (1992).

[157]   V. Bouton, A. R. Allouche, F. Spiegelmann, J. Chevaleyre and M. Aubert Frécon, *Eur. Phys. J. D* **2**, 63 (1998).

[158]   M. E. Garcia, G. M. Pastor and K. H. Bennemann, *Phys. Rev. Lett.* **67**, 1142 (1991).

[159]   R. W. G. Wyckof, *Crystal Structures*, Wiley Interscience, New York, 1968.

[160]   See, for instance, L. Hanley and S. L. Anderson, *J. Phys. Chem.* **91**, 5161 (1987); L. Hanley, J. L. Whitten and S. L. Anderson, *J. Phys. Chem.* **92**, 5803 (1988) and **94**, 2218 (1990); L. Hanley and S. L. Anderson, *J. Chem. Phys.* **89**, 2848 (1988); S. A. Ruatta, L. Hanley and S. L. Anderson, *J. Chem. Phys.* **91**, 226 (1989); P. A. Hintz, S. A. Ruatta and S. L. Anderson, *J. Chem. Phys.* **92**, 292 (1990).

[161]   R. Kaway and J. H. Weare, *J. Chem. Phys.* **95**, 1151 (1991).

[162]   R. Kaway and J. H. Weare, *Chem. Phys. Lett.* **191**, 311 (1992).

[163]   I. Boustani, *Phys. Rev. B* **55**, 16426 (1997); I. Boustani, *Int. J. Quantum Chem.* **52**, 1081 (1994); A. Ricca and C. Bauschlicher, Jr, *Chem. Phys.* **208**, 233 (1996); F. L. Gu, X. Yang, A.-C. Tang, H. Jiao and P. von R. Schleyer, *J. Comp. Chem.* **19**, 203 (1998).

[164]   K. E. Schriver, J. L. Persson, E. C. Honea and R. L. Whetten, *Phys. Rev. Lett.* **64**, 2539 (1990).

[165]   J. Lermé, M. Pellarin, J. L. Vialle, B. Bacguenard and M. Broyer, *Phys. Rev. Lett.* **68**, 2818 (1992).

[166]   D. M. Cox, D. J. Trevor, R. L. Whetten, E. A. Rohlfing and A. Kaldor, *J. Chem. Phys.* **84**, 4651 (1986).

[167]   D. M. Cox, D. J. Trevor, R. L. Whetten and A. Kaldor, *J. Phys. Chem.* **92**, 421 (1988).

[168]   M. F. Jarrold and J. E. Bower, *J. Chem. Phys.* **98**, 2399 (1993).

[169]   G. Ganteför, M. Gausa, K. H. Meiwes-Broer and H. O. Lutz, *Z. Phys. D* **9**, 253 (1988); K. J. Taylor, C. L. Pettiette, M. J. Craycraft, O. Chesnovsky and R. E. Smalley, *Chem. Phys. Lett.* **152**, 347 (1988).

[170]   C. Y. Cha, G. Ganteför and W. Eberhardt, *J. Chem. Phys.* **100** 995 (1994).

[171]   R. O. Jones, *Phys. Rev. Lett.* **67**, 224 (1991).

[172]   R. O. Jones, *J. Chem. Phys.* **99**, 1194 (1993).

[173]   X. G. Gong and E. Tosatti, *Phys. Lett. A* **166**, 369 (1992).

[174]   B. Wilkens, *Nucl. Instr. Meth. A* **236**, 340 (1985); D. L. Barr, *J. Vac. Sci. Technol. B* **5**, 184 (1987); F. G. Rudenauer, W. Steiger, H. Studnicka and P. Pollinger, *Int. J. Mass Spectrom. Ion Processes* **77**, 63 (1989).

[175]   W. A. de Heer, P. Milani and A. Châtelain, *Phys. Rev. Lett.* **65**, 488 (1990); I. M. Billas, J. A. Becker, A. Châtelain and W. A. de Heer, *Phys. Rev. Lett.* **71**, 4067 (1993).

[176]   I. M. L. Billas, A. Châtelain and W. A. de Heer, *J. Magnetism and Magnetic Materials* **168**, 64 (1997).

[177]   D. C. Douglass, J. P. Bucher and L. A. Bloomfield, *Phys. Rev. Lett.* **68**, 1774 (1992).

[178]   A. J. Cox, D. C. Douglass, J. G. Louderback, A. M. Spencer and L. A. Bloomfield, *Z. Phys. D* **26**, 319 (1993).

[179]   A. J. Cox, J. G. Louderback and L. A. Bloomfield, *Phys. Rev. Lett.* **71**, 923 (1993).

[180]   G. Ganteför and W. Eberhardt, *Phys. Rev. Lett.* **76**, 4975 (1996).

[181]   L.-S. Wang, H.-S. Cheng and J. Fan, *Chem. Phys. Lett.* **236**, 57 (1995).

[182]   H. Yoshida, A. Terasaki, K. Kobayashi, M. Tsukada and T. Kondow, *J. Chem. Phys.* **102**, 5960 (1995).

[183]   See, for instance, L.-S. Wang, H. Wu and S. R. Desai, *Phys. Rev. Lett.* **76**, 4853 (1996); J. B. Griffin and P. B. Armentrout, *J. Chem. Phys.* **106**, 4448 (1997).

[184]   The magnetic moment of the Fe, Co and Ni atoms is 4, 3 and 2 μB, respectively, whereas in the bulk the corresponding values are 2.22, 1.27 and 0.66 μB.

[185]   I. M. L. Billas, A. Châtelain and W. A. de Heer, *Science* **265**, 1682 (1994).

[186]   It is worth pointing out that even in magnetic-deflection experiments the magnetic moments are not measured directly but obtained from the observed deflections via the *superparamagnetic model* described in S. H. Khanna and S. Linderoth, *Phys. Rev. Lett.* **67**, 742 (1991).

[187]   See, however, G. M. Pastor, J. Dorantes-Davila, S. Pick and H. Dreyssé, *Phys. Rev. Lett.* **75**, 326 (1995).

[188]   D. C. Matthis, *The Theory of Magnetism I*, Springer Verlag, Berlin, Heidelberg, 1988.

[189]   D. D. Awschalom, M. A. McCord and G. Grinstein, *Phys. Rev. Lett.* **65**, 783 (1990).

[190]   C. Y. Yang, K. H. Johnson, D. R. Salahub, J. Kaspar and R. P. Messmer, *Phys. Rev. B* **24**, 5673 (1981); K. Lee, J. Callaway and S. Dhar, *Phys. Rev. B* **30**, 1724 (1984); K. Lee, J. Callaway, K. Kwonk, R. Tang and A. Ziegler, *Phys. Rev. B* **31**, 1796 (1985).

[191]   G. M. Pastor, J. Dorantes-Davila and K. H. Bennemann, *Phys. Rev. B* **40**, 7642 (1989); G. M. Pastor, R. Hirsch and B. Mühlschlegel, *Phys. Rev. Lett.* **72**, 3879 (1994) and *Phys. Rev. B* **53**, 10382 (1996).

[192]   P. Ballone, P. Milani and W. A. de Heer, *Phys. Rev. B* **44**, 10350 (1991); A. Maiti and L. M. Falicov, *Phys. Rev. B* **48**, 13596 (1993).

[193]   K. Binder, H. Rauch and V. Wildpaner, *J. Phys. Chem. Solids* **31**, 391 (1990).

[194]   H. J. F. Jansen, K. B. Hathaway and A. J. Freeman, *Phys. Rev. B* **30**, 6177 (1985). Recent computations (see, for instance, C. Elsässer, J. Zhu, S. G. Louie, M. Fähnle and C. T. Chan, *J. Phys. Cond. Matter* **10**, 5081 (1998)) show that gradient-corrected schemes are able to predict the magnetic bcc ground state of iron. Given the small energy differences between different phases, however, the result may well depend on the specific gradient-corrected functional.

[195]   M. Weinert, S. Blügel and P. D. Johnson, *Phys. Rev. Lett.* **71**, 4097 (1993).

[196]   M. Castro, C. Jamorski and D. R. Salahub, *Chem. Phys. Lett.* **271**, 133 (1997).

[197]   (a) DFT–LSDA computations for Fe_2 are reported in S. Dhar and N. R. Kestner, *Phys. Rev. A* **38**, 1111 (1988). (b) Configuration interaction results are given in T. Noro, C. Ballard, M. H. Palmer and H. Tatewaki, *J. Chem. Phys.* **100**, 452 (1994).

[198]   J. L. Chen, C. S. Wang, K. A. Jackson and M. R. Pederson, *Phys. Rev. B* **44**, 6558 (1991).

[199]   M. Castro and D. R. Salahub, *Phys. Rev. B* **49**, 11842 (1994).

[200]   M. Castro, *Int. J. Quantum Chem.* **64**, 223 (1997).

[201]   P. Ballone and R. O. Jones, *Chem. Phys. Lett.* **233**, 632 (1995).

[202]   C. Jamorski, A. Martinez, M. Castro and D. R. Salahub, *Phys. Rev. B* **55**, 10905 (1997).

[203]   F. A. Reuse and S. N. Khanna, *Chem. Phys. Lett.* **234**, 77 (1995).

[204]   D. P. Pappas, A. P. Popov, A. N. Anisimov, B. V. Reddy and S. N. Khanna, *Phys. Rev. Lett.* **76**, 4332 (1996).

[205]   B. V. Reddy, S. N. Khanna and B. I. Dunlap, *Phys. Rev. Lett.* **70**, 3323 (1993); Y. Jinlong, F. Toigo and W. Kelin, *Phys. Rev. B* **50**, 7915 (1994).

[206]   D. C. Douglass, J. P. Bucher and L. A. Bloomfield, *Phys. Rev. B* **45**, 6341 (1992).

[207]   H. Grönbeck and A. Rosén, *J. Chem. Phys.* **107**, 10620 (1997).

[208]   A. Kant and B. Strauss, *J. Chem. Phys.* **45**, 3161 (1966).

[209]   B. Delley, A. J. Freeman and D. E. Ellis, *Phys. Rev. Lett.* **50**, 488 (1983).

[210]   H. Cheng and L.-S. Wang, *Phys. Rev. Lett.* **77**, 51 (1996).

[211]   C. A. Baumann, R. J. Van Zee, S. Bhat and W. Weltner, Jr, *J. Chem. Phys.* **78**, 190 (1983).

[212]   S. K. Nayak and P. Jena, *Chem. Phys. Lett.* **289**, 473 (1998).

[213]   B. Piveteau, M.-C. Desjonqueres, A. M. Oles and D. Spanjaard, *Phys. Rev. B* **53**, 9251 (1996).

[214]   Interatomic magnetic interactions in iron have been studied in S. S. Peng and H. J. F. Jansen, *Phys. Rev. B* **43**, 3518 (1991) on the basis of DFT–LSDA computations. The results show that the spin–spin interactions are fairly long-range, and contain sizable many-body interactions.

[215] Z. Xu, F.-S. Xiao, S. K. Purnell, O. Alexeev, S. Kawi, S. E. Deutsch and B. C. Gates, *Nature* **372**, 346 (1994).

[216] Calculations on Ir microclusters have been made within the Hartree–Fock scheme by J.-N. Feng, X.-R. Huang and Z.-S. Li, *Chem. Phys. Lett.* **276**, 334 (1998).

[217] P. Fayet, F. Granzer, G. Hegenbart, E. Moisar, B. Pischel and L. Wöste, *Phys. Rev. Lett.* **55**, 3002 (1985).

[218] *The Physics of Latent Image Formation in Silver Halides*, ed. A. Baldereschi, W. Czaja, E. Tosatti and M. Tosi (World Scientific, Singapore, 1984).

[219] V. Bonačič-Koutecký, L. Cespiva, P. Fantuccii and J. Koutecký, *J. Chem. Phys.* **98**, 7981 (1993).

[220] R. L. Whetten, J. T. Khouri, M. M. Alvarez, S. Murthy, I. Vezmar, Z. L. Wang, P. W. Stephens, C. L. Cleveland, W. D. Luedtke and U. Landman, *Adv. Mater.* **8**, 428 (1996), and references therein.

[221] M. Valden, X. Lai and D. W. Goodman, *Science* **281**, 1647 (1998).

[222] See, for example, the discussion in M. Koerling and J. Haeglund, *Phys. Rev. B* **45**, 13293 (1992).

[223] H. Grönbeck, A. Rosén and W. Andreoni, *Phys. Rev. A* (in press).

[224] J. C. Boettger, *Phys. Rev. B* **57**, 8743 (1998), and references therein.

[225] K. Balasubramanian, P. Yi Feng and M. Z. Liao, *J. Chem. Phys.* **91**, 3561 (1989).

[226] W. Andreoni and J. L. Martins, *Phys. Rev. A* **28**, 3637 (1983) and *Surf. Sci.* **165**, 635 (1985).

[227] R. Santamaria, I. G. Kaplan and O. Novarc, *Chem. Phys. Lett.* **218**, 395 (1994).

[228] O. D. Häberlen, S.-C. Chung, M. Steiner and N. Rösch, *J. Chem. Phys.* **106**, 5189 (1997).

[229] L. Goodwin and D. R. Salahub, *Phys. Rev. A* **47**, R774 (1993).

[230] H. Grönbeck and A. Rosén, *Phys. Rev. B* **54**, 1549 (1996).

[231] H. Grönbeck, A. Rosén and W. Andreoni, *Z. Phys. D* **40**, 206 (1997).

[232] H. Grönbeck and W. Andreoni, *Chem. Phys. Lett.* **269**, 385 (1997).

[233] P. Ballone, W. Andreoni, R. Car and M. Parrinello, *Europhys. Lett.* **8**, 73 (1989).

[234] A. Mañanes, M. P. Iñiguez, M. J. López and J. A. Alonso, *Phys. Rev. B* **42**, 5000 (1990).

[235] S. B. Zhang, M. L. Cohen and M. Y. Chou, *Phys. Rev. B* **36**, 3455 (1987).

[236] See also C. Baldrón and J. A. Alonso, *Physica B* **154**, 73 (1988).

[237] U. Röthlisberger and W. Andreoni, *Chem. Phys. Lett.* **198**, 478 (1992).

[238] M. M. Kappes, M. Schär, C. Yeretzian, U. Heiz, A. Vayloyan and E. Schumacher, *Proc. Intl Symp. on the Physics and Chemistry of Small Clusters*, NATO ASI Series B, Vol. **158**, ed. P. Jena, (Plenum, New York, 1987), p. 263; C. Yeretzian, PhD Thesis, University of Bern, Switzerland (1990).

[239] U. Röthlisberger and W. Andreoni, *Int'l J. Mod. Phys. B* **6**, 3675 (1992).

[240] L. C. Balbas and J. L. Martins, *Phys. Rev. B* **54**, 2937 (1996).

[241] C. Yeretzian, U. Röthlisberger and E. Schumacher, *Chem. Phys. Lett.* **237**, 334 (1995).

[242] H.-P. Cheng, R. N. Barnett and U. Landman, *Phys. Rev. B* **48**, 1820 (1993).

[243] V. Kumar, *Phys. Rev. B* **57**, 8827 (1998).

[244] W. Andreoni and A. Curioni, *Phys. Rev. Lett.* **77**, 834 (1996).

[245] P. Carloni, W. Andreoni and M. Parrinello, *Phys. Rev. Lett.* **79**, 761 (1997).

[246] J. C. Charlier, A. DeVita and R. Car, *Science* **275**, 646 (1997).

[247] R. N. Barnett and U. Landman, *Nature* **387**, 788 (1997).

[248] A. Curioni and W. Andreoni (in preparation).

[249] L. H. Thomas, *Proc. Cambridge Philos. Soc.* **23**, 542 (1927); E. Fermi, *Z. Phys.* **48**, 73 (1928).

[250] E. Smargiassi and P. A. Madden, *Phys. Rev. B* **49**, 5220 (1994).

[251] D. Nehete, V. Shah and D. G. Kanhere, *Phys. Rev. B* **53**, 2126 (1996); A. Dhavale, V. Shah and D. G. Kanhere, *Phys. Rev. B* **57**, 4522 (1998).

[252] These methods are oversimplified models and can be considered as simple approximations to *ab initio* calculations with localized basis functions, using either a fixed, small and thus nonconvergent basis set [253] or a fixed number of matrix elements [254]. Unfortunately they are often presented as *ab initio* MD. This is highly misleading for the non-expert readership.

[253] O. F. Sankey and D. J. Niklewski, *Phys. Rev. B* **40**, 3979 (1989).

[254] G. Seifert, D. Porezag and Th. Frauenheim, *Int. J. Quantum Chem.* **58**, 185 (1996).

[255] M. Gausa, R. Kaschner, H. O. Lutz, G. Seifert and K.-H. Meiwes-Broer, *Chem. Phys. Lett.* **230**, 99 (1994).

[256] M. S. Daw and M. I. Baskes, *Phys. Rev. Lett.* **50**, 1285 (1983) and *Phys. Rev. B* **29**, 6443 (1984).

[257] K. W. Jacobsen, J. K. Nørskov and M. J. Puska, *Phys. Rev. B* **35**, 7423 (1987).

[258] M. E. Tuckerman, G. J. Martyna and B. J. Berne, *J. Chem. Phys.* **93**, 1287 (1990).

[259] B. Hartke, D. A. Gibson and E. A. Carter, *Int. J. Quantum Chem.* **45**, 59 (1993); D. A. Gibson and E. A. Carter, *J. Phys. Chem.* **97**, 13429 (1993).

[260] M. E. Tuckerman and M. Parrinello, *J. Chem. Phys.* **101**, 1316 (1994).

[261] J. Hutter, M. Parrinello and S. Vogel, *J. Chem. Phys.* **101**, 3862 (1994).

[262] P. E. Blöchl and M. Parrinello, *Phys. Rev. B* **45**, 9413 (1992).

[263] S. Nosé, *Mol. Phys.* **52**, 255 (1984); W. G. Hoover, *Phys. Rev. A* **31**, 1695 (1985).

[264] M. E. Tuckerman and M. Parrinello, *J. Chem. Phys.* **101**, 1302 (1994).

[265] A. St-Amant and D. R. Salahub, *Chem. Phys. Lett.* **169**, 387 (1990).

[266] G. B. Bachelet, D. R. Hamann and M. Schlüter, *Phys. Rev. B* **26**, 4199 (1982).

[267] In Table I of Ref. [123] there are two errors in the parameters of the NLXC pseudopotential. These are the values of R_1 and R_{cc}. The correct values are: 1.13228 and 1.2080 respectively (rather than 0.9182 and 0.685227).

[268] R. Stumpf, X. Gonze and M. Scheffler, *Research Report, Fritz-Haber-Institute, Berlin* (1990).

[269] A. Pasquarello, K. Laasonen, R. Car, C. Lee and D. Vanderbilt, *Phys. Rev. Lett.* **69**, 1982 (1992).

[270] J. Andzelm and E. Wimmer, *J. Chem. Phys.* **96**, 1280 (1992).

[271] B. I. Dunlap and N. Rösch, *Adv. Quantum Chem.* **21**, 317 (1990).

[272] B. Delley, *J. Chem. Phys.* **92**, 508 (1990).

4   Dissociation, Fragmentation and Fission of Simple Metal Clusters

CONSTANTINE YANNOULEAS, UZI LANDMAN and
ROBERT N. BARNETT

Georgia Institute of Technology, Atlanta, USA

4.1   INTRODUCTION

Dissociation, fragmentation and fissioning processes underlie physical and chemical phenomena in a variety of finite-size systems, characterized by a wide spectrum of energy scales, nature of interactions, and characteristic spatial and temporal scales. These include nuclear fission [1, 2], unimolecular decay and reactions in atoms and molecules [3], and, more recently, dissociation and fragmentation processes in atomic and molecular clusters [4–6]. Investigations of the energetics, mechanisms, pathways and dynamics of fragmentation processes provide ways and means for explorations of the structure, stability, excitations and dynamics in the many-body finite systems mentioned above, as well as allowing for comprehensive tests of theoretical methodologies and conceptual developments, and have formed active areas of fruitful research endeavors in nuclear physics, and more recently in cluster science.

Under the general title of dissociation and fragmentation [7] processes in metal clusters, one usually distinguishes two classes of phenomena, i.e. (1) dissociation of neutral monomers and/or dimers, and (2) fission. The physical processes in the first class are most often referred to as 'evaporation' of monomers and/or dimers, since they are endothermic processes and are usually induced through laser heating of the cluster. The unimolecular

Metal Clusters. Edited by W. Ekardt
© 1999 John Wiley & Sons Ltd

equations associated with these processes are

$$M_N{}^+ \longrightarrow M_{N-1}{}^+ + M, \tag{1}$$

for monomer separation, and

$$M_N{}^+ \longrightarrow M_{N-2}{}^+ + M_2, \tag{2}$$

for dimer separation (N denotes the number of atoms in the clusters [8]). The parent clusters $M_N{}^+$ have been taken here as being singly ionized, in order to conform with available experimental measurements [4]. Fission on the other hand, is most often an exothermic process and is due to the Coulombic forces associated with excess charges on the cluster. It has been found that the minimum excess charge required to induce fission is 2 elementary units (either positive or negative). In this case the doubly charged parent cluster splits into two singly charged fragments, and the corresponding unimolecular equation can be written as

$$M_N{}^{2\pm} \longrightarrow M_P{}^{1\pm} + M_{N-P}{}^{1\pm}, \quad P = 1, \ldots, [N/2]. \tag{3}$$

It needs to be emphasized that fragmentation through fission involves most often the overcoming of a fission barrier, while monomer and dimer separation are barrierless processes [4].

4.1.1 Metal Cluster Fission and Nuclear Fission: Similarities and Differences

Multiply charged metallic clusters ($M_N{}^{Z+}$) are observable in mass spectra if they exceed a critical size of stability $N_c{}^{Z+}$ (e.g. for $Z = 2$, $N_c{}^{2+} = 27$ for Na and $N_c{}^{2+} = 20$ for K [4, 9]). For clusters with $N > N_c{}^{Z+}$, evaporation of neutral species is the preferred dissociation channel, while, below the critical size, fission into two charged fragments dominates (for $Z = 2$, two singly charged fragments emerge). Nevertheless, at low enough temperature, such $M_N{}^{Z+}(N < N_c{}^{Z+})$ clusters can be metastable above a certain size $N_b{}^{Z+}$, because of the existence of a fission barrier E_b (for $Na_N{}^{2+}$ and $K_N{}^{2+}$, $N_b{}^{2+} = 7$ [10, 11]).

These observations indicate that fission of metal clusters occurs when the repulsive Coulomb forces due to the accumulation of the excess charges overcome the electronic binding (cohesion) of the cluster. This reminds us immediately of the well-studied nuclear fission phenomenon and the celebrated liquid drop model (LDM) according to which the binding nuclear forces are expressed as a sum of volume and surface terms, and the balance between the Coulomb repulsion and the increase in surface area upon volume-conserving deformations allows for an estimate of the stability and fissility of the nucleus [12, 13].

We note that for doubly charged metal clusters with $N \leq 12$ microscopic descriptions of energetics and dynamics of fission, based on first-principles electronic-structure calculations in conjunction with molecular dynamics (MD) simulations, have been performed [10, 11] (see Section 4.3.3.1 for details). Several of the trends exhibited by the microscopic calculations (such as influence of magic numbers, associated with electronic shell-closing, on fission energetics and barrier heights; predominance of an asymmetric fission channel; double-humped fission-barrier shapes; shapes of deforming clusters along the fission trajectory portraying two fragments connected through a stretching neck) suggest that appropriate adaptation of methodologies developed originally in the context of nuclear fission may provide a useful conceptual and calculational framework for studies of systematics and patterns of fission processes in metallic clusters.

In this context, it is useful to comment on the earliest treatments of pertinent nuclear processes, i.e. fission [12, 1] and alpha radioactivity [14, 15, 2]. Adaptation of the simple one-center LDM to charged metallic clusters [5], involving calculation of the Coulomb repulsive energy due to an excess charge localized at the surface, yields a reduced LDM fissility parameter $\xi = (Z^2/N)/(Z^2/N)_{cr}$, where $(Z^2/N)_{cr} = 16\pi r_s^3 \sigma/e^2$ with the surface energy per unit area denoted by σ and r_s being the Wigner–Seitz radius (using bulk r_s and σ values, $(Z^2/N)_{cr} = 0.44$ and 0.39 for K_N^{2+} and Na_N^{2+}, respectively). Accordingly, a cluster is unstable for $\xi > 1$ (implying that for K_N^{2+} with $N \leq 9$ and Na_N^{2+} with $N \leq 10$ barrierless fission should occur) with the most favorable channel being the symmetric one (i.e. when the two fragments have equal masses, which is only approximately true for nuclear fission, and certainly not the case for small metal clusters). For $0.351 < \xi < 1$, the system is metastable (i.e. may fission in a process involving a barrier), and for $0 < \xi < 0.351$ the system is stable.

At the other limit, α-radioactivity, which may be viewed as an extreme case of (super-asymmetric) fission, is commonly described as a process where the fragments are formed (or, as often said, preformed) before the system reaches the top of the barrier (saddle point), and as a result the barrier is mainly Coulombic [2]. We note here that asymmetric emission of heavier nuclei is also known (e.g. ^{223}Ra $\rightarrow$ ^{14}C + ^{209}Pb, which is referred to as 'exotic' or 'cluster' radioactivity [16–18]), and the barriers in these cases resemble the one-humped barrier of alpha radioactivity and do not exhibit modulations due to shell effects [18]. We also remark that such α-radioactivity-type (essentially Coulombic) barriers have been proposed recently [19] for describing the overall shape of the fission barriers in the case of metal clusters.

Although, several aspects of the simple LDM (e.g. competition between Coulomb and surface terms) and the α-particle, Coulombic model (e.g. asymmetric channels and a scission configuration close to the location of the saddle of the multidimensional potential-energy surface) are present in the fission of metal clusters, neither model is adequate in light of the characteristic behavior revealed from the microscopic calculations and experiments. Rather, we find that proper treatments of fission in these systems require consideration of shell effects (for a recent experimental study that demonstrates the importance of shell effects in metal-cluster fission, see Ref. [9b]). While such effects are known to have important consequences in nuclear fission (transforming the one-humped LDM barrier for symmetric fission into a two-humped barrier [20, 2]), their role in the case of metal clusters goes even further. Indeed, as illustrated below (see Section 4.3.3.2) for the case of the magic Na_{10}^{2+} (8 delocalized electrons), shell effects can be the largest contribution to the fission barrier, in particular in instances when the LDM component exhibits no barrier (in this case the LDM fissility $\xi > 1$). In this respect, Na_{10}^{2+} is analogous to the case of superheavy nuclei, which are believed [21] to be stabilized by the shell structure of a major shell closure at $Z_p = 114$, $N_n = 184$ (Z_p is the number of protons and N_n is the number of neutrons; unfortunately such nuclei have not been yet observed or synthesized artificially).

4.1.2   Other Decay Modes in Atomic and Molecular Clusters

In this Chapter, we will concentrate on the unimolecular processes in metal clusters described by Eqs (1–3). However, there is a variety of additional dissociation and fragmentation modes in atomic and molecular clusters (see reviews in Ref. [22]), which have been discovered experimentally or anticipated theoretically; among them we mention:

1. unimolecular fission of triply and higher charged cationic simple metal clusters [6, 23, 24];
2. metastability against electron autodetachment of multiply charged *anionic* atomic clusters [25–27] and fullerenes [26–28];
3. fragmentation of cationic fullerenes via sequential evaporation of carbon dimers [29];
4. ultrarapid fragmentation of rare-gas clusters following excitation (involving excimer formation [30]) or ionization [22];
5. multifragmentation phase transitions according to microcanonical thermodynamics of highly excited atomic clusters [31]; and
6. pathways and dynamics of dissociation and fragmentation of ionized van der Waals and hydrogen-bonded molecular clusters [22, 32].

4.1.3 Organization of the Chapter

In the following, we will present jellium-related theoretical approaches (specifically the shell-correction method (SCM) and variants thereof) appropriate for describing shell effects, energetics and decay pathways of metal-cluster fragmentation processes (both the monomer/dimer dissociation and fission), which were inspired by the many similarities with the physics of shell effects in atomic nuclei (Section 4.2). In Section 4.3, we will compare the experimental trends with the resulting theoretical SCM interpretations, and in addition we will discuss theoretical results from first-principles MD simulations (Section 4.3.3.1). Section 4.4 will discuss some of the latest insights concerning the importance of electronic-entropy and finite-temperature effects. Finally, Section 4.5 will provide a summary.

4.2 THEORY OF SHAPE DEFORMATIONS

In early applications of the jellium model, the shape of metal clusters was assumed in all instances to be spherical [33, 34], but soon it became apparent that the spherical symmetry was too restrictive [35, 36]. Indeed clusters with open electronic shells (between the magic numbers $N_e = 2, 8, 20, 40, 58, 92$, etc.) are subjected to Jahn–Teller distortions [37]. By now it has been well established that a quantitative description of the underlying shell effects and of fragmentation phenomena (as well as of other less complicated phenomena such as ionization and vertical electron detachment) requires a proper description of the deformed shapes of both parent and daughter clusters (of both precursor and final ionic or neutral product in the case of ionization and vertical electron detachment).

A most successful method for describing both deformation and shell effects in simple metal clusters (i.e. those that can be described by the jellium background model) is the SCM, originally developed in the field of nuclear physics [38, 2]. In a series of recent publications [25, 26, 28, 39–45], the SCM was further developed, adapted and applied in the realm of finite-size, condensed-matter nanostructures (i.e. metal clusters [25, 26, 39–43], but also multiply charged fullerenes [28], ^{3}He clusters [44], and metallic nanowires and nanoconstrictions [45]). Additionally, Refs [46–49] have used semiempirical versions (see below) of the SCM to study the shapes of neutral Na clusters [46, 47] and aspects of metal-cluster fission [48, 49].

The SCM derives its justification from the local-density-approximation (LDA) functional theory and has been developed as a two-level method.

At the microscopic level, referred to as the LDA-SCM, the method has been shown to be a nonselfconsistent approximation to the Kohn–Sham (KS)-LDA approach [50]. Apart from computational efficiency, an important physical insight provided by the LDA-SCM is that the total KS-LDA energy $E_{\text{total}}(N)$ (or in another notation $E_{\text{KS}}(N)$) of a finite system of interacting delocalized electrons (or more generally of other fermions, like nucleons or ^{3}He atoms) can be divided into two contributions, i.e.

$$E_{\text{total}}(N) = \tilde{E}(N) + \Delta E_{\text{sh}}(N), \qquad (4)$$

where $\tilde{E}$ is the part that varies smoothly as a function of the system size (e.g. the number, N, of atoms in a metal cluster), while $\Delta E_{\text{sh}}(N)$ is an oscillatory term accounting for the shell effects; this latter arises from the discretization of the electronic states (quantum size effect). $\Delta E_{\text{sh}}(N)$ is usually called a 'shell-correction' in the nuclear [38, 1] and cluster [25, 26] literature.

Starting from the fundamental microscopic separation in Eq. (4), various semiempirical implementations (referred to as SE-SCM, see Section 4.2.2) of such a division consist of different approximate choices and methods for evaluating the two terms contributing to this separation.

As an illustration of the physical content of Eq. (4) (which also serves as a motivating example for the SCM), we show in Figure 4.1 the size-evolutionary pattern of the ionization potentials (IPs) of Na$_N$ clusters, which exhibits odd–even oscillations in the observed

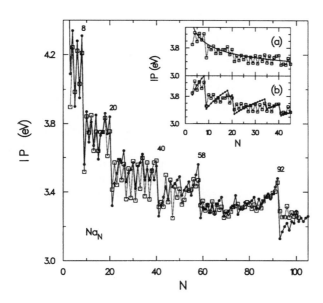

Figure 4.1 IPs of Na$_N$ clusters. Open squares: experimental measurements [51, 52]. Solid circles: theoretical IPs derived from the SCM assuming ellipsoidal (triaxial) deformations. Inset (a): the solid line represents the smooth contribution to the theoretical SCM IPs. Inset (b): the solid circles are the IPs derived from the SCM assuming spherical symmetry

spectrum in addition to the major features (major IP drops) at the magic numbers. Theoretical calculations at three different levels are contrasted to the experimental observations, namely, a smooth description of the pattern (inset (a)), and two levels of shell-corrected descriptions — one assuming spherical symmetry (inset (b)), and the other allowing for triaxial shape deformations (main frame of figure). The progressive improvement of the level of agreement between the experimental [51, 52] and theoretical patterns is evident.

Below, we first outline the microscopic derivation of Eq. (4), and subsequently we proceed with a presentation of the SE-SCM.

4.2.1 Microscopic Foundation of Shell-Correction Methods — the LDA-SCM

The LDA-SCM approach, which has been shown to yield results in excellent agreement with self-consistent KS-LDA calculations [25, 26], is equivalent to a Harris functional [53] approximation ($E_{\text{Harris}}[\rho^{\text{in}}]$, see below) to the KS-LDA total energy [50] ($E_{\text{KS}}[\rho_{\text{KS}}]$), with the input density ρ^{in} obtained through a variational minimization of an extended Thomas–Fermi (ETF) energy functional, $E_{\text{ETF}}[\rho]$.

The property of the nonselfconsistent Harris functional to yield total energies close to the KS-LDA ones is based on the following equality:

$$E_{\text{KS}}[\rho_{\text{KS}}] = E_{\text{Harris}}[\rho^{\text{in}}] + O(\delta\rho^2), \tag{5}$$

where $\delta\rho = \rho_{\text{KS}} - \rho^{\text{in}}$. That is, the KS-LDA energy is, to second order in $\delta\rho$, equal to the Harris energy.

Several recent publications have proven [54–56] the validity of equation (5) in connection with the Harris functional, which is often used in electronic structure calculations of molecules, surfaces and other condensed-matter systems. We note that, in the context of nuclear physics, Strutinsky had earlier proven [38] the validity of Eq. (5), with the difference that he utilized the Hartree–Fock (HF) functional instead of the KS-LDA one. In the nuclear physics literature, the HF version of Eq. (5) is referred to as the 'Strutinsky theorem'.

Usually, in the Harris functional, the *input* density ρ^{in} is taken as a superposition of site densities. Initially [53], the site components of the input density were not optimized. Later [55, 56], it was realized that the results could be improved by variationally adjusting the site components through a *maximization* of the Harris functional itself. However, doing so adds the burden of a matrix diagonalization for obtaining the eigenvalues (see below) at each step of the variation. Our method differs from the Harris approach in that the optimization of the input density is achieved by us through a variational ETF method [57] (which does not require such a step-by-step matrix diagonalization).

The nonselfconsistent Harris functional is given by the following expression,

$$E_{\text{Harris}}[\rho^{\text{in}}] =$$

$$E_{\text{I}} + \sum_{i=1}^{\text{occ}} \varepsilon_i^{\text{out}} - \int \left\{ \tfrac{1}{2} V_{\text{H}}[\rho^{\text{in}}(\boldsymbol{r})] + V_{\text{xc}}[\rho^{\text{in}}(\boldsymbol{r})] \right\} \rho^{\text{in}}(\boldsymbol{r})\, d\boldsymbol{r} + \int \mathcal{E}_{\text{xc}}[\rho^{\text{in}}(\boldsymbol{r})]\, d\boldsymbol{r}, \tag{6}$$

where V_{H} is the Hartree (electronic) repulsive potential, E_{I} is the repulsive electrostatic energy of the ions, and $E_{\text{xc}}[\rho] \equiv \int \mathcal{E}_{\text{xc}}[\rho]\, d\boldsymbol{r}$ is the exchange–correlation (xc) functional [58] (the corresponding xc potential is given as $V_{\text{xc}}(\boldsymbol{r}) \equiv \delta E_{\text{xc}}[\rho]/\delta\rho(\boldsymbol{r})$). $\varepsilon_i^{\text{out}}$ are the eigenvalues (nonselfconsistent) of the single-particle Hamiltonian,

$$\hat{H} = -\frac{\hbar^2}{2m_e}\nabla^2 + V_{in}, \tag{7}$$

with the mean-field potential given by

$$V_{in}[\rho^{in}(r)] = V_H[\rho^{in}(r)] + V_{xc}[\rho^{in}(r)] + V_I(r), \tag{8}$$

$V_I(r)$ being the attractive potential between the electrons and ions.

The ETF-LDA energy functional, $E_{ETF}[\rho]$, is obtained by replacing the kinetic energy term, $T[\rho]$, in the usual LDA functional, namely in the expression

$$E_{LDA}[\rho] = T[\rho] + \int \left\{ \tfrac{1}{2}V_H[\rho(r)] + V_I(r) \right\} \rho(r)\, dr + \int \mathcal{E}_{xc}[\rho(r)]\, dr + E_I, \tag{9}$$

by the ETF kinetic energy, given to the 4th-order gradients as follows [59]:

$$T_{ETF}[\rho] = \int t_{ETF}[\rho]\, dr = \frac{\hbar^2}{2m_e} \int \left\{ \frac{3}{5}(3\pi^2)^{2/3}\rho^{5/3} + \frac{1}{36}\frac{(\nabla\rho)^2}{\rho} + \frac{1}{270}(3\pi^2)^{-2/3}\rho^{1/3} \right.$$

$$\left. \times \left[\frac{1}{3}\left(\frac{\nabla\rho}{\rho}\right)^4 - \frac{9}{8}\left(\frac{\nabla\rho}{\rho}\right)^2 \frac{\Delta\rho}{\rho} + \left(\frac{\Delta\rho}{\rho}\right)^2 \right] \right\} dr. \tag{10}$$

We would like to remind the reader that the KS kinetic energy is of course given by the expression

$$T_{KS}[\rho_{KS}] = \sum_{i=1}^{occ} < \phi_{KS,i}| -\frac{\hbar^2}{2m_e}\nabla^2 |\phi_{KS,i} >, \tag{11}$$

where the single-particle wave functions $\phi_{KS,i}(r)$ are obtained from a self-consistent solution of the KS equations.

The optimal ETF-LDA total energy is obtained by minimization of $E_{ETF}[\rho]$ with respect to the density. In our calculations, we use for the trial densities parameterized profiles $\rho(r; \{\gamma_i\})$ [60, 25, 26] with $\{\gamma_i\}$ as variational parameters (the ETF-LDA optimal density is denoted as $\tilde{\rho}$). The single-particle eigenvalues, $\{\varepsilon_i^{out}\}$, in Eq. (6) are obtained then as the solutions to the single-particle Hamiltonian of Eq. (7) with V_{in} replaced by V_{ETF} (given by Eq. (8) with $\rho^{in}(r)$ replaced by $\tilde{\rho}(r)$). Hereafter, these single-particle eigenvalues will be denoted by $\{\tilde{\varepsilon}_i\}$.

In our approach, the smooth contribution in the separation (4) of the total energy is given by $E_{ETF}[\tilde{\rho}]$, while the shell-correction, ΔE_{sh}, is simply the difference [25, 26]

$$\Delta E_{sh} = E_{Harris}[\tilde{\rho}] - E_{ETF}[\tilde{\rho}]$$

$$= \sum_{i=1}^{occ} \tilde{\varepsilon}_i - \int \tilde{\rho}(r)V_{ETF}(r)\, dr - T_{ETF}[\tilde{\rho}]. \tag{12}$$

4.2.2 Semiempirical Shell-Correction Method (SE-SCM)

4.2.2.1 Methodology

Rather than proceed with the microscopic route, Strutinsky proposed a method for the separation of the total energy into smooth and shell-correction terms (see Eq. (4)) based on an averaging procedure. Accordingly, a smooth part, $\tilde{E}_{sp}$, is extracted out of the sum of the single-particle energies $\sum_i^{occ} \tilde{\varepsilon}_i$ (or $\sum_i^{occ} \varepsilon_i^{out}$, see Eq. (6)) by averaging them through an appropriate procedure. Usually, but not necessarily, one replaces the delta functions in the single-particle density of states by Gaussians or other appropriate weighting functions. As a result, each single-particle level is assigned an averaging occupation number $\tilde{f}_i$, and the smooth part $\tilde{E}_{sp}$ is formally written as

$$\tilde{E}_{sp} = \sum_i \tilde{\varepsilon}_i \tilde{f}_i. \tag{13}$$

Consequently, the Strutinsky shell-correction is given by

$$\Delta E_{sh}^{Str} = \sum_{i=1}^{occ} \tilde{\varepsilon}_i - \tilde{E}_{sp}. \tag{14}$$

The Strutinsky prescription (14) has the practical advantage of using only the single-particle energies $\tilde{\varepsilon}_i$, and not the smooth density $\tilde{\rho}$. Taking advantage of this, the single-particle energies can be taken as those of an external potential that empirically approximates the self-consistent potential of a finite system. In the nuclear case, a modified anisotropic three-dimensional harmonic oscillator has been used successfully to describe the shell-corrections in deformed nuclei [1, 2].

The single-particle smooth part, $\tilde{E}_{sp}$, however, is only one component in the smooth contribution $\tilde{E}[\tilde{\rho}]$, which needs to be added to the shell correction term in order to yield the total energy, i.e.

$$E_{total} \approx \Delta E_{sh}^{Str} + \tilde{E}[\tilde{\rho}]. \tag{15}$$

Strutinsky did not address the question of how to calculate microscopically the smooth part $\tilde{E}$ (which necessarily entails specifying the smooth density $\tilde{\rho}$). Instead he circumvented this question by substituting for $\tilde{E}$ the empirical energies, E_{LDM}, of the nuclear liquid-drop model, namely he suggested that

$$E_{total} \approx \Delta E_{sh}^{Str} + E_{LDM}. \tag{16}$$

In applications of Eq. (16), the single-particle energies involved in the averaging (see Eqs (13) and (14)) are commonly obtained as solutions of a Schrödinger equation with phenomenological one-body potentials. This last approximation has been very successful in describing fission barriers and properties of strongly deformed nuclei using harmonic-oscillator-type or Wood–Saxon empirical potentials.

4.2.2.2 Liquid-Drop Model for Neutral and Charged Deformed Clusters

For neutral clusters, the LDM expresses [60, 48, 5] the *smooth* part, $\tilde{E}$, of the total energy as the sum of three contributions, namely a volume, a surface and a curvature term, i.e.

$$\tilde{E} = E_{vol} + E_{surf} + E_{curv}$$

$$= A_v \int d\tau + \sigma \int dS + A_c \int dS \, \kappa, \tag{17}$$

where $d\tau$ is the volume element and dS is the surface differential element. The local curvature κ is defined by the expression $\kappa = 0.5 \, (R_{max}^{-1} + R_{min}^{-1})$, where R_{max} and R_{min} are the two principal radii of curvature at a local point on the surface of the jellium droplet (of a general shape) that models the cluster. The corresponding coefficients can be determined [25, 26, 60] by fitting the ETF-LDA total energy for spherical shapes (see Section 4.2.1) to the following parameterized expression as a function of the number, N, of atoms in the cluster [61]:

$$E_{\mathrm{ETF}}^{\mathrm{sph}} = \alpha_{\mathrm{v}} N + \alpha_{\mathrm{s}} N^{2/3} + \alpha_{\mathrm{c}} N^{1/3}. \tag{18}$$

The following expressions relate [62] the coefficients A_{v}, σ and A_{c} to the corresponding coefficients, (αs), in Eq. (18):

$$A_{\mathrm{v}} = \frac{3}{4\pi r_{\mathrm{s}}^3} \alpha_{\mathrm{v}}; \quad \sigma = \frac{1}{4\pi r_{\mathrm{s}}^2} \alpha_{\mathrm{s}}; \quad A_{\mathrm{c}} = \frac{1}{4\pi r_{\mathrm{s}}} \alpha_{\mathrm{c}}. \tag{19}$$

In the following, we will focus on the case of clusters with ellipsoidal (triaxial) shapes. In the case of ellipsoidal shapes the areal integral and the integrated curvature can be expressed in closed analytical form with the help of the incomplete elliptic integrals $\mathcal{F}(\psi, k)$ and $\mathcal{E}(\psi, k)$ of the first and second kind [63], respectively. Before writing the formulas, we need to introduce some notation. Volume conservation must be employed, namely

$$a'b'c'/R_0^3 = abc = 1, \tag{20}$$

where R_0 is the radius of a sphere with the same volume ($R_0 = r_{\mathrm{s}} N^{1/3}$ is taken to be the radius of the positive jellium assuming spherical symmetry, r_{s} being the corresponding Wigner–Seitz radius), and $a = a'/R_0$, etc. are the dimensionless semiaxes. The eccentricities are defined through the dimensionless semiaxes as follows:

$$e_1^2 = 1 - (c/a)^2,$$
$$e_2^2 = 1 - (b/a)^2,$$
$$e_3^2 = 1 - (c/b)^2. \tag{21}$$

The semiaxes are chosen so that

$$a \geq b \geq c. \tag{22}$$

With the notation $\sin \psi = e_1$, $k_2 = e_2/e_1$ and $k_3 = e_3/e_1$, the relative (with respect to the spherical shape) surface and curvature energies are given [64] by

$$\frac{E_{\mathrm{surf}}^{\mathrm{ell}}}{E_{\mathrm{surf}}^{\mathrm{sph}}} = \frac{ab}{2} \left[\frac{1 - e_1^2}{e_1} \mathcal{F}(\psi, k_3) + e_1 \mathcal{E}(\psi, k_3) + c^3 \right] \tag{23}$$

and

$$\frac{E_{\mathrm{curv}}^{\mathrm{ell}}}{E_{\mathrm{curv}}^{\mathrm{sph}}} = \frac{bc}{2a} \left\{ 1 + \frac{a^3}{e_1} \left[(1 - e_1^2)\mathcal{F}(\psi, k_2) + e_1^2 \mathcal{E}(\psi, k_2) \right] \right\}. \tag{24}$$

The change in the smooth part of the cluster total energy due to the excess charge $\pm Z$ has been discussed for spherical clusters in Refs [25, 26]. The result may be summarized as

$$\Delta \tilde{E}^{\mathrm{sph}}(Z) = \tilde{E}^{\mathrm{sph}}(Z) - \tilde{E}^{\mathrm{sph}}(0) = \mp WZ + \frac{Z(Z \pm 0.25)e^2}{2(R_0 + \delta)}, \tag{25}$$

where the upper and lower signs correspond to negatively and positively charged states, respectively, W is the work function of the metal, R_0 is the radius of the positive jellium assuming spherical symmetry, and δ is a spillout-type parameter.

To generalize the above results to an ellipsoidal shape, $\phi(R_0 + \delta) = e^2/(R_0 + \delta)$, which is the value of the potential on the surface of a spherical conductor, needs to be replaced by the corresponding expression for the potential on the surface of a conducting ellipsoid. The final result, normalized to the spherical shape, is given by the expression

$$\frac{\Delta \tilde{E}^{\mathrm{ell}}(Z) \pm WZ}{\Delta \tilde{E}^{\mathrm{sph}}(Z) \pm WZ} = \frac{bc}{e_1} \mathcal{F}(\psi, k_2), \tag{26}$$

where the $\pm$ sign in front of WZ corresponds to negatively and positively charged clusters, respectively.

4.2.2.3 The Modified Nilsson Potential for Ellipsoidal Shapes

A natural choice for an external potential to be used for calculating shell-corrections with the Strutinsky method is an anisotropic, three-dimensional oscillator with an additional l^2 angular-momentum term for lifting the harmonic oscillator degeneracies [65]. Such an oscillator model for approximating the total energies of metal clusters, but without separating them into a smooth and a shell-correction part in the spirit of Strutinsky's approach, had been used [52] with some success for calculating relative energy surfaces and deformation shapes of metal clusters. However, this simple harmonic oscillator model had serious limitations, since (i) the total energies were calculated by the expression $\frac{3}{4} \sum_i^{\mathrm{occ}} \tilde{\varepsilon}_i$, and thus did not compare with the total energies obtained from the KS-LDA approach, and (ii) the model could not be extended to the case of charged (cationic or anionic) clusters. Thus absolute ionization potentials, electron affinities and fission energetics could not be calculated in this model. Alternatively, in our approach, we are making only a limited use of the external oscillator potential in calculating a modified Strutinsky shell-correction. Total energies are evaluated by adding this shell-correction to the smooth LDM energies (which incorporate xc contributions, since the LDM coefficients are extracted via a comparison with total ETF-LDA energies, or they are taken from experimental values).

In particular, a modified Nilsson Hamiltonian appropriate for metal clusters [35, 36] is given by

$$H_{\mathrm{N}} = H_0 + U_0 \hbar \omega_0 (l^2 - \langle l^2 \rangle_n), \tag{27}$$

where H_0 is the Hamiltonian for a three-dimensional anisotropic oscillator, namely

$$H_0 = -\frac{\hbar^2}{2m_{\mathrm{e}}} \Delta + \frac{m_{\mathrm{e}}}{2} (\omega_1^2 x^2 + \omega_2^2 y^2 + \omega_3^2 z^2)$$

$$= \sum_{k=1}^{3} \left(a_k^\dagger a_k + \frac{1}{2} \right) \hbar \omega_k. \tag{28}$$

U_0 in Eq. (27) is a dimensionless parameter, which for occupied states may depend on the effective principal quantum number $n = n_1 + n_2 + n_3$ associated with the major shells of any spherical oscillator, (n_1, n_2, n_3) being the quantum numbers specifying the single-particle levels of the Hamiltonian H_0 (for clusters comprising up to 100 valence electrons, only a weak dependence on n is found, see Table I in Ref. [40a]). U_0 vanishes for values of n higher than the corresponding value of the last partially (or fully) filled major shell with reference to the spherical limit.

$l^2 = \sum_{k=1}^{3} l_k^2$ is a 'stretched' angular momentum which scales to the ellipsoidal shape and is defined as follows,

$$l_3^2 \equiv (q_1 p_2 - q_2 p_1)^2, \tag{29}$$

(with similarly obtained expressions for l_1 and l_2 via a cyclic permutation of indices), where the stretched position and momentum coordinates are defined via the corresponding natural coordinates, q_k^{nat} and p_k^{nat}, as follows,

$$q_k \equiv q_k^{\mathrm{nat}} (m_e \omega_k / \hbar)^{1/2} = \frac{a_k^\dagger + a_k}{\sqrt{2}}, \quad (k = 1, 2, 3), \tag{30}$$

$$p_k \equiv p_k^{\mathrm{nat}} (1/\hbar m_e \omega_k)^{1/2} = i \frac{a_k^\dagger - a_k}{\sqrt{2}}, \quad (k = 1, 2, 3). \tag{31}$$

The stretched l^2 is not a properly defined angular-momentum operator, but has the advantageous property that it does not mix deformed states that correspond to spherical major shells with different principal quantum numbers $n = n_1 + n_2 + n_3$ (see the Appendix in Ref. [40a] for the expression of the matrix elements of l^2).

The subtraction of the term $\langle l^2 \rangle_n = n(n+3)/2$, where $\langle \ \rangle_n$ denotes the expectation value taken over the nth-major shell in spherical symmetry, guarantees that the average separation between major oscillator shells is not affected as a result of the lifting of the degeneracy.

The oscillator frequencies can be related to the principal semiaxes a', b' and c' (see, Eq. (20)) via the volume-conservation constraint and the requirement that the surface of the cluster is an equipotential one, namely

$$\omega_1 a' = \omega_2 b' = \omega_3 c' = \omega_0 R_0, \tag{32}$$

where the frequency ω_0 for the spherical shape (with radius R_0) was taken according to Ref. [35] to be

$$\hbar\omega_0(N) = \frac{49 \text{ eV bohr}^2}{r_s^2 N^{1/3}} \left[1 + \frac{t}{r_s N^{1/3}} \right]^{-2}. \tag{33}$$

Since in this paper we consider solely monovalent elements, N in Eq. (33) is the number of atoms for the family of clusters $M_N^{Z\pm}$, r_s is the Wigner–Seitz radius expressed in atomic units, and t denotes the electronic spillout for the neutral cluster according to Ref. [35].

4.2.2.4 Shell-Correction and Averaging of Single-Particle Spectra for the Modified Nilsson Potential

Usually $\tilde{E}_{\mathrm{sp}}$ (see Eqs (13) and (14)) is calculated numerically [66]. However, a variation of the numerical Strutinsky averaging method consists in using the semiclassical partition function and in expanding it in powers of $\hbar^2$. With this method, for the case of an anisotropic, fully triaxial oscillator, one finds [1, 67] an analytical result, namely [68]

$$\tilde{E}_{\rm sp}^{\rm osc} = \hbar(\omega_1\omega_2\omega_3)^{1/3}\left(\frac{1}{4}(3N_{\rm e})^{4/3} + \frac{1}{24}\frac{\omega_1^2 + \omega_2^2 + \omega_3^2}{(\omega_1\omega_2\omega_3)^{2/3}}(3N_{\rm e})^{2/3}\right), \qquad (34)$$

where $N_{\rm e}$ denotes the number of delocalized valence electrons in the cluster.

In the present work, expression (34) (as modified below) will be substituted for the average part $\tilde{E}_{\rm sp}$ in Eq. (14), while the sum $\sum_i^{\rm occ}\tilde{\varepsilon}_i$ will be calculated numerically by specifying the occupied single-particle states of the modified Nilsson oscillator represented by the Hamiltonian (27).

In the case of an isotropic oscillator, not only the smooth contribution, $\tilde{E}_{\rm sp}^{\rm osc}$, but also the Strutinsky shell correction (14) can be specified analytically [1] with the result

$$\Delta E_{{\rm sh},0}^{\rm Str}(x) = \frac{1}{24}\hbar\omega_0(3N_{\rm e})^{2/3}[-1 + 12x(1 - x)], \qquad (35)$$

where x is the fractional filling of the highest partially filled harmonic-oscillator major shell. For a filled shell ($x = 0$ or 1), $\Delta E_{{\rm sh},0}^{\rm Str}(0) = -\frac{1}{24}\hbar\omega_0(3N_{\rm e})^{2/3}$, instead of the essentially vanishing value as in the case of the ETF-LDA-defined shell-correction (cf. Figure 1 of Ref. [40a]). To adjust for this discrepancy, we add $-\Delta E_{{\rm sh},0}^{\rm Str}(0)$ to $\Delta E_{\rm sh}^{\rm Str}$ calculated through Eq. (14) for the case of open-shell, as well as closed-shell clusters.

4.2.2.5 Overall Procedure

We are now in a position to summarize the calculational procedure for the SE-SCM in the case of deformed clusters, which consists of the following steps:

1. Parameterize results of ETF-LDA calculations for spherical neutral jellia according to Eq. (18).
2. Use the above parameterization (assuming that parameters per differential element of volume, surface and integrated curvature are shape-independent) in Eq. (17) to calculate the liquid-drop energy associated with neutral clusters, and then add to it the charging energy according to Eq. (26) to determine the total LDM energy $\tilde{E}$ (available experimental values for σ and W can also be used).
3. Use Eqs (27) and (28) for a given deformation (i.e. a', b', c', or equivalently ω_1, ω_2, ω_3, see Eq. (32)) to solve for the single-particle spectrum ($\tilde{\varepsilon}_i$).
4. Evaluate the average, $\tilde{E}_{\rm sp}$, of the single-particle spectrum according to Eq. (34) and subsequent remarks.
5. Use the results of steps 3 and 4 above to calculate the shell correction $\Delta E_{\rm sh}^{\rm Str}$ according to Eq. (14).
6. Finally, calculate the total energy $E_{\rm total}$ as the sum of the liquid-drop contribution (step 2) and the shell-correction (step 5), namely $E_{\rm total} = \tilde{E} + \Delta E_{\rm sh}^{\rm Str}$.

The optimal ellipsoidal geometries for a given cluster $M_N{}^{Z\pm}$, either neutral or charged, are determined by systematically varying the distortion (namely, the parameters a and b) in order to locate the global minimum of the total energy $E_{\rm total}(N, Z)$ (for the global minima and equilibrium shapes of neutral Na_N clusters according to the ellipsoidal model in the range $3 \le N \le 60$, see Figure 22 of Ref. [40a]).

4.2.2.6   Asymmetric Two-Center Oscillator Model for Fission

Naturally, the one-center Nilsson oscillator is not the most appropriate empirical potential for describing binary fission, which involves the gradual emergence of two separate fragments. A better choice is the asymmetric two-center oscillator model (ATCOM). According to the ATCOM approach, the single-particle levels, associated with both the initial one-fragment parent and the separated daughters emerging from binary cluster fission, are determined by the following single-particle Hamiltonian [69, 70]:

$$H = T + \tfrac{1}{2}m_e\omega_{\rho i}^2\rho^2 + \tfrac{1}{2}m_e\omega_{zi}^2(z-z_i)^2 + V_{\text{neck}}(z) + U(l_i^2), \qquad (36)$$

where $i = 1$ for $z < 0$ (left) and $i = 2$ for $z > 0$ (right).

This Hamiltonian is axially symmetric along the z axis. ρ denotes the cylindrical coordinate perpendicular to the symmetry axis [71]. The shapes described by this Hamiltonian are those of two semispheroids (either prolate or oblate) connected by a smooth neck (which is specified by the term $V_{\text{neck}}(z)$). $z_1 < 0$ and $z_2 > 0$ are the centers of these semispheroids. For the smooth neck, the following 4th-order expression [70] was adopted, namely

$$V_{\text{neck}}(z) = \tfrac{1}{2}m_e\xi_i\omega_{zi}^2(z-z_i)^4\theta(|z|-|z_i|), \qquad (37)$$

where $\theta(x) = 0$ for $x > 0$ and $\theta(x) = 1$ for $x < 0$ and $\xi_i = -1/2z_i^2$.

The frequency $\omega_{\rho i}$ in Eq. (36) must be z-dependent in order to interpolate smoothly between the values $\omega_{\rho i}^\circ$ of the lateral frequencies associated with the left ($i = 1$) and right ($i = 2$) semispheroids, which are not equal in asymmetric cases. The frequencies $\omega_{\rho i}^\circ$ ($i = 1, 2$) characterize the lateral harmonic potentials associated with the two semispheroids outside the neck region. In the implementation of such an interpolation, we follow Ref. [70].

The angular-momentum-dependent term $U(l_i^2)$, where l_1 and l_2 are pseudoangular momenta with respect to the left and right centers z_1 and z_2, is a direct generalization of the corresponding term familiar from the one-center Nilsson potential (e.g. see Ref. [40a]). Its function is to lift the usual harmonic-oscillator degeneracies for different angular momenta, that is, for a spherical shape the 1d–2s degeneracy is properly lifted into a 1d shell that is lower than the 2s shell (for the parameters entering into this term, which ensure a proper transition from the case of the fissioning cluster to that of the separated two fragments, we have followed Ref. [70]).

The cluster shapes associated with the spatial-coordinate-dependent single-particle potential $V(\rho, z)$ in the Hamiltonian (36) (i.e., the second, third and fourth terms) are determined by the assumption that the cluster surface coincides with an equipotential surface of value V_0, namely, from the relation $V(\rho, z) = V_0$. Subsequently, one solves for ρ and derives the cluster shape $\rho = \rho(z)$. For the proper value of V_0, we take the one associated with a spherical shape containing the same number of atoms, N, as the parent cluster, namely, $V_0 = \tfrac{1}{2}m_e\omega_0^2R^2$, where $\hbar\omega_0 = 49r_s^{-2}N^{-1/3}$ eV, $R = r_sN^{1/3}$, and r_s is the Wigner–Seitz radius in atomic units (monovalent metals have been assumed). Volume conservation is implemented by requiring that the volume enclosed by the fissioning cluster surface (even after separation) remains equal to $4\pi R^3/3$.

The cluster shape in this parameterization is specified by four independent parameters. We take them to be: the separation $d = z_2 - z_1$ of the semispheroids, the asymmetry ratio $q_{\text{as}} = \omega_{\rho 2}^\circ/\omega_{\rho 1}^\circ$, and the deformation ratios for the left (1) and right (2) semispheroids $q_i = \omega_{zi}/\omega_{\rho i}^\circ$ ($i = 1, 2$).

The single-particle levels of the Hamiltonian in Eq. (36) are obtained by numerical diagonalization in a basis consisting of the eigenstates of the following auxiliary Hamiltonian:

$$H_0 = T + \tfrac{1}{2} m_e \overline{\omega}_\rho^2 \rho^2 + \tfrac{1}{2} m_e \omega_{zi}^2 (z - z_i)^2, \tag{38}$$

where $\overline{\omega}_\rho$ is the arithmetic average of $\omega_{\rho 1}^\circ$ and $\omega_{\rho 2}^\circ$. The eigenvalue problem specified by the auxiliary Hamiltonian (38) is separable in the cylindrical variables ρ and z. The general solutions in ρ are those of a two-dimensional oscillator, while in z they can be expressed through the parabolic cylinder functions [72]. The matching conditions at $z = 0$ for the left and right domains yield the z-eigenvalues and the associated eigenfunctions [69].

Having obtained the single-particle spectra, the empirical shell correction (in the spirit of Strutinsky's method [38]), ΔE_{sh}^{Str}, is determined from Eq. (14).

The single-particle average, E_{av}^{Str} (i.e. $\tilde{E}_{sp}$ in Eq. (14)), is calculated [73] through an $\hbar$ expansion of the semiclassical partition function introduced by Wigner and Kirkwood (see references in Ref. [73]). For general-shape potentials, this last method amounts [73] to eliminating the semiclassical Fermi energy $\tilde{\lambda}$ from the set of the following two equations:

$$N_e = \frac{1}{3\pi^2} \left(\frac{2m_e}{\hbar^2} \right)^{3/2} \int^{r_{\tilde{\lambda}}} d\mathbf{r} \left[(\tilde{\lambda} - V)^{3/2} - \frac{1}{16} \frac{\hbar^2}{2m_e} (\tilde{\lambda} - V)^{-1/2} \nabla^2 V \right], \tag{39}$$

and

$$
\begin{aligned}
E_{av}^{Str} = \frac{1}{3\pi^2} \left(\frac{2m_e}{\hbar^2} \right)^{3/2} \int^{r_{\tilde{\lambda}}} d\mathbf{r} \Bigg(&\left[\frac{3}{5} (\tilde{\lambda} - V)^{5/2} + V (\tilde{\lambda} - V)^{3/2} \right] \\
&+ \frac{1}{16} \frac{\hbar^2}{2m_e} [(\tilde{\lambda} - V)^{1/2} \nabla^2 V - V (\tilde{\lambda} - V)^{-1/2} \nabla^2 V] \Bigg),
\end{aligned}
\tag{40}
$$

where N_e is the total number of delocalized valence electrons, and $V(\rho, z)$ is the potential in the single-particle Hamiltonian of Eq. (36). The domain of integration is demarcated by the classical turning point $r_{\tilde{\lambda}}$, such that $V(r_{\tilde{\lambda}}) = \tilde{\lambda}$.

Finally, from the liquid-drop-model contributions, we retain the two most important ones, namely the surface contribution and the Coulomb repulsion. To determine the surface contribution, we calculate numerically the area of the surface of the fissioning cluster shape, $\rho = \rho(z)$, and multiply it by a surface-tension coefficient specified via an ETF-LDA calculation for spherical jellia [25, 39] (or even from experimental values). The Coulomb repulsion is calculated numerically using the assumption of a classical conductor, namely the excess 2 units of positive charge are assumed to be distributed over the surface of the fissioning cluster, and in addition each of the fragments carries one unit of charge upon separation (for a more elaborate application of the LDM to triaxially deformed ground states of neutral and charged metal clusters described via a one-center shape parameterization, see our discussion in Section 4.2.2.2 in connection with Eqs (17–26) and Ref. [40]).

As a result, the total energy E_{total} for a specific fission configuration is given by

$$E_{total} = E_{LDM} + \Delta E_{sh}^{Str} = E_S + E_C + \Delta E_{sh}^{Str}, \tag{41}$$

where E_S and E_C are the surface and Coulomb terms, respectively.

4.3 EXPERIMENTAL TRENDS AND THEORETICAL INTERPRETATION

In the following, we describe applications of the SE-SCM approach to systematic investigations of the effects of shape deformations on the energetics of fragmentation processes of metal clusters [26, 40, 41], and to studies of deformations and barriers in fission of charged metal clusters [43]. We mention that, in addition, Strutinsky calculations using phenomenological potentials have been reported for the case of neutral sodium clusters assuming axial symmetry in Refs [46, 47, 74], and for the case of fission in Refs [48, 49].

4.3.1 Electronic Shell Effects in Monomer and Dimer Separation Energies

Monomer and dimer separation energies associated with the unimolecular reactions $K_N^+ \rightarrow K_{N-1}^+ + K$, $K_N^+ \rightarrow K_{N-2}^+ + K_2$, and $Na_N^+ \rightarrow Na_{N-1}^+ + Na$ can be calculated as follows:

$$D_{1,N}^+ = E_{\text{total}}(\mathcal{Z} = +1, N-1) + E_{\text{total}}(\mathcal{Z} = 0, N = 1) - E_{\text{total}}(\mathcal{Z} = +1, N), \quad (42)$$

and

$$D_{2,N}^+ = E_{\text{total}}(\mathcal{Z} = +1, N-2) + E_{\text{total}}(\mathcal{Z} = 0, N = 2) - E_{\text{total}}(\mathcal{Z} = +1, N), \quad (43)$$

where $\mathcal{Z} = \pm Z$ (Z being the excess positive or negative charge in absolute units).

The theoretical results for $D_{1,N}^+$ and $D_{2,N}^+$ for potassium are displayed in Figure 4.2 and Figure 4.3, respectively, and are compared to the experimental measurements [75]. The theoretical and experimental [76] results for $D_{1,N}^+$ in the case of sodium are displayed in Figure 4.4 (bottom panel). An inspection of all three figures leads to the same conclusion as in the case of IPs and electron affinities (see Figure 4.1 and Ref. [40a]), i.e. that results obtained via calculations restricted to spherical shapes compare rather poorly with the experiment, that improvement is evident when spheroidal (axially symmetric) deformations are considered, and that the agreement between theory and experiment becomes detailed when triaxiality (i.e. ellipsoidal shapes) is taken into consideration. The feature of the appearance of strong odd–even alternations for $N = 12–15$ together with a well-defined quartet in the range $N = 16–19$ is present in the experimental monomer separation energies of both potassium and sodium clusters, and theoretically it can be accounted for only after the inclusion of triaxial deformations.

We note that in the case of dimer separation energies (Figure 4.3) the odd–even alternations cancel out. Parents with closed shells or subshells correspond to maxima, while daughters with closed shells or subshells are associated with minima (e.g. the triplets $N = 9–11$, or $N = 15–17$).

We also include for comparison results obtained by KS-LDA calculations [77] for deformed Na_N clusters restricted to spheroidal (axial) symmetry (Figure 4.4, top panel). As expected, except for very small clusters ($N < 9$), these results do not exhibit odd–even oscillations. In addition, significant discrepancies between the calculated and experimental results are evident, particularly pertaining to the amplitude of oscillations at shell and subshell closures.

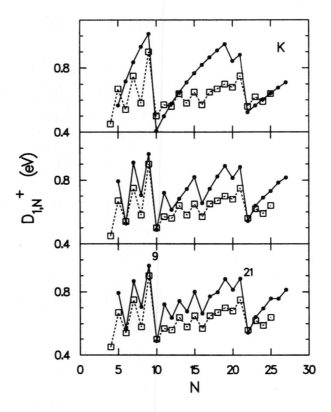

Figure 4.2 Monomer separation energies, $D_{1,N}^+$ (see Eq. (42)), from singly cationic K_N^+ clusters in the range $5 \leq N \leq 27$. Solid dots: theoretical results derived from the SE-SCM method. Open squares: experimental measurements from Ref. [75]. Top panel: the spherical model compared to experimental data. Middle panel: the spheroidal (axially symmetric) model compared to experimental data. Lower panel: the ellipsoidal (triaxial) model compared to experimental data

4.3.2 Electronic Shell Effects in Fission Energetics

Fission of doubly charged metal clusters, $M_N^{2\pm}$, has attracted considerable attention in the last few years. LDA calculations for fission energetics have usually been restricted to spherical jellia for both parent and daughters [78, 19], with the exception of molecular-dynamical calculations for sodium [10] and potassium [11] clusters with $N \leq 12$. We present here systematic calculations for the dissociation energies $\Delta_{N,P}$ of the fission processes $K_N^{2+} \rightarrow K_P^+ + K_{N-P}^+$, as a function of the fission channels P.

We have calculated the dissociation energies

$$\Delta_{N,P} = E_{\text{total}}(\mathcal{Z} = +1, P) + E_{\text{total}}(\mathcal{Z} = +1, N - P) - E_{\text{total}}(\mathcal{Z} = +2, N), \quad (44)$$

for the cases of parent clusters having $N = 26, 23, 18$ and 15 potassium atoms, and compared them with experimental results [79]. The theoretical calculations compared to the experimental results are displayed in Figures 4.5–8 for $N = 26, 23, 18, 15$, respectively. Again, while consideration of spheroidal shapes improves greatly the agreement between theory and experiment over the spherical model, fully detailed correspondence

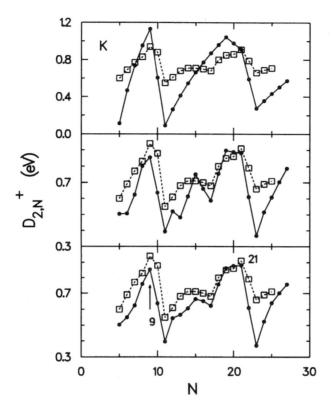

Figure 4.3 Dimer separation energies, $D_{2,N}{}^+$ (see Eq. (43)), from singly cationic $K_N{}^+$ clusters in the range $5 \leq N \leq 27$. Solid dots: theoretical results derived from the SE-SCM method. Open squares: experimental measurements from Ref. [75]. Top panel: the spherical model compared to experimental data. Middle panel: the spheroidal model compared to experimental data. Lower panel: the ellipsoidal model compared to experimental data

is achieved only upon allowing for triaxial-shape deformations (notice the improvement in the range $P = 12–14$ for $N = 26$, and in the range $P = 10–13$ for $N = 23$). In the cases $N = 18$ and $N = 15$ (Figures 4.7 and 4.8), the biaxial and triaxial results are essentially identical, since no fragment with more than nine electrons is involved. We note that the magic fragments $K_3{}^+$ and $K_9{}^+$ always correspond to strong minima, and that for $N = 18$ the channel associated with the double magic fragments ($K_9{}^+$, $K_9{}^+$) is clearly the favored one over the other magic channel with $K_3{}^+$, in agreement with the experimental analysis.

Finally, we carried out calculations of dissociation energies, Δ_f^{pos} and Δ_f^{neg}, of the most favored fission channels over the whole range up to $N = 100$ atoms for the cases of doubly charged cationic and anionic sodium clusters, respectively. The triaxial results compared to the spherical-jellium calculations according to the LDA-SCM method [25] are displayed in Figures 4.9 and 4.10. In both cases, the main difference from the spherical jellium is a strong suppression of the local minima, indicating that the critical size for exothermic fission is significantly smaller than $N = 100$ (about $N = 30$), as indeed has been observed experimentally for hot cationic alkali-metal clusters [79] (the spherical-jellium results

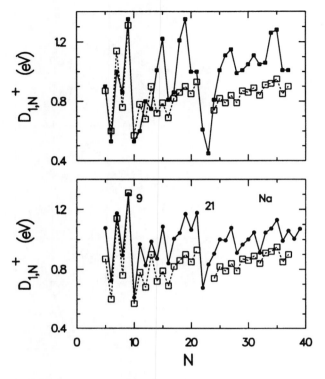

Figure 4.4   Monomer separation energies, $D_{1,N}^+$ (see Eq. (42)), from singly cationic Na_N^+ clusters in the range $5 \leq N \leq 39$. Open squares: Experimental measurements from Ref. [76]. Solid dots (bottom panel): theoretical results derived from the SE-SCM method in the case of triaxial deformations. Solid squares (top panel): theoretical results according to the KS-LDA spheroidal calculations of Ref. [77]

clearly are not compatible with the emergence of such experimental critical sizes in the size range $N \leq 100$).

4.3.3   Electronic Shell Effects in Fission Barriers and Fission Dynamics of Metal Clusters

In this section, we focus our discussion on recent trends in studies of binary fission processes in doubly charged metal clusters.

4.3.3.1   Molecular-Dynamics Studies of Fission

Before discussing applications of the SE-SCM (and variants thereof) to the description of cluster fission, we note that for atomic and molecular clusters microscopic descriptions of energetics and dynamics of fission processes, based on modern electronic structure calculations in conjunction with molecular dynamics simulations (where the classical trajectories of the ions, moving on the concurrently calculated Born–Oppenheimer (BO) electronic potential-energy surface, are obtained via integration of the Newtonian

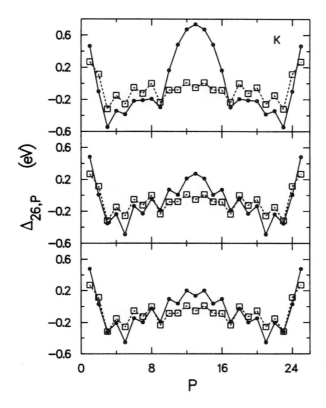

Figure 4.5 Fission dissociation energies, $\Delta_{26,P}$ (see Eq. (44)), for the doubly cationic K_{26}^{2+} cluster as a function of the fission channels P. Solid dots: theoretical results derived from the SE-SCM method. Open squares: experimental measurements from Ref. [79]. Top panel: the spherical model compared to experimental data. Middle panel: the spheroidal model compared to experimental data. Lower panel: the ellipsoidal model compared to experimental data

equations of motion), are possible and have been performed [10, 11] using the BO-local-spin-density(LSD)-functional-MD method [80]. Such calculations, using norm-conserving nonlocal pseudopotentials and self-consistent solutions of the KS-LSD equations [10, 11], applied to small sodium [10] and potassium [11] clusters, revealed several important trends (Figures 4.11–13).

(i)   The energetically favorable fission channel for such doubly charged clusters is the asymmetric one, $M_N^{2+} \rightarrow M_{N-3}^+ + M_3^+$, containing a 'magic' daughter M_3^+ (M = Na, K), i.e. $\Delta_{N,P} = E(M_{N-P}^+) + E(M_P^+) - E(M_N^{2+})$ is smallest for $P = 3$.

(ii)   Fission of clusters with $N \geq N_b^{2+}$, where $N_b^{2+} = 7$, involves barriers, whose magnitudes reflect the closed-shell stability of the parent cluster (i.e. E_b for $N = 10$ is particularly high), exhibiting a double-humped barrier shape [see Figures 4.11 and 4.13(a)].

(iii) The eventual fission products may already be distinguishable (i.e. preformed) at a rather early stage of the fission process (on the top of the exit barrier for Na_{10}^{2+}, see Figure 4.12, or prior to the exit barrier for K_{12}^{2+}, see Figure 4.13), and the electronic binding between the two fragments is long-range in nature.

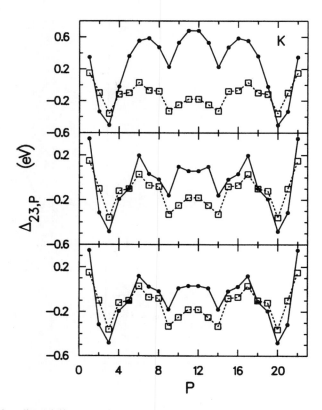

Figure 4.6 Fission dissociation energies, $\Delta_{23,P}$ (see Eq. (44)), for the doubly cationic K_{23}^{2+} cluster as a function of the fission channels P. Solid dots: theoretical results derived from the SE-SCM method. Open squares: experimental measurements from Ref. [79]. Top panel: the spherical model compared to experimental data. Middle panel: the spheroidal model compared to experimental data. Lower panel: the ellipsoidal model compared to experimental data

(iv)   The kinetic energy release $\mathcal{E}_r$ in the favorable channel obtained via dynamic simulations was found to be given by $\mathcal{E}_r \approx E_b + |\Delta_{N,3}|$, and the results are in correspondence with experimental measurements [11] for K_N^{2+} ($5 \leq N \leq 12$).

Furthermore, in agreement with experimental findings, the emerging fragments are vibrationally excited, with the heating of the internal nuclear degrees of freedom of the fission products in the exit channel originating from dynamical conversion of potential into internal kinetic energy (see K_{9+}^{int} in Figure 4.13(b)).

4.3.3.2   SE-SCM Interpretation of Fissioning Processes

The method we adopt in this section for further studying metal-cluster fission is the SE-SCM described in Section 4.2.2.6 (see also Ref. [43]).

As discussed above (see Section 4.2.2), in the SE-SCM method we need to introduce appropriate empirical potentials. As will become apparent from our results, one-center potentials (like the one-center modified, anisotropic harmonic oscillator) are not adequate

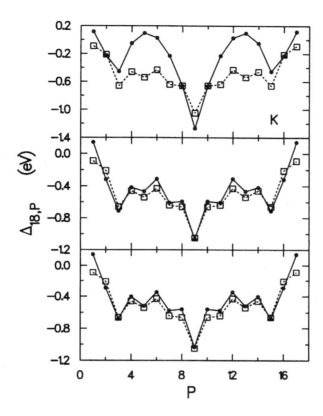

Figure 4.7 Fission dissociation energies, $\Delta_{18,P}$ (see Eq. (44)), for the doubly cationic K_{18}^{2+} cluster as a function of the fission channels P. Solid dots: theoretical results derived from the SE-SCM method. Open squares: experimental measurements from Ref. [79]. Top panel: the spherical model compared to experimental data. Middle panel: the spheroidal model compared to experimental data. Lower panel: the ellipsoidal model compared to experimental data

for describing shell effects in the fission of small metal clusters; rather, a two-center potential is required. Indeed, the empirical potentials should be able to simulate the fragmentation of the initial parent cluster towards a variety of asymptotic daughter-cluster shapes, e.g. two spheres in the case of double magic fragments, a sphere and a spheroid in the case of a single magic fragment, or two spheroids in a more general case. In the case of metal clusters, asymmetric channels are most common, and thus a meaningful and flexible description of the asymmetry is of primary concern. We found [43] that such a required degree of flexibility can be provided via the shape parameterization of the asymmetric two-center-oscillator shell model (ATCOSM) introduced earlier in nuclear fission [69] (see Section 4.2.2.6).

In addition to the present shape parameterization [43], other two-center shape parameterizations (mainly in connection with KS-LDA jellium calculations) have been used [81–83] in studies of metal-cluster fission. They can be grouped into two categories, namely, the two-intersected-spheres jellium [81, 84], and the variable-necking-in parameterizations [82, 83]. In the latter group, Ref. [82] accounts for various necking-in situations by using the 'funny-hills' parameterization [85], while Ref. [83] describes the necking-in

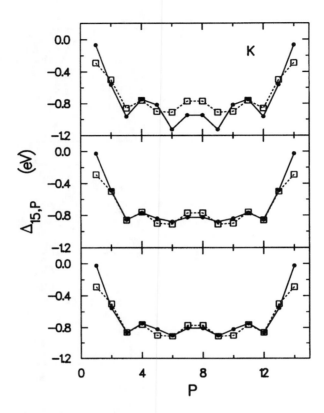

Figure 4.8 Fission dissociation energies, $\Delta_{15,P}$ (see Eq. (44)), for the doubly cationic K_{15}^{2+} cluster as a function of the fission channels P. Solid dots: theoretical results derived from the SE-SCM method. Open squares: experimental measurements from Ref. [79]. Top panel: the spherical model compared to experimental data. Middle panel: the spheroidal model compared to experimental data. Lower panel: the ellipsoidal model compared to experimental data

by connecting two spheres smoothly through a quadratic surface. The limitation of these other parameterizations is that they are not flexible enough to account for the majority of the effects generated by the shell structure of the parent and daughters, which in general do not have spherical, but deformed (independently from each other), shapes. An example is offered by the case of the parent Na_{18}^{2+}, which has a metastable oblate ground state, and thus cannot be described by any one of the above parameterizations. We wish to emphasize again that one of the conclusions of the present work is that the shell structures of the (independently deformed) parent and daughters are the dominant factors specifying the fission barriers, and thus parameterizations [81–83] with restricted final fragment (or parent) shapes are deficient in accounting for some of the most important features governing metal-cluster fission.

As a demonstration of our method, we present results for two different parents, namely Na_{10}^{2+} and Na_{18}^{2+}.

Figure 4.14 presents results for the channel $Na_{10}^{2+} \rightarrow Na_7^+ + Na_3^+$ for three different cases, namely, when the larger fragment Na_7^+ is oblate (left column), spherical (middle column) or prolate (right column). From our one-center analysis, we find, as expected, that

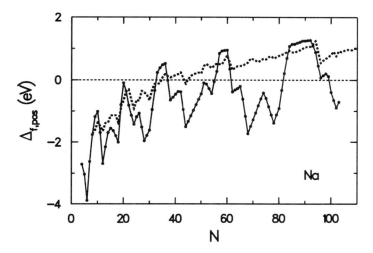

Figure 4.9   Solid dots: LDA-SCM results for the dissociation energies Δ_f^{pos} for the most favorable fission channel for doubly charged cationic parents Na_N^{2+} when the spherical jellium is used. The influence of triaxial deformation effects (calculated with the SE-SCM approach) is shown by the thick dashed line

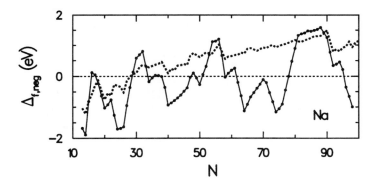

Figure 4.10   Solid dots: LDA-SCM results for the dissociation energies Δ_f^{neg} for the most favorable fission channel for doubly charged anionic parents Na_N^{2-} when the spherical jellium is used. The influence of triaxial deformation effects (calculated with the SE-SCM approach) is shown by the thick dashed line

Na_7^+ (with six electrons) has an oblate global minimum and a higher-in-energy prolate local minimum. In the two-center analysis, we have calculated the fission pathways so that the emerging fragments correspond to possible deformed one-center minima. It is apparent that the most favored channel (i.e. having the lowest barrier, see the solid line in the bottom panels) will yield an oblate Na_7^+ (left column in Figure 4.14), in agreement with the expectations from the one-center energetics analysis.

The middle panels exhibit the decomposition of the total barrier into the three components of surface, Coulomb and shell-correction terms (see Eq. (41)), which are denoted by an upper dashed curve, a lower dashed curve and a solid line, respectively. The total

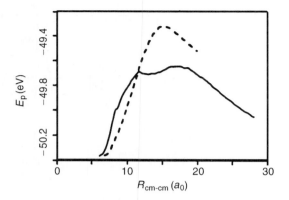

Figure 4.11 Molecular dynamics results for potential energy plotted against distance (in atomic units) between the centers of mass for the fragmentation of Na_{10}^{2+} into Na_7^+ and Na_3^+ (solid) and Na_9^+ and Na^+ (dashed), obtained via constrained minimization of the LSD ground-state energy of the system [10]

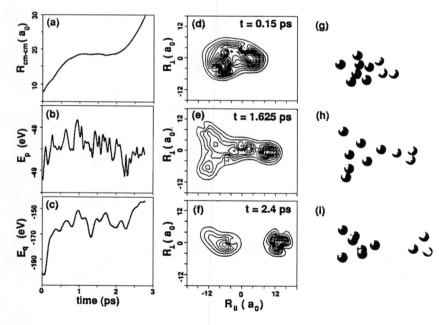

Figure 4.12 Fragmentation dynamics of Na_{10}^{2+} from first-principles Born–Oppenheimer local-spin-density functional molecular-dynamics simulations [10]. (a)–(c) Center-of-mass distance between the eventual fission products ($R_{c.m.-c.m.}$), total potential energy (E_p), and the electronic contribution E_q to E_p, against time. (d)–(f) Contours of the total electronic charge distribution at selected times calculated in the phane containing the two centers of mass. The $R_{\parallel}$ axis is parallel to $R_{c.m.-c.m.}$ (g)–(i) Cluster configurations for the times given in (d)–(f). Dark and light balls represent ions in the large and small fragments, respectively. Energy, distance and time in units of eV, bohr (a_0) and ps, respectively

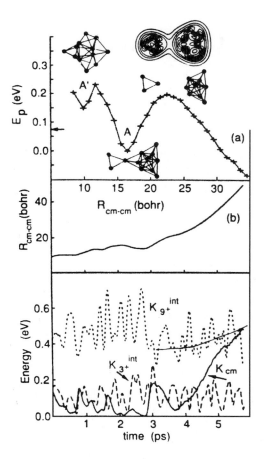

Figure 4.13   (a) Potential energy of K_{12}^{2+} fissioning in the favorable channel ($K_3^+ + K_9^+$) versus the interfragment distance $R_{c.m.-c.m.}$ obtained via constrained minimization. The origin of the E_p scale is set at the optimal pre-barrier configuration (A). For large $R_{c.m.-c.m.}$, $E_p = -0.9$ eV, i.e. Δ_3. Also included are cluster configurations of K_{12}^{2+} corresponding to: a compact isomer (A') (the energy of the optimal compact isomer found is denoted by an arrow); the optimal bound configuration (A); the structure on top of the exit-channel barrier for which contours of the total electronic charge density, ρ, are shown [11]

(b) Time evolution of $R_{c.m.-c.m.}$, the internal vibrational kinetic energies of the fragments (K_{3+}^{int} and K_{9+}^{int}) and the sum of the fragments' translational kinetic energies (K_{cm}) obtained via a BO-LSD-MD simulation starting from ionization ($t = 0$) of a K_{12}^+ cluster at 500 K. A line is drawn in K_{9+}^{int} (for $t \geq 3$ ps) to guide the eye, illustrating heating of the internal vibrational degrees of freedom of the departing fragment

LDM contribution (surface plus Coulomb) is also exhibited in the bottom panels (dashed lines).

It can be seen that the LDM barrier is either absent or very small, and that the total barrier is due almost exclusively to electronic shell effects. The total barrier has a double-humped structure, with the outer hump corresponding to the LDM saddle point, which also happens to be the scission point (indicated by an empty vertical arrow). The inner hump coincides with the peak of the shell-effect term, and is associated with the rearrangement

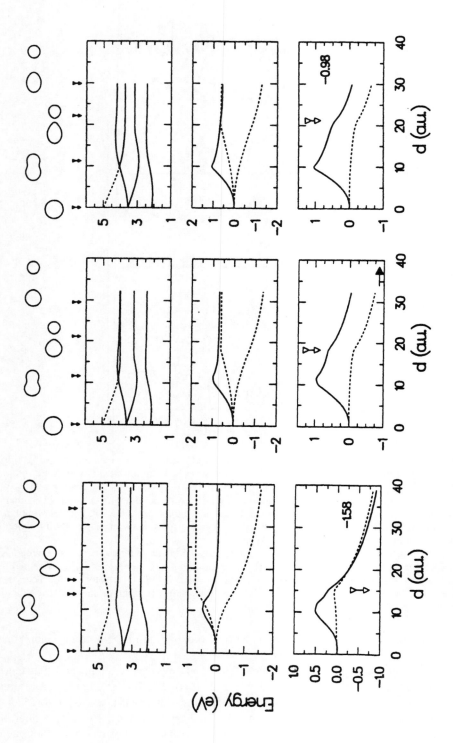

of single-particle levels from the initial spherical parent to a molecular configuration resembling an Na_7^+ attached to an Na_3^+. Such molecular configurations (discovered earlier in first-principles MD simulations [10, 11] of fission of charged metal clusters, as well as in studies of fusion of neutral clusters [86]) are a natural precursor of full fragment separation and complete fission, and naturally they give rise to the notion of preformation of the emerging fragments [10, 11].

Figure 4.15(a) displays the ATCOSM results for the symmetric channel $Na_{18}^{2+} \rightarrow 2Na_9^+$, when, for illustrative purposes, the parent is assumed to be spherical at $d = 0$ (this channel is favored compared to that of the trimer [43] from both energetics and barrier considerations; for small clusters, this is the only case where a channel other than that of the trimer is the most favored one). The top panel of Figure 4.15(a) describes the evolution of the single-particle spectra. The spherical ordering 1s, 1p, 1d, 2s, etc., for the parent at $d = 0$ is clearly discernible. With increasing separation distance, the levels exhibit several crossings, and, after the scission point, they naturally regroup to a new ordering associated with the spherical Na_9^+ products (at the end of the fission process, the levels are doubly degenerate compared to the initial configuration, since there are two Na_9^+ fragments). It is seen that the ATCOSM leads to an oscillator energy (i.e. the gap between two populated major shells exhibited at the right-hand end of the figure) of 1.47 eV for each Na_9^+ fragment in agreement with the value expected from the one-center model (the 1s state of Na_9^+ lies at 2.21 eV; in the case of the initial spherical Na_{18}^{2+} ($d = 0$), the oscillator energy corresponding to the gap between major shells is 1.17 eV, and the corresponding 1s state lies at 1.75 eV).

From the middle panel of Figure 4.15(a), we observe that the shell correction (solid line) contributes a net gain in energy of about 1.6 eV upon dissociation into two Na_9^+ fragments. This gain is larger than the increase in energy (i.e. positive energy change) due to the surface term, which saturates at a value of about 1 eV after the scission point at $d \approx 23$ a.u. The total energy is displayed in the bottom panel of Figure 4.15(a) (solid line) along with the LDM barrier (dashed line). Even though distorted (when compared to the cases of Figure 4.14), the total barrier still exhibits a two-peak structure, the inner peak arising from the hump in the shell-correction, and the outer peak arising from the point of saturation of the surface term (this last point coincides again with the scission point,

---

Figure 4.14 ATCOSM results for the asymmetric channel $Na_{10}^{2+} \rightarrow Na_7^+ + Na_3^+$. The final configuration of Na_3^+ is spherical. For the heavier fragment Na_7^+, we present results associated with three different final shape configurations, namely, oblate ((o,s); left), spherical ((s,s); middle), and prolate ((p,s); right). The ratio of shorter over longer axis is 0.555 for the oblate case and 0.75 for the prolate case.

Bottom panel: LDM energy (surface plus Coulomb, dashed curve) and total potential energy (LDM plus shell-corrections, solid curve) as a function of fragment separation d. The empty vertical arrow marks the scission point. The zero of energy is taken at $d = 0$. A number (-1.58 eV or -0.98 eV), or a horizontal solid arrow, denotes the corresponding dissociation energy.

Middle panel: shell-correction contribution (solid curve), surface contribution (upper dashed curve), and Coulomb contribution (lower dashed curve) to the total energy, as a function of fragment separation d.

Top panel: single-particle spectra as a function of fragment separation d. The occupied (fully or partially) levels are denoted with solid lines. The unoccupied levels are denoted with dashed lines. On top of the figure, four snapshots of the evolving cluster shapes are displayed. The solid vertical arrows mark the corresponding fragment separations. Observe that the doorway molecular configurations correspond to the second snapshot from the left. Notice the change in energy scale for the middle and bottom panels, as one passes from (o,s) to (s,s) and (p,s) final configurations

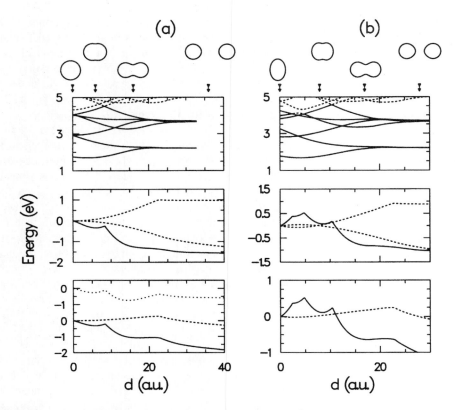

Figure 4.15 ATCOSM results for the symmetric channel $Na_{18}^{2+} \rightarrow 2Na_9^+$, when the initial parent shape is assumed (a) spherical, and (b) oblate (with a shorter over longer axis ratio equal to 0.699). Panel distribution and other notations and conventions are the same as in Figure 4.14. The top dotted line in the bottom panel of (a) represents the total energy without the Coulomb contribution. Observe that the doorway molecular configurations correspond to the third snapshot from the left. Notice that the zero of all energies is taken at $d = 0$

as well as with the saddle of the LDM barrier). An inner local minimum is located at $d \approx 8$ a.u., and corresponds to a compact prolate shape of the parent (see second drawing from the left at the top of Figure 4.15(a)), while a second deeper minimum appears at $d \approx 18$ a.u., corresponding to a superdeformed shape of a molecular configuration of two Na_9^+ clusters tied up together (preformation of fragments, see third drawing from the left at the top of Figure 4.15(a)). The inner barrier separating the compact prolate configuration from the superdeformed molecular configuration arises from the rearrangement of the single-particle levels during the transition from the initially assumed spherical Na_{18}^{2+} configuration to that of the supermolecule $Na_9^+ + Na_9^+$. We note that the barrier separating the molecular configuration from complete fission is very weak, being less than 0.1 eV.

The top dotted line of the bottom panel displays the total energy in the case when the Coulomb contribution is neglected. This curve mimics the total energy for the fusion of two neutral Na_8 clusters, namely the total energy for the reaction $2Na_8 \rightarrow Na_{16}$. Overall, we find good agreement with a KS-LDA calculation for this fusion process (see Figure 4.1

of Ref. [86]). We further note that the superdeformed minimum for the neutral Na_{16} cluster is deeper than that in the case of the doubly charged Na_{18}^{2+} cluster. Naturally, this is due to the absence of the Coulomb term.

The natural way for producing experimentally the metastable Na_{18}^{2+} cluster is by ionization of the stable singly charged Na_{18}^{+} cluster. Since this latter cluster contains 17 electrons and has a deformed oblate ground state [40a], it is not likely that the initial configuration of Na_{18}^{2+} will be spherical or prolate as was assumed for illustrative purposes in Figure 4.15(a). Most likely, the initial configuration for Na_{18}^{2+} will be that of the oblate Na_{18}^{+}. To study the effect that such an oblate initial configuration has on the fission barrier, we display in Figure 4.15(b) ATCOSM results for the pathway for the symmetric fission channel, starting from an oblate shape of Na_{18}^{2+}, proceeding to a compact prolate shape, and then to full separation between the fragments via a superdeformed molecular configuration. We observe that additional potential humps (in the range 2 a.u. $\leq d \leq 6$ a.u.), associated with the shape transition from the oblate to the compact prolate shape, do develop. Concerning the total energies, the additional innermost humps result in the emergence of a significant fission barrier of about 0.52 eV for the favored symmetric channel (see $d \approx 5$ a.u. in Figure 4.15(b)).

From the above analysis, we conclude that considerations of the energy pathways leading from the parent to preformation configurations (i.e. the inner-barrier hump, or humps) together with the subsequent separation processes are most important for proper elucidation of the mechanisms of metal-cluster fission processes. This corroborates earlier results obtained via first-principles MD simulations [10, 11] pertaining to the energetics and dynamical evolution of fission processes, and emphasizes that focusing exclusively [81, 83] on the separation process between the preformed state and the ultimate fission products provides a rather incomplete description of fission phenomena in metal clusters. It is anticipated that, with the use of emerging fast spectroscopies [87], experimental probing of the detailed dynamics of such fission processes could be achieved.

4.4 INFLUENCE OF ELECTRONIC ENTROPY ON SHELL EFFECTS

In the previous sections, we showed that consideration of triaxial (ellipsoidal) shapes in the framework of the SCM leads to overall substantial systematic improvement in the agreement between theory and experimental observations pertaining to the major and the fine structure of the size-evolutionary patterns associated with the energetics of fragmentation processes (monomer/dimer dissociation energies and fission energetics) and ionization.

The theoretical methods and discussion of deformation effects in the previous sections were restricted to zero temperature. However, the experiments are necessarily made with clusters at finite temperatures, a fact that strongly motivates the development of finite-temperature theoretical approaches.

Due to the difficulty of the subject, to date only a few finite-temperature theoretical studies of metal clusters have been performed. In this section, we discuss briefly some of the conclusions of a recent SCM study [41] regarding the importance of thermal effects. The theoretical details pertaining to this finite-temperature (FT) SE-SCM will not be elaborated here, but they can be found in the aforementioned reference.

The main conclusion of Ref. [41] was that, in conjunction with deformation effects, electronic-entropy effects in the size-evolutionary patterns of relatively small (as small as 20 atoms) simple-metal clusters become already prominent at moderate temperatures. At smaller sizes, electronic-entropy effects are less prominent, but they can still be discernible. As an example, we present in Figure 4.16 the monomer separation energies of K_N^+ clusters for two temperatures ($T = 10$ K and $T = 300$ K), along with the available experimental measurements [75] (open squares) in the size range $N = 4$–23. First notice that the $T = 10$ K results are practically indistinguishable from the $T = 0$ K results presented in Figure 4.2. Compared to the $T = 10$ K results, the theoretical results at $T = 300$ K are in better agreement with the experimental ones due to an attenuation of the amplitude of the alternations (e.g. notice the favorable reduction in the size of the drops at $N = 9, 15$ and 21). This amplitude attenuation, however, is moderate, and it is remarkable that the $T = 300$ K SCM results in this size range preserve in detail the same relative pattern as the $T = 0$ K ones (in particular, the well-defined odd–even oscillations in the range $N = 4$–15 and the ascending quartet at $N = 16$–19 followed by a dip at $N = 20$).

As a further example, the theoretical IPs of K_N clusters in the size range $3 \leq N \leq 102$ for three temperatures, $T = 10$ K, 300 K and 500 K, are displayed in Figure 4.17, and are compared with the experimental measurements [36] (open squares; the experimental uncertainties are 0.06 eV for $N \leq 30$ and 0.03 eV for $N > 30$). As was the case with our earlier $T = 0$ K results [40], the $T = 10$ K theoretical results exhibit the following two characteristics. (i) Above $N = 21$, there is a pronounced fine structure between major-shell closures that is not present in the experimental measurements. (ii) There are steps at the major-shell closures that are much larger than the experimental ones, i.e. three-to-five times for $N = 40, 58$ and 92, and two-to-three times for $N = 8$ and 20 (this needs to be

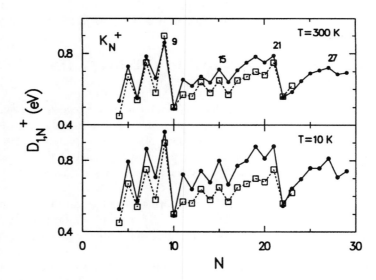

Figure 4.16 Monomer separation energies of K_N^+ clusters at two temperatures, $T = 10$ K and 300 K. Solid dots: theoretical FT-SE-SCM results. Open squares: experimental measurements [75]. To facilitate comparison, the SE-SCM results at the higher temperature have been shifted by 0.07 eV, so that the theoretical curves at both temperatures refer to the same point at $N = 10$

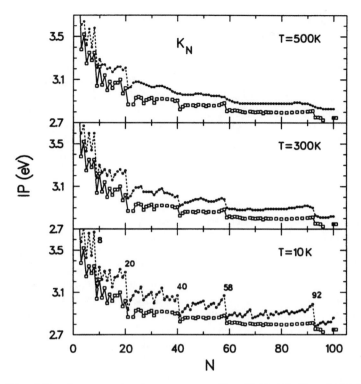

Figure 4.17 IPs of K_N clusters at three temperatures, $T = 10$ K, 300 K and 500 K. Solid dots: theoretical FT-SE-SCM results. Open squares: experimental measurements [36]

contrasted to the experimental IPs for cold Na_N clusters, which are in overall good agreement with our $T = 0$ K SE-SCM results regarding both characteristics (see Figure 4.1)).

The agreement between theory and experiment is significantly improved at $T = 300$ K. Indeed, in comparison with the lower-temperature calculations, the $T = 300$ K results exhibit the following remarkable changes. (i) Above $N = 21$, the previously sharp fine-structure features are smeared out, and, as a result, the theoretical curve follows closely the mild modulations of the experimental profile. In the size range $N = 21-34$, three rounded, hump-like formations (ending to the right at the subshell closures at $N = 26$, 30 and 34) survive in very good agreement with the experiment (the sizes of the drops at $N = 26$, 30 and 34 are comparable to the experimental ones [88]). (ii) The sizes of the IP drops at $N = 20$, 40, 58 and 92 are reduced drastically and are now comparable to the experimental ones. In the size range $N \leq 20$, the modifications are not as dramatic. Indeed, one can clearly see that the pattern of odd–even alternations remains well defined, but with a moderate attenuation in amplitude, again in excellent agreement with the experimental observation.

For $T = 500$ K, the smearing out of the shell structure associated with the calculated results progresses even further, obliterating the agreement between theory and experiment. Specifically, the steps at the subshell closures at $N = 26$ and 30, as well as at the major-shell closures at $N = 40$, 58 and 92 are rounded and smeared out over several clusters (an analogous behavior has been observed in the logarithmic abundance spectra of hot, singly cationic, copper, silver and gold clusters [89]). At the same time, however, the odd–even alternation remains well defined for $N \leq 8$. We further notice that, while some residue of fine structure survives in the range $N = 9-15$, the odd–even alternations there are essentially absent (certain experimental measurements [90] of the IPs of hot Na_N clusters appear to conform to this trend).

The influence of the electronic entropy on the height of fission barriers has not been studied as yet, but it will undoubtedly be the subject of future research in metal-cluster physics. In any case, based on the results of this section, it is natural to conjecture that electronic-entropy effects will tend to quench the barrier heights, especially in the case of larger multiply charged clusters.

4.5 SUMMARY

In this chapter, we have elucidated certain issues pertaining to evaporation and fission processes of metallic clusters, focusing on electronic shell effects and their importance in determining the energetics, structure, pathways and dynamical mechanisms of dissociation and fragmentation in these systems, and have outlined and demonstrated various theoretical approaches currently used in investigations of cluster fragmentation phenomena, ranging from microscopic first-principles electronic-structure calculations coupled with molecular-dynamics simulations to adaptation of models of a more phenomenological nature that originated in studies of atomic nuclei. In this respect, a recurrent theme in this exposition has been the crucial importance of deformation and electronic-entropy (temperature) effects, as well as their treatment with the help of shell-correction methods.

By drawing analogies, as well as considering differences, between certain aspects of nuclear fission and nuclear radioactivity phenomena and atomic (metallic) cluster fission processes, we have attempted to provide a unifying conceptual framework for discussion of the physical principles underlying modes of cluster fission (i.e. importance of deformations, shell effects originating from fragments and parent, asymmetric and symmetric fission, single- and double-humped barriers, fissioning cluster shapes and dynamical aspects, such as the time scale of fission processes, kinetic energy release and dynamical energy redistribution among the fission products).

We conclude by commenting on some experimental and theoretical issues in cluster fission that remain as future challenges (limiting ourselves to metallic clusters). These include: fission dynamics of multiply charged large metal clusters [6, 23, 24]; systematic investigations of temperature effects on modes of cluster fission, and ternary, and higher multifragmentation processes; time-resolved spectroscopy of fission processes and of fission isomers; spin effects in fission; tunnelling processes and corresponding lifetimes in subbarrier fission modes of clusters of light elements, e.g. lithium; and fission processes of nonsimple metal clusters.

4.6 ACKNOWLEDGMENTS

This research was supported by a grant from the US Department of Energy (Grant No. FG05-86ER45234). Calculations were performed on CRAY computers at the Supercomputer Center at Livermore, California, and the Georgia Institute of Technology Center for Computational Materials Science.

4.7 REFERENCES AND NOTES

[1] Å. Bohr and B. R. Mottelson, *Nuclear Structure, Vol. II*, Benjamin, Reading, MA, 1975.

[2] M. A. Preston and R. K. Bhaduri, *Structure of the Nucleus*, Addison-Wesley, London, 1975.

[3] W. Frost, *Theory of Unimolecular Reactions*, Academic, New York, 1973.

[4] C. Bréchignac, Ph. Cahuzac, F. Carlier, M. de Frutos, J. Leygnier, J. Ph. Roux and A. Sarfati, *Comments At. Mol. Phys.* **31**, 361 (1995), and references therein.

[5] W. A. Saunders, *Phys. Rev. A* **46**, 7028 (1992).

[6] T. P. Martin, U. Näher, H. Göhlich and T. Lange, *Chem. Phys. Lett.* **196**, 113 (1992); U. Näher, H. Göhlich, T. Lange and T. P. Martin, *Phys. Rev. Lett.* **68**, 3416 (1992).

[7] The usage of the term 'fragmentation' here should not be confused with the 'fragmentation of the oscillator strength' (also referred to as 'Landau damping' or 'Landau fragmentation'), which is a phenomenon associated with the profile of the photoabsorption cross-section of metal clusters, and it may lead to the broadening of the photoabsorption profiles and/or to the appearance of a multipeak profile. It was first described using the matrix-RPA/LDA version of linear response for the case of neutral Na_{20} in C. Yannouleas, R. A. Broglia, M. Brack and P.-F. Bortignon, *Phys. Rev. Lett.* **63**, 255 (1989). For other metal-cluster species and sizes, see also C. Yannouleas and R. A. Broglia, *Europhys. Lett.* **15**, 843 (1991); C. Yannouleas and R. A. Broglia, *Phys. Rev. A* **44**, 5793 (1991); C. Yannouleas, *Chem. Phys. Lett.* **193**, 587 (1992); C. Yannouleas, P. Jena and S. N. Khanna, *Phys. Rev. B* **46**, 9751 (1992); C. Yannouleas, E. Vigezzi and R. A. Broglia, *Phys. Rev. B* **47**, 9849 (1993); C. Yannouleas, F. Catara and N. Van Giai, *Phys. Rev. B* **51**, 4569 (1995); C. Yannouleas, *Phys. Rev. B* **58**, 6748 (1998).

[8] For monovalent elements (Na, K, Cu, etc.), N is also equal to the number N_e of delocalized valence electrons.

[9] (a) C. Bréchignac, Ph. Cahuzac, F. Carlier and M. de Frutos, *Phys. Rev. Lett.* **64**, 2893 (1990); (b) C. Bréchignac, Ph. Cahuzac, F. Carlier and M. de Frutos, *Phys. Rev. B* **49**, 2825 (1994).

[10] R. N. Barnett, U. Landman and G. Rajagopal, *Phys. Rev. Lett.* **67**, 3058 (1991).

[11] C. Bréchignac, Ph. Cahuzac, F. Carlier, M. de Frutos, R. N. Barnett and U. Landman, *Phys. Rev. Lett.* **72**, 1636 (1994).

[12] N. Bohr and J. A. Wheeler, *Phys. Rev.* **56**, 426 (1939).

[13] J. R. Nix and W. J. Swiatecki, *Nucl. Phys.* **71**, 1 (1965).

[14] G. Gamow, *Z. Phys.* **51**, 204 (1928).

[15] E. U. Condon and R. W. Gurney, *Nature* **122**, 439 (1928).

[16] A. Sandulescu, D. N. Poenaru, and W. Greiner, *Sov. J. Part. Nucl.* **11**, 528 (1980).

[17] P. B. Price, *Ann. Rev. Nucl. Part. Sci.* **39**, 19 (1989).

[18] For a theoretical review, see W. Greiner, M. Ivascu, D. N. Poenaru and A. Sandulescu, 'Cluster radioactivities', in *Treatise on Heavy-Ion Science*, ed. D. A. Bromley, Plenum, New York, 1989, Vol. 8, p. 641.

[19] J. M. López, J. A. Alonso, F. Garcias and M. Barranco, *Ann. Physik* (Leipzig) **1**, 270 (1992).

[20] S. G. Nilsson, C.-F. Tsang, A. Sobiczewski, Z. Szymanski, S. Wycech, C. Gustafson, I.-L. Lamm, P. Moller and B. Nilsson, *Nucl. Phys.* **A131**, 1 (1969).

[21] W. D. Myers and W. J. Swiatecki, *Nucl. Phys.* **81**, 1 (1966).

[22] T. D. Märk and O. Echt, in *Clusters of Atoms and Molecules II*, ed. H. Haberland (Springer-Verlag, Berlin, 1994), ch. 2.6; and O. Echt and T. D. Märk, ch. 2.7.

[23] C. Bréchignac, Ph. Cahuzac, F. Carlier, M. de Frutos, N. Kebaili, J. Leygnier, A. Sarfati and V. M. Akulin, in *Large Clusters of Atoms and Molecules*, ed. T. P. Martin (Kluwer, Dordrecht, 1996), p. 315.

[24] U. Näher, S. Frank, N. Malinowski, U. Zimmermann and T. P. Martin, *Z. Phys. D* **31**, 191 (1994).

[25] C. Yannouleas and U. Landman, *Phys. Rev. B* **48**, 8376 (1993); *Chem. Phys. Lett.* **210**, 437 (1993).

[26] C. Yannouleas and U. Landman, in *Large Clusters of Atoms and Molecules*, ed. T. P. Martin (Kluwer, Dordrecht, 1996), p. 131.

[27] M. K. Scheller, R. N. Compton and L. S. Cederbaum, *Science* **270**, 1160 (1995).

[28] C. Yannouleas and U. Landman, *Chem. Phys. Lett.* **217**, 175 (1994).
[29] P. Scheier, B. Dunser, R. Worgotter, D. Muigg, S. Matt, O. Echt, M. Foltin and T. D. Märk, *Phys. Rev. Lett.* **77**, 2654 (1996).
[30] D. Scharf, J. Jortner and U. Landman, *J. Chem. Phys.* **88**, 4273 (1988).
[31] D. H. E. Gross, M. E. Madjet and O. Schapiro, *Z. Phys. D* **39**, 75 (1997).
[32] R. N. Barnett and U. Landman, *J. Phys. Chem.* **99**, 17305 (1995).
[33] W. Ekardt, *Phys. Rev. B* **29**, 1558 (1984); *Phys. Rev. B* **31**, 6360 (1985).
[34] D. E. Beck, *Solid State Commun.* **49**, 381 (1984).
[35] K. L. Clemenger, *Phys. Rev. B* **32**, 1359 (1985); PhD Dissertation, University of California, Berkeley, 1985.
[36] W. A. Saunders, PhD Dissertation, University of California, Berkeley, 1986; W. A. Saunders, K. Clemenger, W. A. de Heer and W. D. Knight, *Phys. Rev. B* **32**, 1366 (1985).
[37] H. A. Jahn and E. Teller, *Proc. R. Soc. London Ser. A* **161**, 220 (1937).
[38] V. M. Strutinsky, *Nucl. Phys. A* **95**, 420 (1967); *Nucl. Phys. A* **122**, 1 (1968).
[39] R. N. Barnett, C. Yannouleas and U. Landman, *Z. Phys. D* **26**, 119 (1993).
[40] (a) C. Yannouleas and U. Landman, *Phys. Rev. B* **51**, 1902 (1995); (b) *J. Chem. Phys.* **107**, 1032 (1997).
[41] C. Yannouleas and U. Landman, *Phys. Rev. Lett.* **78**, 1424 (1997).
[42] C. Yannouleas and U. Landman, *J. Phys. Chem. A* (Letter) **102**, 2505 (1998).
[43] C. Yannouleas and U. Landman, *J. Phys. Chem.* (Letter) **99**, 14577 (1995); C. Yannouleas, R. N. Barnett and U. Landman, *Comments At. Mol. Phys.* **31**, 445 (1995).
[44] C. Yannouleas and U. Landman, *J. Chem. Phys.* **105**, 8734 (1996); *Phys. Rev. B* **54**, 7690 (1996).
[45] C. Yannouleas and U. Landman, *J. Phys. Chem. B* (Letter) **101**, 5780 (1997); C. Yannouleas, E. N. Bogachek and U. Landman, *Phys. Rev. B* **57**, 4872 (1998).
[46] S. M. Reimann, M. Brack and K. Hansen, *Z. Phys. D* **28**, 235 (1993).
[47] S. Frauendorf and V. V. Pashkevich, *Ann. der Physik* **5**, 34 (1996).
[48] H. Koizumi, S. Sugano and Y. Ishii, *Z. Phys. D* **28**, 223 (1993); M. Nakamura, Y. Ishii, A. Tamura and S. Sugano, *Phys. Rev. A* **42**, 2267 (1990).
[49] A. Vieira and C. Fiolhais, *Z. Phys. D* **37**, 269 (1996).
[50] W. Kohn and L. J. Sham, *Phys. Rev.* **140**, A1133 (1965).
[51] M. L. Homer, E. C. Honea, J. L. Persson and R. L. Whetten (unpublished).
[52] W. A. De Heer, *Rev. Mod. Phys.* **65**, 611 (1993).
[53] J. Harris, *Phys. Rev. B* **31**, 1770 (1985).
[54] W. M. C. Foulkes and R. Haydock, *Phys. Rev. B* **39**, 12 520 (1989).
[55] M. W. Finnis, *J. Phys.: Condens. Matter* **2**, 331 (1990).
[56] E. Zaremba, *J. Phys.: Condens. Matter* **2**, 2479 (1990).
[57] The choice of the ETF energy functional as a vehicle for optimization of the input density is a rather natural one. Indeed, in several attempts (such as the integral formulation of density functional theory (W. Yang, *Phys. Rev. A* **38**, 5494 (1988)), or through the use of generalized nonlocal kinetic-energy functionals (L.-W. Wang and M. P. Teter, *Phys. Rev. B* **45**, 13 196 (1992))) to construct theories based on the Hohenberg–Kohn density functional theorem (P. Hohenberg and W. Kohn, *Phys. Rev.* **136**, B684 (1964)) directly, that is without the use of orbitals, the ETF functional (see Wang and Teter above) (with the kinetic functional expanded in density gradients), or the TF density (see, Yang above) appear as a limiting case, or lowest level of approximation, respectively.
[58] Here we use the Gunnarsson–Lundqvist xc functional (see O. Gunnarsson and B. I. Lundqvist, *Phys. Rev. B* **13**, 4274 (1976)).
[59] C. H. Hodges, *Can. J. Phys.* **51**, 1428 (1973).
[60] M. Brack, *Phys. Rev. B* **39**, 3533 (1989).
[61] Here, we consider clusters of monovalent elements (Na, K, Cu, etc.). For polyvalent elements, N in Eq. (18) must be replaced by Nv, where v is the valency.
[62] For materials with high electronic densities, it is known that the usual LDA fails to provide adequate values for the surface tension. Therefore, in Ref. [40] (see also Ref. [28] for the case of fullerenes), in the case of Cu and Li clusters, we have carried an ETF calculation

using the *stabilized-jellium-LDA* (SJ-LDA) functional, which yields substantially improved values for the surface tension. It is also well known that both the simple-jellium LDA and the stabilized-jellium LDA in the case of plane-surface calculations yield only approximate values for the work function. In our method, we can easily overcome these LDA discrepancies by taking the work function and the surface tension from experimental observations. This latter procedure has the additional advantage that contributions from the atomic d electrons in the case of noble-metal clusters (i.e. Ag, Au, Cu) are automatically included in the SE-SCM (see Refs [41, 42]).

[63] I. S. Gradshteyn and I. M. Ryzhik, *Table of Integrals, Series, and Products*, Academic, New York, 1980, ch. 8.11.

[64] R. W. Hasse and W. D. Myers, *Geometrical Relationships of Macroscopic Nuclear Physics*, Springer-Verlag, Berlin, 1980 ch. 6.5.

[65] S. G. Nilsson, *K. Danske Vidensk. Selsk. Mat.-Fys. Medd.* **29** (1955), no. 16.

[66] J. R. Nix, *Annu. Rev. Nucl. Part. Sci.* **22**, 65 (1972).

[67] R. K. Bhaduri and C. K. Ross, *Phys. Rev. Lett.* **27**, 606 (1971).

[68] The perturbation $l^2 - \langle l^2 \rangle_n$ in the Hamiltonian (27) influences the shell-correction ΔE_{sh}^{Str}, but not the average, $\tilde{E}_{sp}$, of the single-particle spectrum, since $U_0 = 0$ for all shells with an effective quantum number n higher than the minimum number required for accomodating N_e electrons (see Ref. [1], pp. 598 ff.).

[69] J. Maruhn and W. Greiner, *Z. Phys.* **251**, 431 (1972).

[70] M. G. Mustafa, U. Mosel and H. W. Schmitt, *Phys. Rev. C* **7**, 1519 (1973).

[71] This meaning of the variable ρ should not be confused with the meaning of ρ as a particle density in a density functional (see Section 4.2.1).

[72] *Handbook of Mathematical Functions*, ed. M. Abramowitz and I. A. Stegun, Dover, New York, 1965.

[73] B. K. Jennings, *Nucl. Phys. A* **207**, 538 (1973); B. K. Jennings, R. K. Bhadhuri and M. Brack, *Phys. Rev. Lett.* **34**, 228 (1975).

[74] A. Bulgac and C. Lewenkopf, *Phys. Rev. Lett.* **71**, 4130 (1993).

[75] C. Bréchignac, Ph. Cahuzac, F. Carlier, M. de Frutos and J. Leygnier, *J. Chem. Phys.* **93**, 7449 (1990).

[76] C. Bréchignac, Ph. Cahuzac, J. Leygnier and J. Weiner, *J. Chem. Phys.* **90**, 1492 (1989).

[77] Z. Penzar and W. Ekardt, *Z. Phys. D* **17**, 69 (1990).

[78] M. P. Iñiguez, J. A. Alonso, M. A. Aller and L. C. Balbás, *Phys. Rev. B* **34**, 2152 (1986).

[79] C. Bréchignac, Ph. Cahuzac, F. Carlier, J. Leygnier and A. Sarfati, *Phys. Rev. B* **44**, 11 386 (1991).

[80] For a detailed description of the BO-LSD-MD method, see R. N. Barnett and U. Landman, *Phys. Rev. B* **48**, 2081 (1993).

[81] F. Garcías, J. A. Alonso, M. Barranco, J. M. López, A. Mañanes and J. Németh, *Z. Phys. D.* **31**, 275 (1994).

[82] H. Koizumi and S. Sugano, *Phys. Rev. A* **51**, R886 (1995).

[83] A. Rigo, F. Garcías, J. A. Alonso, J. M. López, M. Barranco, A. Mañanes and J. Németh, *Surf. Rev. and Letters* **3**, 617 (1996).

[84] The two-intersected-spheres jellium has also been used for describing the fusion of two neutral magic clusters (see Ref. [86]).

[85] In this three-variables parameterization, the B parameter controls the necking-in, the C parameter controls the distance, and the α parameter controls the asymmetry, leaving no freedom for the shapes of the parent or the emerging fragments to be varied. In particular, both fragments remain simultaneously either prolate-like or oblate-like, while final spherical shapes are excluded altogether. The weaknesses of the 'funny hills' parameterization with respect to metal-cluster fission have been discussed in Ref. [24].

[86] O. Knospe, R. Schmidt, E. Engel, U. R. Schmitt, R. M. Dreizler and H. O. Lutz, *Phys. Lett. A* **183**, 332 (1993).

[87] S. Wolf, G. Sommerer, S. Rutz, E. Schreiber, T. Leisner, L. Wöste and R. S. Berry, *Phys. Rev. Lett.* **74**, 4177 (1995).

[88] Notice that experimental measurements at $N = 33$ and 35 have not been obtained (see Ref. [36]).

[89] I. Katakuse, T. Ichihara, Y. Fujita, T. Matsuo, T. Sakurai and H. Matsuda, *Int. J. Mass. Spectrom. Ion Processes* **67**, 229 (1985).

[90] M. M. Kappes, M. Schär, U. Röthlisberger, C. Yeretzian and E. Schmacher, *Chem. Phys. Lett.* **143**, 251 (1988).

5 Optical and Thermal Properties of Sodium Clusters

HELLMUT HABERLAND

Universität Freiburg

5.1 HISTORY AND MOTIVATION

The Victoria and Albert Museum in London has on display a glass drinking cup that was manufactured during the later years of the Roman Empire [1]. Its shining colors of red and yellow are an early and very beautiful example of an artistic and commercial application of cluster science. It is also a fine example of how the properties of finely dispersed matter change with size. The 'golden' color of Au changes if the gold particles become small enough.

The glass makers of ancient days did not know, of course, what produced the shining colors, but they knew from experiment that if metal salts were mixed into the hot glass it would result in a brilliant coloration. Metal atoms become mobile in the hot glass, and if they meet during their diffusive motion they will stick together and form what we call today a 'cluster' [2]. The process was employed over the centuries for making colored glass, and is still in use today. It was only in 1908 that Gustav Mie [3] showed that the beautiful colors are due to resonantly enhanced light-scattering. A microscopic understanding of the process had to wait for the theory pioneered by Ekardt [4] and the experiments of the Knight group [5]. These seminal works gave new impetus to the field of metal cluster science in 1984 — the results can be judged from this book.

The glass-coloring 'experiments' have been performed with gold, silver, nickel and other metals, which are much more difficult to handle theoretically than the alkalis. Among the latter, sodium is the best representative of the nearly free electron gas or jellium model which forms the basic assumption of some of the articles found here. Therefore this review is restricted to sodium clusters, and more specifically to their optical and thermal properties.

Metal Clusters. Edited by W. Ekardt
© 1999 John Wiley & Sons Ltd

At first, the assumptions of the jellium model seem outrageous to somebody with an atomic or molecular physics background, like the present author. How can electrons be noninteracting, if they are confined to a density of a thousand times that of air at atmospheric pressure? Just from Coulomb's law their mutual repulsion is several electron volts. But, as is explained in books on solid-state physics, although the electrons themselves are strongly interacting, the jellium model is in fact a very good one for the heavier alkalis. The solution of this problem is not to treat the strongly interacting electrons by themselves. Instead the interaction of one electron with all its neighboring ones is considered, and this new entity — the electron and its polarization cloud — is called a 'dressed electron' or a 'quasiparticle'. It can behave — under happy circumstances as in sodium — like a free electron. Why and when the interaction of the electrons with the ions can be neglected is explained in the contribution by W. Ekardt and J. Pacheco in this book.

The present chapter is not written for a complete beginner — more introductory articles can be found in [6–9]. The reader is assumed to be somewhat familiar with the basic notions of the field, like magic numbers, evaporative ensemble, the main features of jellium theory and so on.

5.2 EXPERIMENT

For the neutral sodium dimer and trimer, beautifully resolved optical spectra are available, which are in excellent agreement with quantum-chemistry-type calculations [10, 11]. The early experiments on larger neutral clusters [12] gave many beautiful results but had one serious drawback. At best it is very difficult to mass-select neutral clusters, so that one has nearly always to work with a broad distribution of masses. In some cases it was possible, nevertheless, to deduce optical data [5, 12].

Due to this problem, the experiments are done nearly exclusively today on charged clusters, which were pioneered for the alkalis by the Bréchignac group [13, 14]. Only those details of the experiments will be given here that are necessary to understand the results. Many further details can be found elsewhere [9, 13–17].

Figure 5.1 shows the principle of the experiment. Cluster ions are produced, thermalized and mass-selected in a first time-of-flight mass spectrometer. A packet of clusters of one size is irradiated with photons of variable energy. These induce an electronic transition which decays rapidly into vibrational excitation. If the cluster is small or hot enough the increase in the cluster temperature (δT) is sufficient to evaporate at least one atom. The overall process studied can thus be written:

$$\text{Na}_n{}^+(T_0) + \hbar\omega \longrightarrow \text{Na}_n{}^+(T_0 + \delta T) \longrightarrow \text{Na}_{n-s}{}^+ + s\text{Na}. \tag{1}$$

The charged photofragments are separated in a second time-of-flight mass spectrometer and detected. The intensity on the selected mass ($\text{Na}_n{}^+$ in Eq. (1)) is measured with (I) and without (I_0) laser interaction. From the Lambert–Beer law one has for a good geometric overlap of cluster and laser beam:

$$I/I_0 = \exp(-\sigma\phi\tau), \tag{2}$$

where σ is the cross-section for photofragmentation, and ϕ and τ are the laser fluence (photons per second and area) and pulse length, respectively. It is easy to measure the product $\phi\tau$ with a calibrated light detector, so that one can obtain an *absolute* value of σ from Eq. (2) just by measuring the ratio of two intensities.

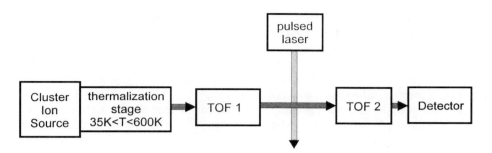

Figure 5.1 Schematic of the experimental set-up. Clusters are produced in a cluster ion source, thermalized and mass-selected in a first time-of-flight (TOF) mass spectrometer. They are irradiated by photons from a pulsed dye laser, and the resultant mass spectrum is measured in a second TOF. From the decrease of the originally selected mass one can calculate via Eq. (2) the optical cross-sections. The internal energy U of the cluster leaving the thermalization stage, which is needed in the thermodynamic experiments is obtained from an analysis of the photofragmentation pattern

If for each photon at least one atom is ejected one can equate the measured photofragmentation to the desired photoabsorption cross-section. It is still possible to extract a cross-section if two photons are necessary for ejection of an atom [16, 17].

5.2.1 Variable-Temperature Cluster Source

A schematic of the cluster ion source is given in Figure 5.2. The cluster ions are produced by evaporating sodium into a helium atmosphere of 105 K and about 70 Pa. Charged clusters are produced by running a weak electric gas discharge in the aggregation chamber. The cluster formation is similar to cloud formation in nature or track formation in cloud chambers. The clustering process is terminated when the clusters are carried out of the aggregation chamber by the continuous helium gas stream.

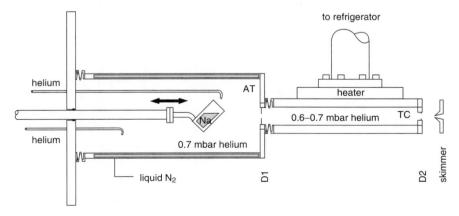

Figure 5.2 Source for producing thermalized cluster ions. Sodium atoms are evaporated, and aggregate in a 10 cm diameter aggregation tube (AT) which is cooled by liquid nitrogen. A hollow cathode discharge (not shown) is used to supply charged species. The clusters are swept via a diaphragm (D1) into the thermalization chamber (TC), whose temperature can be varied. The clusters pass through a second diaphragm (D2) and the skimmer into the acceleration stage of the first TOF as shown in Figure 5.1

After passing through diaphragm D1 the cluster ions enter a thermalization chamber (4 cm diameter, 30 cm long), whose temperature can be varied between 50 K and 700 K. For the 35 K experiment a shorter tube was employed [16]. The helium buffer gas transfers this temperature to the clusters. The clusters undergo about 10^5 to 10^6 collisions with the helium atoms before leaving the thermalization tube. Detailed analysis shows that a shorter aggregation tube would be sufficient [18, 19].

At thermal equilibrium, clusters and gas do not exchange energy on the average, which leads to a *canonical* distribution of internal energies. Care is taken that this distribution is preserved during the flight of the clusters through the machine. Estimates show that the 300 K background radiation does not influence the cluster temperature.

5.2.2   Temperature Distributions

For a comparison with theory it is necessary to know the distribution of internal energies of the clusters before the photoabsorption. For all temperatures well below that of the evaporative ensemble [20–22], the clusters are in thermal equilibrium with the heat bath of the thermalization chamber (TC) of Figure 5.2. In other words: the temperature of this chamber is, by definition, also the temperature of the ensemble of clusters, which gives a *canonical* distribution of internal energies. A single cluster taken alone is a very small system, and consequently its internal energy fluctuates strongly — as long as the cluster is in the heat bath. After leaving it, the internal energy of each single cluster is fixed, with the ensemble still retaining the canonical distribution.

Modeling the cluster by a system of coupled harmonic oscillators, one obtains a Poisson distribution for the microcanonical temperatures T in the cluster ensemble [23],

$$P(T) = \lambda^{s-1} \exp(-\lambda)/\Gamma(s-1) \quad \text{with} \quad \lambda = (s-1)T/T_{\text{bath}} \tag{3}$$

with the Γ-function, the number of coupled oscillators, $s = 3n - 6$, and the temperature of the thermalization chamber, T_{bath}. The Poisson distributions (see Figure 5.3) go over into Gaussians for large n. Only in the limit of arbitrarily large n does one obtain a

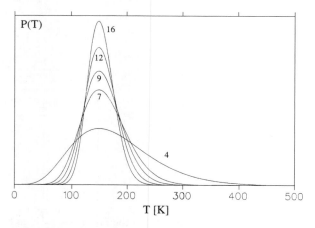

Figure 5.3   Thermal distributions for clusters of 4, 7, 9, 12 and 16 atoms in equilibrium with a heat bath at 150 K, assuming a model of coupled harmonic oscillators. The real distributions of sodium clusters can be expected to be sharper than shown here, as anharmonicities and melting have been neglected

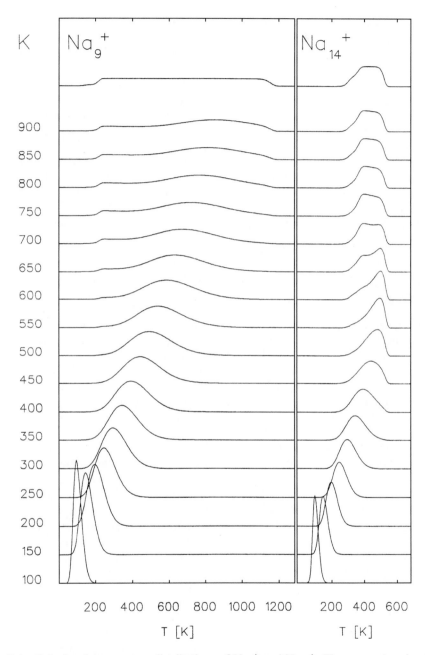

Figure 5.4 Calculated temperature distributions of Na_9^+ and Na_{14}^+. The parameter given on the left hand side is the temperature of the thermalization cell, or T_{bath} of Eq. (3). For low T_{bath} one observes the distributions seen in Figure 5.3. These shift to higher values, and then remain nearly fixed when the cluster becomes so hot that it loses atoms by evaporation. The difference between $n = 9$ and 14 is that the former has a much higher binding energy, thus making it more resistant to evaporation. Note that there is insufficient intensity to measure optical data, once the clusters start evaporating. This happens at 447 K for $n = 9$, and at 309 K for $n = 14$

$P(T) \sim \delta(T - T_{bath})$, i.e. the same temperature for all clusters. The real thermal distribution of sodium clusters can be expected to be narrower than those given by the harmonic oscillator model, because the width of the distribution depends inversely on the heat capacity [23]. At high temperature, the heat capacity of real sodium ($\sim 0.6 k_B$ per degree of freedom) is higher than the Dulong–Petit value of $k_B/2$. Also, melting of the cluster raises the heat capacity [24, 25] and leads to a further reduction of the width of the thermal distribution near the melting point. Note that the distributions have been discussed here in terms of temperature. If instead the internal energy is taken as the independent variable, the situation is different. The width is larger near the melting point!

At the highest temperatures used in this experiment, the clusters can no longer be thermalized by the heat bath as they cool continuously by evaporation. The temperature distribution is thus that of an evaporative ensemble [20–22]. Figure 5.4 shows temperature distributions calculated for $n = 9$ and 15, which were obtained by using an appropriate combination of the two concepts [19]. Note that isomerization and melting have been neglected in the calculation for Figure 5.4.

5.3   TEMPERATURE DEPENDENCE OF THE OPTICAL RESPONSE

Photoabsorption cross-sections are shown in Figures 5.5–7, where the absolute cross-section, divided by the number of valence electrons, is plotted against the photon energy [26]. Note that each data point represents an independent, *absolute* determination of the cross-section. The temperature, indicated on the vertical scale, is changed in steps of 20 to 30 K. Clusters having a higher temperature than T_{evap} cannot be measured in this experiment, as they fragment before they reach the laser interaction region. The high-temperature data agree with results of earlier experiments (where available), i.e. for $n = 9$ and 21 [14] and for $n = 14$, 15 and 16 [27].

The spectra of the clusters from $n = 4$ to 16 show rather sharp features at low temperatures, which broaden with increasing temperature. However, this broadening is not accompanied by an overall shift; the mean of the spectrum is nearly temperature-independent, as discussed below in Section 5.4.3.4. Also, there is no qualitative change between 35 and 105 K, i.e. the number of peaks in the spectra remains constant. Contrary to this, large changes are seen between 105 K and the highest possible temperature. For Na_9^+, the low-temperature double peak becomes one broad peak at higher temperature, and for Na_{11}^+ six well-resolved lines transform into two broad humps.

In Figure 5.8 the dependence of the average line width for Na_4^+ and Na_5^+ is plotted as a function of cluster temperature. The line width increases with the square root of the temperature. For the smallest cluster, Na_4^+, one very small individual resonance was measured. A comparison of this resonance at 35 K and 100 K is given in Figure 5.9. The line width increases by a factor of two upon the threefold increase of temperature, whereas the oscillator strength in this resonance remains unchanged at $(5 \pm 1) \cdot 10^{-3}$ per 3s electron. This is the narrowest resonance ever found in alkali clusters, and could therefore serve as a stringent test for theory.

No special feature can be seen in the optical spectra which could be interpreted as a melting transition. There are many changes of the spectra, but all 'transitions' happen very gradually with temperature. This does not imply that no melting occurs. The transition might be just too broad in these small clusters to be observable, as discussed below in Section 5.7.2.

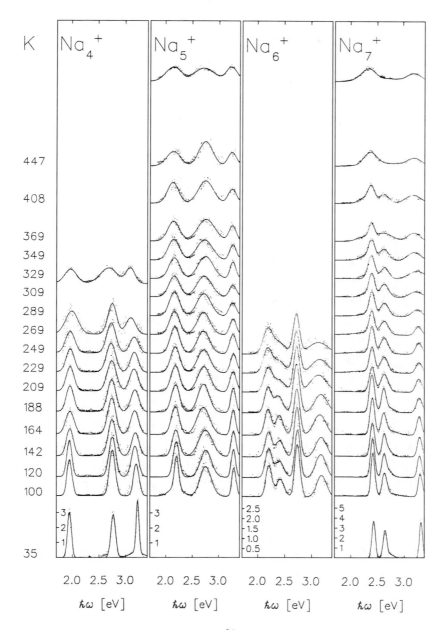

Figure 5.5 Photoabsorption cross-sections (in $\mathring{A}^2$ per valence electron) for Na_n^+, $n = 4$ to 7, as a function of photon energy. The absolute value of the cross-section scale is indicated in the lower left corner of each column. The temperature in kelvins is indicated on the left-hand side. At low temperatures, rather sharp peaks are seen which broaden with increasing temperature. The highest temperature data for each cluster size correspond to the evaporative ensemble [20]. The value of T_{evap} is a direct measure of the dissociation energy. For example, Na_6^+ is so weakly bound that it decays above 250 K within the time window of the experiment. Clusters with an even number of valence electrons have a higher binding energy and can thus be measured to much higher temperatures

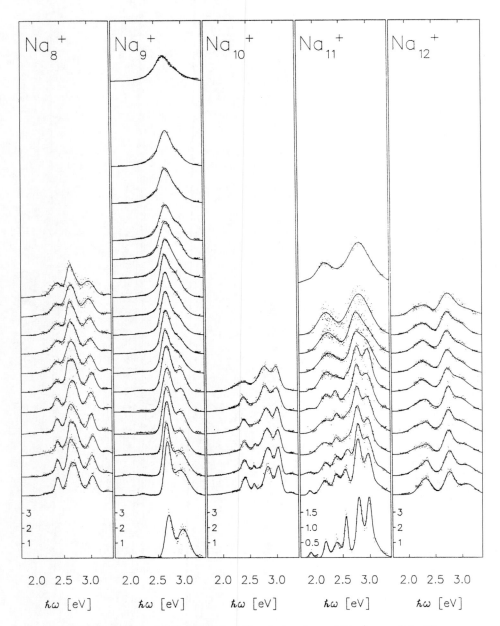

Figure 5.6   As in Figure 5.5 but for cluster sizes of Na$_n^+$, $n = 8$ to 12

5.3.1   Low-Temperature Spectra

At 35 K, the atoms in the cluster make mainly small-amplitude motions [28, 29]. The cluster thus has a definite geometric structure and can be considered to be a large molecule, and the powerful methods of quantum chemistry can be used [30–32]. For $n \leq 6$ and for $n = 9$, the agreement between experiment and theory is good [30]. For all other clusters

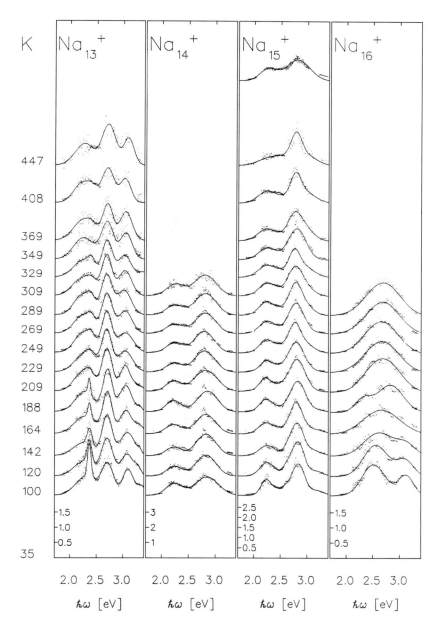

Figure 5.7 As in Figure 5.5 but for cluster sizes of Na_n^+, $n = 13$ to 16

the calculations show the main experimental features, but small details are not reproduced. The main conclusion [17, 30] from an analysis of the spectra is: the geometry is
essential for understanding electronic properties at low temperature. A peak in the optical
spectra corresponds to an electronic transition between the ground and some excited state
of this molecule. One should therefore observe vibrational (and possibly even rotational)

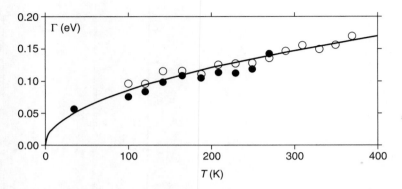

Figure 5.8   Average line width Γ of Na_4^+ and Na_5^+ as a function of cluster temperature

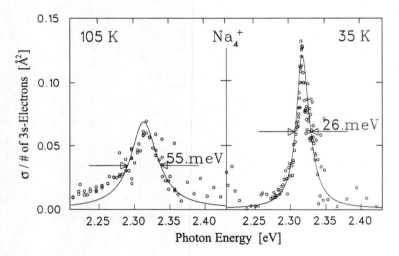

Figure 5.9   Comparison of a narrow line of Na_4^+ at two temperatures. The line shape is better described by a Lorentzian (solid line) than a Gaussian, in contrast to the major resonances in Figures 5.5 to 5.7. The line width increases strongly with temperature, while the oscillator strength remains unchanged

structure on these lines. This conclusion is corroborated by the experimental observation (see Figure 5.8) that the mean line widths for Na_4^+ and Na_5^+ are compatible with a, as predicted for a thermal average over several vibrational states [33]. For sodium, no vibrational structure has been observed so far, but for small Li_n^+ clusters some electronic resonances show a fine structure which could be due to an only partially resolved vibrational structure [34].

5.3.2   High-Temperature Spectra

At the highest temperature, the positions of the atoms are no longer fixed. The atoms perform random motions around a local minimum until they move to a neighboring site.

See Figure 33 of Ref. [31] for a calculation on Na_{20} at 340 and 640 K which shows this behavior. Under these conditions, at which the cluster can be considered as molten, the peak positions and intensities are for $n = 7$ and $n \geq 9$ in good agreement with the predictions of the jellium model [16, 35].

The peak positions of the hard-wall jellium model are slightly blue-shifted compared to the experimental data [8, 27, 36–39]. Agreement can be obtained by using a soft-wall jellium model [16, 36], or, more correctly, by pseudopotential perturbation theory [38, 39].

It is surprising that the spectrum of Na_7^+, a cluster with only six valence electrons, can be well described by the jellium model at a high temperature. Thus, even for this small size, the geometry of the hot cluster is determined by the cloud of valence electrons which finds its shape by minimizing the energy. The atoms, which are mobile at high temperature, distribute themselves following the potential given by the cloud of electrons. It is not possible to verify this for $n = 6$ and 8, as the atomic binding energies are so low that these clusters evaporate before the spectrum could become jellium-like. The spectra of the even smaller ones, $n = 4$ and 5, show at high temperature the same structure with three peaks as at low temperature, whereas a jellium calculation predicts only two peaks [36].

5.4   SIZE DEPENDENCE OF THE OPTICAL RESPONSE

Figure 5.10 shows photoabsorption cross-sections measured at a temperature of 105 K in the size range of 3 to 64 atoms per cluster. The data agree within experimental accuracy with earlier measurements for Na_{21}^+ made by the Orsay group [13] (only the small dip in the spectrum was not seen here) and with the data of the Copenhagen group [27] for $n = 14$ to 48.

For the small cluster sizes ($n = 3–9$), single, well-separated resonances are observed, which could be well fitted by Gaussians. As discussed above (see Section 5.3.1), the peaks are interpreted as vibrationally broadened electronic transitions of an Na_n^+ *molecule* [17, 30, 40].

The number of electronic lines increases with cluster size and in the size range $n = 10$ to 15 the lines begin to overlap. For even bigger clusters, single electronic resonances can no longer be resolved. Instead broad peaks are seen which can be characterized by one, two or three Lorentzian peaks, which were accordingly fitted by

$$\sigma(E) = \frac{\hbar e^2}{m_e c \varepsilon_0} \sum_{i=1}^{3} \frac{f_i E^2 \Gamma_i}{(E^2 - E_i^2)^2 + (E \Gamma_i)^2}, \tag{4}$$

where m_e is the bare electronic mass and $E = \hbar \omega$ the photon energy. The oscillator strength f_i, the peak positions E_i and the widths of the single resonances Γ_i were used as fit parameters, with the restriction that the oscillator strengths of all the resonances be identical ($f_1 = f_2 = f_3$). When two resonances were sufficient to describe the spectrum, $f_1 = 2f_2$ was demanded. Fit curves are included in Figure 5.10. Numerical values are given in Ref. [41]. The peak positions are indicated by the vertical bars in Figure 5.10 and are shown in Figure 5.12a.

The larger the cluster size, the stronger is an additional broad absorption on the high-energy side of the optical spectrum. This can have one of two origins: (1) it could

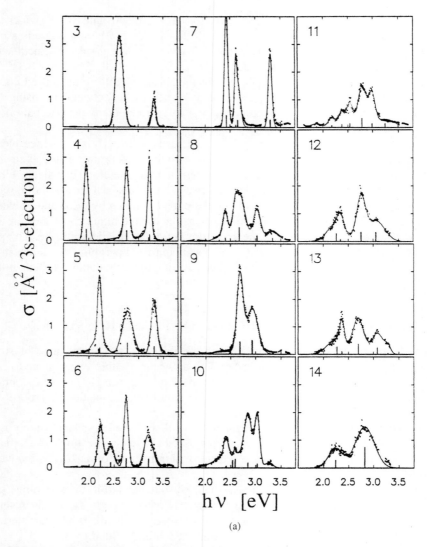

(a)

Figure 5.10 (a to d). Photoabsorption cross-section as a function of the photon energy. Plotted is the absolute value per valence electron for all cluster sizes between $n = 3$ and $n = 64$. The last spectrum, marked 'Mie', has been calculated from Eq. (6)

be the beginning of the volume plasmon modes [4, 8, 9], or (2) it could be the microscopic analog of the bulk interband transition (see Section 5.6.1 below). The data on the high-energy side cannot be described by a Lorentzian or Gaussian peak shape and thus were neglected for the fit to Eq. (4).

5.4.1 Peak Positions

In the jellium model, one has a closed electronic shell for $n = 8, 20, 40, 58, \ldots$ valence electrons. This gives a spherical shape for the cluster, and one dominant line in the

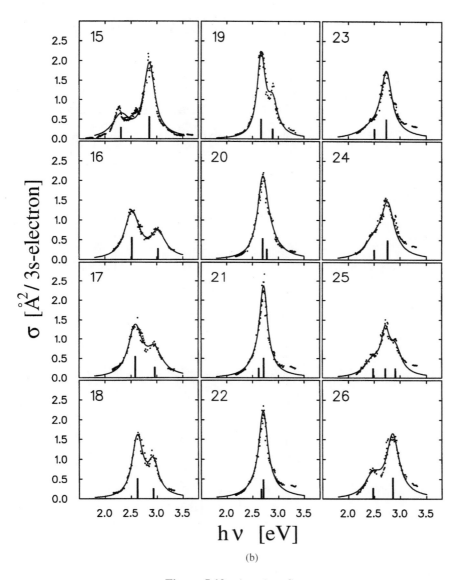

Figure 5.10 (*continued*)

spectrum. The most general form for nonspherical clusters is a spheroid, whose three axes are all different. The data suggest that often two axes are identical (or differ not too much from each other). This gives either a prolate (cigar-like) or oblate (disc-like) deformation. If two of the principal axes are identical, the resonances along them are degenerate in energy and the optical peak is thus of about double intensity. As the resonance parallel to the longer axis has a lower frequency, one can deduce the cluster's shape from a simple inspection of the optical spectrum. An example is shown in Figure 5.11. For the spherical symmetric Na_{21}^+ we have one single peak, and a change-over from an oblate geometry for smaller to a prolate one at larger cluster sizes.

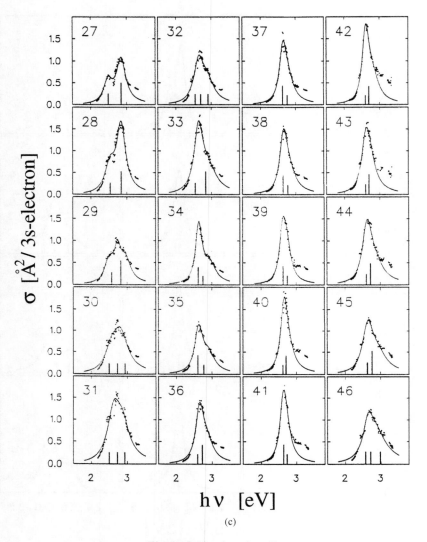

(c)

Figure 5.10   (*continued*)

The location of the peak maxima up to $n = 66$ are given in Figure 5.12(a). If the intense peak (marked by a full dot) has a higher frequency, we have a prolate structure, and vice versa. There is a change from oblate to prolate geometry at each closed shell, and also a change in between closed shells. For some clusters one could also make a fit with three peaks of equal intensity, corresponding to a triaxial structure. These are indicated by three small open dots in Figure 5.12(a). Triaxial shapes have been calculated for neutral sodium clusters for $N = 11$–13, 23–25, 61–65, ... valence electrons [8, 42–44]. The experimental spectrum shows a clear sign for $N = 12$ electrons, and just an indication of a possible triaxial shape for $N = 24$, 28–31, 44–48 valence electrons. The $N = 12$ splitting must be more pronounced than the calculated one, where it was concluded that even small thermal fluctuations will erase the threefold peaks.

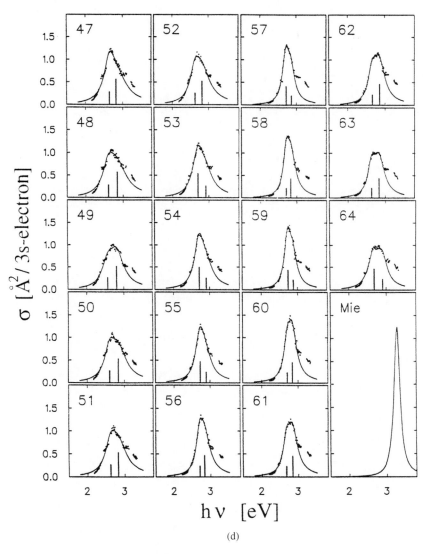

(d)

Figure 5.10   (*continued*)

The Nillson–Clemenger model [45] predicts that the ratio of the two axes R_1 and R_2 of a deformed cluster equals the ratio of the resonance energies. The connection between the energetic splitting of the resonances and the deformation is given by

$$\Delta = \frac{\Delta R}{\overline{R}} = 2\frac{|R_1 - R_2|}{R_1 + R_2} = 2\frac{|E_1 - E_2|}{E_1 + E_2}.$$

(5)

The result is shown in Figure. 5.12(b). The deformation Δ is smallest — but not zero — for spherical clusters. Many authors have derived results similar to Eq. (5), as discussed in detail in Ref. [27].

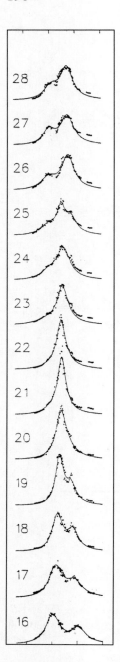

2.0 2.5 3.0 3.5

$\hbar\omega$ [eV]

Figure 5.11 Cluster sizes between 16 and 20 are oblate, above $n = 23$ they are prolate. The spherical shape of the Na_{21}^+ cluster leads to the narrowest line

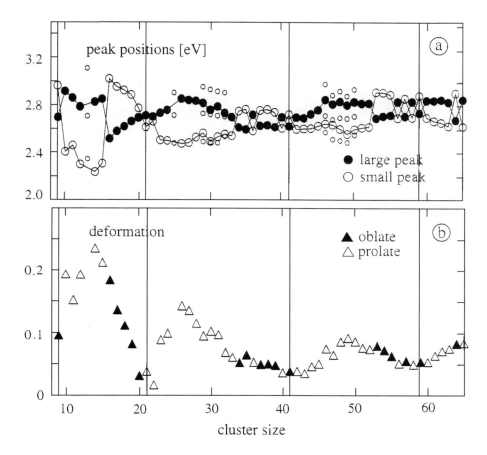

Figure 5.12 Peak energies as a function of cluster size are shown in the upper figure, (a). The energies have been obtained by a Lorentz fit to the data (see Eq. (4)). The position of the intense peak is given by a full point, the position of the less intense peak by a large open circle. Three open circles are shown where a three-peak fit is possible. Closed electronic shells are indicated by the vertical lines. The lower figure, (b), shows the deformation parameter Δ of Eq. (5). Full triangles indicate oblate, open ones prolate deformations

5.4.2 Asymptotic Limit

The classical equation for light absorption by a small sphere of radius R reads [2, 46]:

$$\sigma(\hbar\omega) = \frac{4\pi\omega}{c} R^3 \mathrm{Im}\left(\frac{\varepsilon(\omega) - 1}{\varepsilon(\omega) + 2}\right). \tag{6}$$

Here Im stands for the imaginary part, and $\varepsilon(\omega)$ is the complex dielectric function. Using the experimental $\varepsilon(\omega)$ for sodium [47] one obtains the absorption profile given as the last curve in Figure 5.10(d) (marked Na_{Mie}). The calculated Mie resonance is narrower than the cluster plasmon peaks, but has nearly the same oscillator strength per electron. It has a maximum at $E_{Mie} = 3.27$ eV, which differs from the pure jellium result of 3.41 eV. The small but significant difference of 140 meV is attributed to core polarization effects, i.e. the interaction of the 3s electrons with the much more strongly bound core electrons [48]. This effect is neglected in the jellium model, but could be incorporated by an effective electron mass. The full width at half maximum is 0.173 eV which corresponds to $\Gamma_0 = 0.19$ eV which is identical to the variance of $\delta = 0.19$ (for a definition of δ see Eq. (12)). The oscillator strength between 1.5 and 3.7 eV is $f = 0.77$ per electron. These asymptotic values are included in Figures 5.13, 5.14, 5.17 and 5.18.

5.4.3 Sum Rules and Moments

Independent of the specific form of the optical response, there is additional valuable information in the integrals of the spectra [8, 37]. The moment M_k is defined [49] as

$$M_0 = \int_0^\infty \sigma(E)\,\mathrm{d}E \quad \text{for} \quad k = 0, \qquad M_k = \int_0^\infty E^k \sigma(E)\,\mathrm{d}E / M_0 \quad \text{for} \quad k > 0 \qquad (7)$$

Experimental information for several moments will now be discussed.

5.4.3.1 Oscillator Strength

For $k = 0$ one obtains the Thomas–Reiche–Kuhn sum rule for the oscillator strength f, which states that the zeroth moment is proportional to the number $N(e^-)$ of electrons.

$$f = \frac{2m_e c \varepsilon_0}{\pi \hbar e^2} M_0 = \frac{0.911}{\text{Å}^2\,\text{eV}} \int \sigma \,\mathrm{d}E = N(e^-). \qquad (8)$$

This is an exact equation giving $f(\text{total}) = 7$ per Na atom if the integral is extended over all photon energies. In the experiment $\sigma(E)$ was measured only between 1.5 and 3.7 eV, so the integral can only be extended over this finite energy interval. In all the integrals of this section these finite limits are implied.

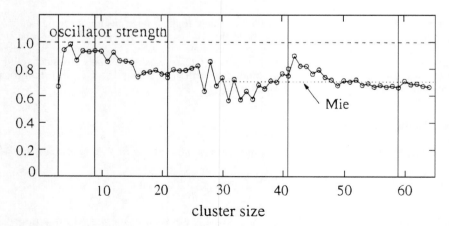

Figure 5.13 Oscillator strength per valence electron between 1.5 and 3.7 eV as a function of cluster size, calculated by Eq. (8). A value of only 0.77 is obtained for the Mie plasmon in this limited energy range, due to the cutoff imposed by the upper limit of the photon energy used. Note that starting with Na_{26}^+ and going up to at least Na_{60}^+ all odd cluster sizes show a significantly higher oscillator strength than the even ones. The estimated error for the smaller clusters is below 10%, resulting mainly from the fact that the laser beam profile is not completely flat

Under certain conditions one can decompose the integral of Eq. (8) and obtain the oscillator strength due to just one electron. For sodium, with its $1s^2 2s^2 2p^6 3s$ electronic structure, one writes:

$$f(\text{total}) = f(1s, 2s, 2p) + f(3s). \tag{9}$$

The approximation $f(3s) = 1$ is very good for sodium, as (i) the optical spectrum due to the 3s electrons is well separated from that of the other electrons, and (ii) the interaction of the 3s electrons with the ions can be well approximated by a local pseudopotential [50].

From a jellium calculation one gets exactly $f = 1$ per 3s electron for sodium. Experimentally, one sees something very similar, showing how good the nearly free electron assumption is for sodium. Also the experimental f-values are nearly independent of temperature, as could be expected.

Integrating the experimental spectra one obtains for $n \geq 15$ that more than 85% of the maximal oscillator strength contributes to the experimental data. The results are given in Figure 5.13. One can expect that the remaining 15% are hidden in the energy range above 3.7 eV, the highest energy used in this experiment. It can be seen from Figure 5.10 that the cross-sections are not zero at the highest photon energies used experimentally. This is in agreement with the fact that the oscillator strength for the limiting Mie plasmon is also only $f(R \rightarrow \infty) = 0.77$ per valence electron if one takes into account the limited photon range of this experiment.

This discussion does not apply to the very small sizes. For $n = 2$ and 3 the photon reaches states which do not dissociate in the time scale of the experiment; i.e. the dimer and trimer have only a probability of about 1/3 and 2/3 to dissociate, respectively. This can be understood in term of symmetry arguments as discussed elsewhere [17].

5.4.3.2 Mean Transition Energy

The first moment, M_1, is equal to the mean transition energy:

$$E_0 = M_1. \tag{10}$$

Experimental values are shown in Figure 5.14(a). As could be expected, the stronger bound electrons in the closed-shell clusters have a higher mean resonance energy compared to the electrons in open-shell clusters of similar size. The mean transition energy is nearly independent of temperature as discussed below.

5.4.3.3 Polarizability

The polarizability α is proportional to the negative second moment:

$$\alpha = \frac{e^2 \hbar^2}{m} M_{-2}. \tag{11}$$

Experimental results are shown in Figure 5.14(b). They also show some structure due to electronic shell-closings. Included in the figure are directly measured polarizabilities of the Knight group for neutral sodium clusters [51]. For $N(e^-) \geq 17$ the two polarizabilities agree. Only for smaller clusters is the α (neutral cluster) higher than that for charged ones. This is physically plausible. The extra charge leads to a stronger bond of the electrons, which are thus less perturbable by an external electric field. The smaller the cluster

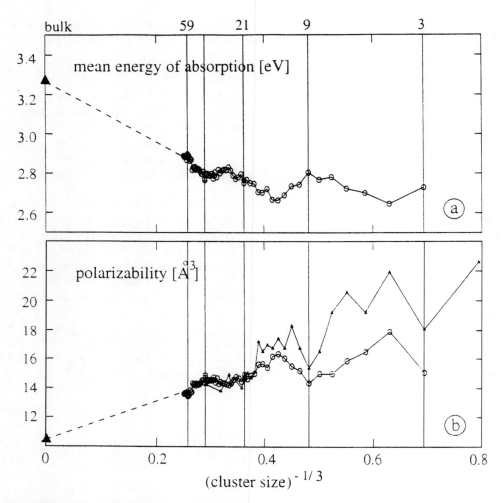

Figure 5.14 (a) Mean energies of the optical spectra. Closed electronic shells are indicated by vertical lines. (b) Polarizability per valence electron of neutral (solid dots) and charged (open circles) clusters. The plot for neutral clusters has been shifted by one mass, so that clusters with the same number of valence electrons are vertically aligned

becomes, the more pronounced becomes this effect. For the extrapolation to $R \to \infty$ one again expects a smooth R^{-1} dependence.

5.4.3.4   *Temperature Dependence*

The temperature dependence of the moments has been calculated from the data given in Figure 5.10. The oscillator strength turns out to be nearly temperature-independent, a surprising result if one looks how the spectra change, but in agreement with the spirit of the local pseudopotential extension of the jellium model [38, 39].

Figure 5.15 shows the temperature dependence of three cluster properties averaged over the optical response. Plotted are (1) the mean transition energy (E_0, see Eq. (10)), (2) the

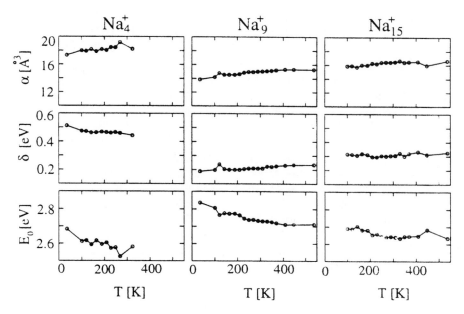

Figure 5.15 Temperature dependence of spectrally averaged properties for the cluster sizes $n = 4$, 9 and 15, i.e. the center of gravity of a spectrum, E_0, the root-mean-square deviation, δ, and the polarizability α. Note that the zero is suppressed for E_0 and α

polarizability (α, Eq. (11)), and (3) the variance δ around E_0, which is defined by

$$\delta^2 = \int (E - E_0)^2 \sigma(E)\, dE / M_0$$

$$= M_2 - M_1^2. \tag{12}$$

The variance is a model-independent measure of the width of the spectrum. The surprising result from Figure 5.15 is that, although the overall appearance of the spectra changes dramatically with temperature, the integral values are nearly temperature-independent. This is valid for the other cluster sizes shown in Figures 5.5–7 [19].

Overall, one notices a slight decrease in E_0 and a weak increase in δ and α with increasing temperature. The behavior of E_0 and α can be understood by the thermal expansion of sodium [16]. The thermal expansion causes a lower charge density, resulting in a lower resonance energy. This shift of the resonance to lower energies results in an increase of the integral in Eq. (7) for $k = -2$, which gives via Eq. (11) a higher polarizability. The slightly larger δ can be rationalized by the fact that more isomers are visited when the cluster is liquid.

5.5 CHARGE DEPENDENCE OF THE OPTICAL RESPONSE

The optical response of three double-charged clusters has been measured [52], i.e. for Na_{22}^{2+}, Na_{42}^{2+} and Na_{60}^{2+}. All these clusters have an even number of atoms, which poses an experimental problem. Any combination of electric and magnetic fields selects masses with the same mass-to-charge ratio. So the clusters Na_{11}^+ and Na_{22}^{2+} can normally not be resolved. This problem was solved in Ref. [52], by first accelerating a singly charged cluster and then photoionizing it.

The clusters studied have 20, 40 and 58 valence electrons, respectively, and are spherical in the jellium model. Figure 5.16 shows a comparison of the absorption cross-section for neutral, singly and doubly charged clusters with 40 valence electrons. The data for neutral Na_{40} is from Ref. [53], and the solid lines have been calculated in Ref. [54].

The spectra become narrower and more blue-shifted the higher the charge state. This is physically expected. For a higher charge the electrons are more strongly bound. This leads to a reduced spillout of the electron cloud, giving a blue-shift of the optical response, as the plasmon frequency is proportional to the square root of the electronic charge density. The narrower line and thus increased peak height has been calculated in Ref. [54], where the term 'concentration of oscillator strength' was used.

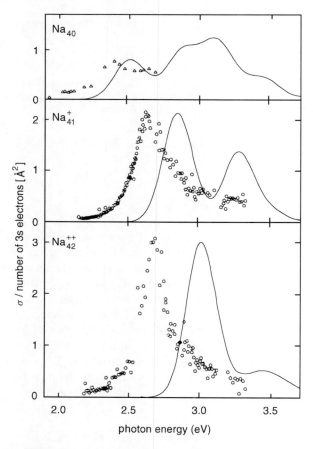

Figure 5.16 Photoabsorption cross-section of neutral/singly and doubly charged clusters, each having 40 valence electrons. The solid line is from Ref. [54], whose results have been averaged over the experimental line width of 0.25 eV. The resonances are more blue-shifted and narrower for the higher charge states

5.6   LINE SHAPE

In contrast to the peak positions, the width of the spectrum is not yet well understood. The only general agreement seems to be that the total width of a spectrum cannot be due to a lifetime effect. This is very evident from the discussion above and an inspection of Figure 5.12(b), which shows that the geometric deformations of a single cluster has a large contribution to the overall line width.

5.6.1   Width of the Bulk Mie Plasmon

The dipolar response of a large sodium sphere can be calculated from the experimental bulk dielectric function as discussed in the context of Eq. (6). The calculated width is $\Gamma_0 = 0.19$ eV. This asymptotic 'experimental' width is due to a structure in the dielectric function which is caused by an interband transition [55]. The collective oscillation can decay by exciting a single electron to a higher electronic band. The same mechanism occurs in the damping of the bulk plasmon. In this case, the width can be well correlated with the strength of the pseudopotential (see Figure 9 of Ref. [48]).

The word 'interband' is defined only for a crystalline lattice. The question thus arises how can this concept be generalized to a finite system? In the bulk, the k value of the alkali interband transition is not so far from the edge of the Brillouin zone [55]. At the very edge of the Brillouin zone, an interband transition can be discussed in real space, that is, electronic charge is taken from the position of the nuclei to positions between the nuclei [55]. This concept can easily be applied to a finite system. Clearly, in a jellium-type calculation this transition cannot occur. There are no atoms from which the electrons could scatter and thus no band structure. It is therefore plausible that all these calculations yield $\Gamma_0 = 0$.

5.6.2   Single Line Width and Lifetime of the Resonance

It has been discussed above that, save for the peak at the high-energy side, the experimental peaks are well represented by Lorentzians. Figure 5.17 shows the mean Γ_i value (as

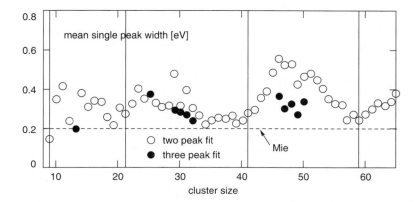

Figure 5.17   The mean single peak width is shown as a function of the cluster size. Note that all spherical clusters and the Mie plasmon have a line width of about 0.2 eV

defined by Eq. (4)) for each mass. The result is surprising. There is no systematic shift with cluster size, if one disregards the shell effects. Also, the value for the Mie plasmon lies in the same range. As the Γ values are obtained from a Lorentz fit, one is tempted to interpret the width as being due to a lifetime τ:

$$\Gamma = \hbar/\tau. \qquad (13)$$

With $\hbar = 0.656$ fs·eV one obtains lifetimes of the order of 1.5 to 3.5 fs for a line width of 0.2 to 0.4 eV.

There exists only one measurement of the lifetime for a free sodium cluster ion that gives for the lifetime of the collective resonance of Na_{93}^+ a value of $\delta t = 10$ to 20 fs [56]. This would correspond to a lifetime broadening of only $\Gamma = 33$ to 66 meV. From this one can conclude that (1) the plasmon lifetime contribution to the width of one single peak is small, and (2) the near-Lorentzian shape of a single line is not due to the lifetime.

The time range obtained for the plasmon lifetime of a free cluster is corroborated by two studies of the plasmon lifetime of large sodium clusters deposited on dielectric surfaces. One obtains a value between 10 and 15 fs, after correcting for inhomogeneous broadening due to the cluster-size distribution, which is unavoidable for deposited clusters [57]. The other experiment obtains a cluster-size-dependent lifetime of 2 to 10 fs, without performing this correction [58].

5.6.3   Total Width and Scaling Laws

The size dependence of the total widths has been studied by several theoretical groups [8, 37, 40, 59–61]. The jellium model was used nearly exclusively. In this case, the interband decay of the plasmon discussed above is not possible, and the collective plasmon oscillations can only decay by exciting a single electron from the same band — a process that has been termed 'Landau damping'. For sufficiently large clusters this gives (Equation 9.7 of Ref. [37]) a width like

$$\Gamma = C_1 \upsilon_{\text{Fermi}}/R, \qquad (14)$$

where υ_{Fermi} is the Fermi velocity (1.07×10^6 m/s for bulk sodium), R is the cluster radius and C_1 is a constant varying for spherical clusters between 0.55 [64] and 0.75 [60]. In the dipolar bulk limit one obtains a vanishing width from Eq. (14), i.e. $\Gamma(R \rightarrow \infty) = 0$.

Figure 5.18 shows the total experimental width of the optical response as defined in Eq. (12). A better representation of the experimental data might therefore be

$$\Gamma = \Gamma_0 + C_1 \upsilon_{\text{Fermi}}/R. \qquad (15)$$

This equation has the expected size dependence and the correct experimental limit for large clusters. The data point to a value that is much smaller than that given by Eq. (15) if the C_1 values cited above are used. It can be concluded from this discrepancy that if there is some applicability of Eq. (15) to real clusters it becomes applicable only for larger ones than studied so far. Note, also, that if Eqs (14) and (15) were approximately valid for large clusters, the width Γ has a maximum as a function of cluster size. This has indeed been observed theoretically [61, 65].

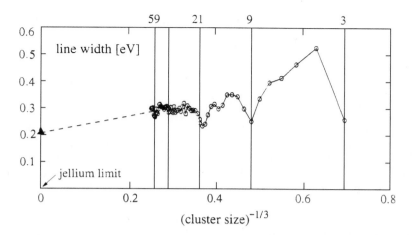

Figure 5.18   Mean-square deviation δ of the optical spectra as defined by Eq. (12)

Another approach is the broadening of the line due to thermal width at temperature T. This leads (see p. 180 of [37]) to a line shape as

$$\Gamma = C_2\sqrt{T}/R. \qquad (16)$$

This indeed gives the experimental result for Na_{21}^+ at 100 K, showing that the measured width has probably a sizable contribution from thermal excitations. One problem is that for a thermal average one would expect a Gaussian line shape, but a Lorentzian gives a better fit to the data. Treating thermal disorder by a random matrix model [66], good agreement has also been obtained for the width of Na_{21}^+.

In principle, one would expect that all the broadening effects add up, leading to rather broad lines, which has not been observed experimentally. One has to conclude that a satisfactory understanding of the line shapes of the plasmon peaks is still eluding us.

5.7   THERMODYNAMIC PROPERTIES

Phase transitions in bulk material have been studied for a very long time. For microscopically small particles [18, 67, 68], one observes that they differ in three main aspects from their bulk counterparts. (1) The melting point decreases with decreasing particle size, which is mainly due to the large percentage of atoms on the surface; here atoms have fewer nearest neighbors and are thus more weakly bound and less constrained in their thermal motion [68]. (2) The latent heat of fusion is reduced, which is also mainly due to the reduced coordination number, and (3) the finite number of particles causes the phase transition to be spread out over a finite temperature range. It has recently become possible to study the solid-to-liquid phase transition in mass-selected free clusters [18, 25, 69–71].

5.7.1   Caloric Curve of a Free Cluster

It is easy to prepare clusters with some known temperature T using the cluster source shown in Figure 5.2. It has recently become possible to determine the mean value of the internal energy U, as explained elsewhere [25, 69, 70, 72]. The curve

$$U = U(T) \tag{17}$$

is called the 'caloric curve' and its derivative, $c(T) = \partial U / \partial T$, the 'heat capacity' or 'specific heat'. The machine for the thermodynamic experiments is the same as that for the optical absorption experiments (see Figure 5.1). The basic measuring process is also given by Eq. (1) with the only difference being that many photons are usually necessary before the larger clusters studied for thermodynamic experiments start to evaporate atoms. The value of U can be obtained from the pattern of evaporated atoms.

It is important to remember, in this context, that the clusters leave the thermalization stage with a *canonical* distribution of internal energies. Afterwards they do not make any collisions and have to be treated microcanonically.

Figure 5.19 shows the results for the heat capacity of Na_{139}^+. The δ-function of the bulk result has become a broad maximum, shifted to a lower temperature by 104 K. Away from the maximum, the data agree with the bulk result. Three data can be read off this curve: (1) the melting point, as given by the maximum of the curve; (2) the latent heat of fusion, as given by the integral under the maximum; and (3) the width of the phase transition. The data for the melting points are plotted in Figure 5.20. Further results can be found in Ref. [72].

5.7.2 Size Dependence of the Melting Point

The reduction in the melting point has been studied intensively for small particles and clusters on a surface. One observes typically a linear reduction of the melting point as a

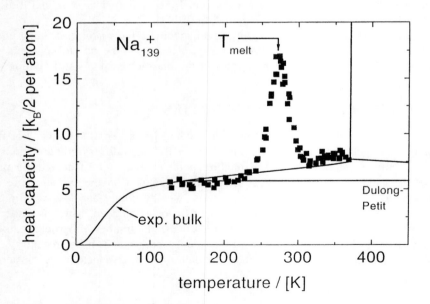

Figure 5.19 The heat capacity of Na_{139}^+ is plotted against the temperature of the cluster thermalization stage. The experimental value for bulk sodium has a δ-function at the melting point of 371 K. The open dots give the experimental results. The δ-function has become a broad peak, whose maximum is taken to be the melting point

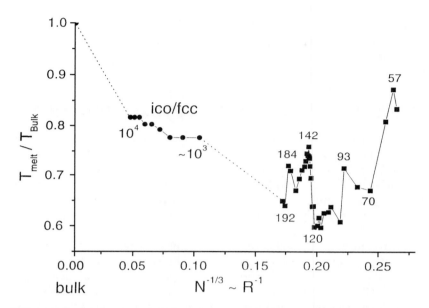

Figure 5.20 Melting points as a function of the inverse cluster radius. The data from Ref. [18] for larger clusters are included. The behavior is far from that expected from the many studies on supported clusters, which all find a linear decrease from the bulk value in such a plot

function of the inverse cluster radius [67]. The data presented here also show a reduction of the melting point, typically by about 30% compared to the bulk value. The actual size variation, on the other hand, shows a very different behavior. One sees large variations of up to ±50 K, with maxima and minima whose positions do not correlate well with any known number of special stability or instability.

There are two pronounced maxima in Figure 5.20, one near 57 and the other at 142 atoms per cluster. One is tempted to correlate these numbers with shell-closings, where the energetic difference between the ground state and the first excited state is largest [8, 9, 18]. Icosahedral shell-closings occur at 13, 55, 147, 309, ... atoms, while electronic shell-closings happen for positively charged sodium clusters for ..., 41, 59, 93, 139, 199, 255, 339, ... atoms. For none of these values is there a maximum in the melting point. But, both maxima in T_{melt} are bracketed by an electronic *and* an icosahedral shell-closing.

Icosahedral shell-closings are known to occur for clusters, which can be described by geometrical ball-stacking models, like e.g. the rare gases. It would be surprising if sodium with its soft interaction potential fell into this class. A possible way around this problem can be found in Ref. [73], where an icosahedral precursor was calculated in the melting of gold clusters, although the ground state has a different symmetry.

The data for the larger clusters (10^3 to 10^4 atoms) have been obtained by the T. P. Martin group [18]. From the structure in their mass spectra it is deduced that these large clusters have either icosahedral or fcc symmetry. Bulk sodium is bcc, on the other hand, so that a structural phase transition must occur between $n = 10^4$ and the bulk. An alternative interpretation would be: large sodium clusters near their melting point show the same icosahedral precursor as calculated in Ref. [73] for gold clusters. In this case the ground state could already be bcc-like.

The cluster sizes 55, 57 and 61 have a higher melting temperature than those in the 10^3 to 10^4 range. This is a totally unexpected and not understood result. In fact, it seems that the mean cluster melting temperature increases with decreasing cluster site for n below ~ 90 atoms.

5.8 UNSOLVED PROBLEMS AND FUTURE DIRECTIONS

Although rather good overall understanding has been reached, there still exist some unsolved problems. Experimentally one would like to go to much lower temperatures than the present 35 K, but no idea is available today how to measure the optical response large, cold clusters. Electron diffraction experiments on mass-selected clusters will soon be available [74], and similar results for inner shell excitation will probably also be done in the near future, especially when the new generation of free-electron lasers becomes available.

There are a number of theoretical problems remaining (as seen with the eyes of an experimentalist). (1) The physical origin of the shoulder seen in the optical response on the high-energy side is not understood. Some speculations have been offered above (Section 5.6.1). (2) As discussed in detail, the width/lifetime problem is far from being clear. Pure jellium calculations will not be helpful in this context, as they do not capture the correct asymptotic limit. (3) A more fundamental question is whether the asymptotic limit discussed above is the correct one. It has been calculated from the bulk dielectric constant, which clusters containing even thousands of atoms might not yet have reached. (4) The wild size dependence of the melting points is still a surprise. Also, the conjecture that the icosahedral intermediate observed for gold clusters might play a role for sodium is unchecked. These four problems are related to ground-state properties. A fifth problem relates to higher excited plasmon states which have recently become accessible due to the development of femtosecond lasers. First results are available, both experimentally [56–58] and theoretically [75].

5.9 SUMMARY

The data discussed here give a fairly complete map of the optical response of sodium cluster ions. The response has been studied as a function of three parameters: temperature, size and charge of the cluster. The overall result is that good agreement with the jellium model is obtained. Exceptions are the splitting of the resonances at low temperature and the lifetime/width problem, both of which are influenced by the detailed arrangement of the atoms. The thermodynamic experiments are too new, so that no satisfactory understanding has yet evolved.

5.10 ACKNOWLEDGMENTS

Clearly the data presented are the work of a whole team of people. Contributors have been (in alphabetical order): Christoph Ellert, Thomas Hippler, Bernd von Issendorff, Werner Kronmüller, Robert Kusche, Walter Orlik, Thomas Reiners, Ralph Schlipper Martin Schmidt and Christina Schmitt. Financial support came from the Deutsche Forschungsgemeinschaft through SFB 276.

5.11 REFERENCES AND NOTES

[1] The Lycurgos goblet was manufactured in the 4th century BC. The glass contains large Au and Ag clusters, whose absorption and scattering properties give it a brilliant red in transmission, and a yellow/green tinge in reflection. See also Ref. [2] and U. Kreibig and M. Quinten, in Chapter 3.5 of Vol. 2 of Ref. [76].

[2] G. F. Bohren and D. R. Huffman *Absorption and Scattering of Light from Small Particles*, Wiley, New York, 1983.

[3] G. Mie, *Ann. d. Physik* **25**, 377 (1908).

[4] W. E. Ekardt, *Phys. Rev.* **29**, 1558 (1984); *Phys. Rev. B* **31**, 6360 (1985).

[5] W. D. Knight, K. Clemenger, W. A. de Heer, W. A. Saunders, M. Y. Chou and M. L. Cohen, *Phys. Rev. Lett.* **52**, 2141 (1984).

[6] See articles in Vols 1 and 2 of Ref. [76].

[7] M. Brack, *Sci. Am.* **12**, 32, (1997).

[8] M. Brack, *Rev. Mod. Phys.* **65**, 677, (1993).

[9] W. A. de Heer, *Rev. Mod. Phys.* **65**, 611 (1993).

[10] H.-A. Eckel, J.-M. Gress and W. Demtröder, *J. Chem. Phys.* **98**, 135 (1993).

[11] H. von Busch, Vas Dev, H.-A. Eckel, S. Kasahara, J. Wang, W. Demtröder, P. Sebald and W. D. Meyer, *Phys. Rev. lett.* **81**, 4584 (1998).

[12] C. Wang, S. Pollack, T. Dahlseid, G. M. Koretsky and M. M. Kappes, *J. Chem. Phys.* **96**, 7931 (1992); T. Baumert, R. Thalweiser, V. Weiss and G. Gerber, *Z. Phys. D* **26**, 131 (1993); H. Fallgren, K. M. Brown and T. P. Martin, *Z. Phys. D* **19**, 81 (1991).

[13] C. Bréchignac, Ph. Cahuzac, F. Carlier, M. de Frutos and J. Leygnier, *Chem. Phys. Lett.* **189**, 28 (1992).

[14] C. Bréchignac and Ph. Cahuzac, *Comments on Atomic and Molecular Physics* **31**, 215 (1995), and C. Bréchignac in Vol. 1 of Ref. [76].

[15] H. Haberland, H. Kornmeier, Ch. Ludewigt, A. Risch and M. Schmidt, *Rev. Sci. Instrum.* **62**, 2621 (1991).

[16] M. Schmidt, Ch. Ellert, W. Kronmüller and H. Haberland, *Phys. Rev. B*, accepted for publication.

[17] Ch. Ellert, M. Schmidt, Th. Reiners and H. Haberland, *Z. Phys. D* **39**, 217 (1997).

[18] T. P. Martin, *Shells of Atoms*, Phys. Rep. 273, 199 (96).

[19] Martin Schmidt, PhD thesis, Freiburg, 1996.

[20] C. E. Klots, *Z. Phys. D* **5**, 83 (1987), and *Nature* **90**, 1492 (1989).

[21] M. F. Jarrold, p. 163 of Vol. 1 of Ref. [76].

[22] C. Bréchignac, Ph. Cahuzac, J. Leygnier and J. Weiner, *J. Chem. Phys.* **90**, 1493 (1989).

[23] L. D. Landau and E. M. Lifshitz, *Statistical Physics, Vol. 5*, Pergamon Press, London.

[24] M. Schmidt, R. Kusche, W. Kronmüller, B. von Issendorff and H. Haberland, *Phys. Rev. Lett.* **79**, 99 (1997).

[25] M. Schmidt, R. Kusche and B. von Issendorff, *Nature* **393**, 238 (1998).

[26] The numerical values of the cross-sections are available at http://www.fmf.uni-freiburg.de/cluster/haberland.html

[27] J. Borggreeen, P. Chowdhury, N. Kebaïli, L. Lundsberg-Nielsen, K. Lützenkirchen, M. B. Nielsen, J. Pedersen and H. D. Rasmussen, *Phys. Rev. B* **48**, 17507 (1993).

[28] A. Bulgac, *Phys. Rev. B* **48**, 2721 (1993).

[29] R. Poteau, F. Spiegelmann and P. Labastie, *Z. Phys. D* **30**, 57 (1994).

[30] V. Bonačić-Koutecký, J. Pittner, C. Fuchs, P. Fantucci, M. F. Guest and J. Koutecký, *J. Chem. Phys.* **104**, 1427 (1996), and V. Bonačić-Koutecký, P. Fantucci, and J. Koutecký in Vol. 1 of Ref. [76]. See also Chapter 2 in this book.

[31] U. Röthlisberger and W. Andreoni, *J. Chem. Phys.* **94**, 8129 (1991). See also Chapter 3 in this book.

[32] K. Mishima, K. Yamashita and A. Bandrauk, *J. Phys. Chem. A* **102**, 3157 (1998).

[33] G. F. Bertsch and D. Tomanek, *Phys. Rev. B* **40**, 2749 (1989).

[34] Christoph Ellert, PhD, Freiburg, 1996, and to be published.

[35] Th. Reiners, Ch. Ellert, M. Schmidt and H. Haberland, *Phys. Rev. Lett.* **74**, 1558 (1995).

[36] S. Kasperl, C. Kohl and P.-G. Reinhard, *Phys. Lett. A* **206**, 81 (1995).

[37] G. F. Bertsch and R. A. Broglia *Oscillations in Finite Quantum Systems*, Cambridge University Press, 1994.

[38] W. D. Schöne, W. Ekardt and J. M. Pacheco, *Phys. Rev. B* **50**, 11078 (1994).

[39] J. M. Pacheco and W. D. Schöne, *Phys. Rev. Lett.* **79**, 4989 (1997).

[40] G. Bertsch and D. Tománek, *Chem. Phys. Lett.* **205**, 521 (1993).

[41] M. Schmidt and H. Haberland, *Eur. Phys. J. D*, accepted for publication.

[42] C. Yannouleas and U. Landman, *Phys. Rev. B* **51**, 1902 (1995).

[43] S. Frauendorf and V. V. Pashkevich, *Annalen der Physik* **5**, 34 (1996).

[44] S. M. Reimann, S. Frauendorf and M. Brack, *Z. Phys. D* **34**, 125 (1995).

[45] K. Clemenger, *Phys. Rev. B* **32**, 1359 (1985).

[46] W. Ekardt and Z. Penzar, *Phys. Rev. B* **43**, 1322 (1991).

[47] N. V. Smith, *Phys. Rev.* **183**, 634 (1969).

[48] A. von Felde, J. Sprösser-Prou and J. Fink, *Phys. Rev. B* **40**, 10181 (1989).

[49] A slightly different definition of the moment is often used in theoretical papers. The kth moment as defined here is proportional to the $(k + 1)$th moment of the imaginary part of the dynamical polarizability as defined by Eq. (4.11) of [8].

[50] C. Guet, *Comments At. Mol. Phys.* **31**, 305 (1995).

[51] W. D. Knight, K. Clemenger, W. A. de Heer and W. A. Saunders, *Phys. Rev. B* **31**, 2539 (1985).

[52] Ch. Schmitt, Ch. Ellert, M. Schmidt and H. Haberland, *Z. Phys. D* **42**, 145 (1997).

[53] W. Selby, V. Kresin, V. Masui, J. Vollmer, M. Châtelain and W. D. Knight, *Phys. Rev. Lett.* **59**, 1805, (1987); W. Selby, V. Kresin, V. Masui, J. Vollmer, W. A. de Heer, A. Scheidemann and W. D. Knight, *Phys. Rev. B* **43**, 4565 (1991).

[54] C. Yannouleas and U. Landman, *Phys. Rev. B* **43**, 8374 (1993).

[55] N. W. Ashcroft and N. D. Mermin, *Solid State Physics*, Saunders College, Philadelphia, 1976, see Figures 15.9 and 15.10.

[56] R. Schlipper, R. Kusche, B. von Issendorff and H. Haberland, *Phys. Rev. Lett.* **80**, 1194 (1998).

[57] M. Simon, F. Träger, A. Assion, B. Lang, S. Voll and G. Gerber, *Chem. Phys. Lett.* accepted for publication.

[58] J.-H. Klein-Wiele, P. Simon and H.-G. Rubahn, *Phys. Rev. Lett.* **80**, 45 (1998).

[59] R. A. Broglia and J. M. Pacheco, *Phys. Rev. B* **62**, 1400 (1989).

[60] C. Yannouleas and R. A. Broglia, *Ann. of Phys.* **217**, 105 (1992).

[61] C. Yannouleas, *Phys. Rev. B* **58**, 6748 (1998).

[62] Th. Reiners, Ch. Ellert, M. Schmidt and H. Haberland, *Phys. Rev. Lett.* **74**, 1558 (1995).

[63] Ch. Ellert, M. Schmidt, Th. Reiners and H. Haberland, *Z. Phys. D* **39**, 317 (1997).

[64] A. Liebsch, *Phys. Rev. B* **36**, 7378 (1987).

[65] J. Babst and P. G. Reinhard, *Z. Phys. D* **42**, 209 (1997).

[66] V. M. Akulin, C. Bréchignac and A. Sarfati, *Phys. Rev. B* **55**, 1372 (1997).

[67] Ph. Buffat and J. P. Borel, *Phys. Rev. A* **13**, 2287 (1976).

[68] R. S. Berry, *Scientific American*, p. 50, August 1990.

[69] G. Bertsch, *Science* **277**, 1619 (1997).

[70] M. Schmidt, R. Kusche, W. Kronmüller, B. von Issendorff and H. Haberland, *Phys. Rev. Lett.* **79**, 99 (1997).

[71] R. S. Berry, *Nature* **393**, 212 (1998).

[72] R. Kusche, Th. Hippler, M. Schmidt, B. von Issendorff and H. Haberland, *Eur. Phys. J.D.*, accepted for publication. Proceedings of the 9th ISSPIC, Lausanne, 1998).

[73] C. L. Cleveland, W. D. Luedtke and U. Landman, *Phys. Rev. Lett.* **81**, 2036 (1998).

[74] J. Parks, *Eur. J. Phys. D.* (Proceedings of the 9th ISSPIC, Lausanne, 1998).

[75] P. G. Reinhard, *Eur. J. Phys. D.* (Proceedings of the 9th ISSPIC, Lausanne, 1998).

[76] H. Haberland, ed., *Clusters of Atoms and Molecules I and II*, Springer Series in Chemical Physics, Vols 52 and 56.

6 Magnetic Properties of Transition-Metal Clusters

G. M. PASTOR

Université Paul Sabatier, Toulouse

and

K. H. BENNEMANN

Freie Universität Berlin

6.1 INTRODUCTION

Magnetism is a classical area of physics, and from antiquity to the present day has been a subject of great concern in human knowledge. The driving forces for magnetism are the interactions between electrons. The interplay between electronic kinetic energy and correlations controls the short-range and long-range ordering of electronic spins. Spin–orbit interactions cause magnetic anisotropy and determine, together with dipole–dipole interactions, the axis of easy magnetization. All these properties are affected by the lattice or atomic structure of the transition metal (TM). In small clusters, atomic structures may be realized that do not occur in the bulk metal and as a consequence new magnetic behaviors may appear. Furthermore, for increasing cluster size the changing atomic environment affects itinerancy, electronic correlations and spin–orbit coupling, leading to a close interdependence between magnetism and cluster structure. This offers many interesting possibilities for novel magnetic properties.

The magnetism of transition-metal clusters is of fundamental interest since atomic and bulk TM magnetism are very different. In the atoms magnetism is due to electrons

Metal Clusters. Edited by W. Ekardt
© 1999 John Wiley & Sons Ltd

localized in atomic orbitals while in a macroscopic TM solid the electrons responsible for magnetism are itinerant, conducting d electrons. Therefore, from a general point of view transition-metal clusters may serve as a testing ground for studying as a function of the dimension of the system the transition from local-moment or Heisenberg-like magnetism to itinerant magnetism. As the cluster size increases, the atomic-like magnetic properties of small Cr, Fe or Ni clusters get more and more similar to those observed in the solid. Taking into account that the magnetic moments of isolated atoms are larger than in the bulk, one may expect that the local magnetic moments $\mu(i)$ in small TM clusters should be larger than in the corresponding solids. Moreover, the magnetic order may change with respect to the bulk, particularly if the structure is different (for fcc- or bcc-like clusters). Also the Curie temperature changes if the number of nearest and next-nearest neighbors is different than in the bulk. Small clusters may be magnetic and then lose their local moments and magnetization as the cluster size increases. Interesting candidates for such a behavior are V_N, Rh_N and Pd_N clusters, for example. Small Mn_N and Cr_N clusters should clearly reflect the interplay between magnetic order and atomic structure. This is generally expected for transition-metal systems showing antiferromagnetic order in the bulk. The different magnetic anisotropy and easy axis couples the magnetization with vibrations and rotations of free clusters and this can be also observed when these clusters are embedded in a matrix or deposited on a metallic or inert-crystal surface.

Both theoretical and experimental research on magnetic clusters is difficult. Since magnetism depends sensitively on the dimension of the system, atomic structure and interatomic distances, well-defined and characterizable clusters are necessary. Up to now, mainly Stern–Gerlach deflection experiments have been performed to analyze the magnetic properties of free clusters [1–3]. Theory needs a reliable and sensitive treatment of the electronic correlations due to the relatively small magnetic contributions to the total energy. Typically, density-functional calculations using the local spin density approximation [4–9] and self-consistent tight-binding calculations [10–12] are performed to determine the magnetic properties like local magnetic moments μ_i, magnetization and magnetic anisotropy [13]. These studies yield quite generally that the local magnetic moment μ_i, their order and orientation depend on the atomic positions within the cluster and vary from the interior to the cluster surface [10].

Theorists were the first to study the magnetism of clusters. Salahub *et al.* determined the magnetic moments of small 3d TM clusters using the X-α approximation [4]. It was predicted that the magnetic moment per atom $\overline{\mu}_N$ in small bcc-like Fe_N clusters should be $\overline{\mu}_9 = 2.89~\mu_B$ and $\overline{\mu}_{15} = 2.67~\mu_B$. Both values are considerably larger than the bulk moment $\mu_b = 2.2~\mu_B$. These results were confirmed qualitatively by the local-spin-density calculations of Lee, Callaway and co-workers who obtained $\overline{\mu}_9 = 2.89~\mu_B$ and $\overline{\mu}_{15} = 2.93~\mu_B$ [5]. An alternative to the *ab initio* approach was provided by the self-consistent tight-binding (SCTB) studies on Cr, Fe and Ni clusters ($N \leq 51$) [10]. This method yielded results in good agreement with *ab initio* calculations ($\overline{\mu}(Fe_N) \simeq 3.0~\mu_B$) and, taking advantage of the flexibility of the parameterized, minimal-basis approximation, it explored a variety of cluster structures and sizes that remain even nowadays inaccessible for first principles techniques. The main predictions of these theoretical studies [4, 5, 10] were confirmed afterwards by Stern–Gerlach measurements. A couple of years later, the theoretical research on cluster magnetism was considerably boosted by remarkable experimental findings [2, 3]. An important number of *ab initio* calculations on magnetic

TM clusters were then performed, in most cases within the framework of the local spin density approximation [6–9, 11–13].

The first experiments on the magnetic properties of free TM clusters were performed by Cox *et al.* [1]. As in all the other experimental studies that followed, the Stern–Gerlach (SG) deflection is used to determine the magnetic properties. Three main steps are involved in the experiment. (i) A beam of neutral clusters containing a more or less broad distribution of sizes is produced, typically using a laser vaporization source. (ii) The neutral clusters pass through an SG magnet where they are deflected along the field-gradient direction. The deflection depends on the value of the projection of the net cluster magnetization onto the magnetic field H. Clusters whose magnetization is parallel (antiparallel) to H are deflected in the direction of increasing (decreasing) field. (iii) Finally, the clusters are ionized in order to be detected after mass selection. Cox *et al.* observed that the beam intensity at the beam axis is depleted upon switching on the SG magnetic field. This was the first experimental indication that small Fe_N clusters are magnetic ($2 \leq N \leq 17$) [1]. The measurements were interpreted assuming that the ordered magnetic clusters should be deflected in the external magnetic field H essentially like atoms carrying a large magnetic moment $N\overline{\mu}_N$ and thus with equal probability in the direction of increasing and decreasing field, depending on whether $\overline{\mu}_N$ is parallel or antiparallel to H. However, later experiments [2, 3] revealed that this assumption was incorrect and that performing only on-axis detection (ionization) is a strong limitation. Consequently, measuring only the on-axis depletion factor is insufficient for determining $\overline{\mu}_N$.

The next important experimental progress was achieved by de Heer, Milani and Châtelain [2]. They performed SG deflection experiments on Fe_N ($15 \leq N \leq 650$) and measured the cluster intensity all along the direction of the field gradient, which is perpendicular to the beam axis. In this way they discovered that the clusters deflect uniquely in the direction of *increasing* field, showing a somewhat broad spatial distribution. This implies that the magnetization of *isolated* Fe clusters tends to align parallel to an external magnetic field, a remarkable effect which was not expected [2]. In this work, the average deflection D was related directly to the average magnetic moment per atom $\overline{\mu}$ by using the relation $\overline{\mu} \propto Dmv^2/H$. However, note that actually the average magnetization of the cluster magnetic field should be related to D. The values of $\overline{\mu}$ derived in this way increase with increasing magnetic filed H and are much *smaller* than the bulk atomic moments μ_b. De Heer and co-workers already recognized that the cluster magnetic moments determined in such a way are a lower bound for the magnitude of the average magnetic moments of the cluster [4, 5, 10]. Due to the assumption $\overline{\mu} \propto Dmv^2/H$ the comparison between theory and experiment seemed quite controversial, especially concerning the temperature dependence of D. It was not until the work of Bucher, Douglass and Bloomfield [3] that the relation between the measured deflections D and the moment $\overline{\mu}_N$ of a cluster of N atoms became clear.

Bucher *et al.* performed experiments on Co_N clusters ($20 \leq N \leq 200$) and showed that the observed small average deflections could be interpreted as the result of the relaxation of the magnetization of superparamagnetic clusters in the direction of H [3]. Theoretical analysis by Khanna and Linderoth [14] and by Jensen, Mukherjee and Bennemann [15] supported this. Similar relaxation is also observed in supported magnetic nanoparticles. Assuming that superparamagnetic clusters undergo relaxation, Bucher *et al.* inferred the value of the intrinsic magnetic moment per Co atom, $\overline{\mu} = (2.1 \pm 0.2)\ \mu_B$, which is larger than the bulk moment, in qualitative agreement with existing calculations [4, 5, 10]. The superparamagnetic regime opened the possibility of a series of systematic experimental studies [16–21]. Rh_N clusters were found to have a rather large magnetization [18]. This was the first experimental observation of a transition from nonmagnetic to magnetic behavior upon reduction of the size of a system. Theoretical results already indicated this possibility [11, 22, 23]. Experiments on clusters of other nonmagnetic TMs have not yet yielded measurable magnetic SG deflection [17] and thus no magnetism has been measured for V_N, Cr_N and Pd_N.

Billas *et al.* determined the magnetization of Fe_N clusters as a function of size and temperature for $25 \leq N \leq 700$ and $100 \text{ K} \leq T \leq 900 \text{ K}$ [19]. They obtained that the low-temperature average moment $\overline{\mu}_N$ increases with decreasing cluster size, exhibiting some oscillations as a function of N and reaching a value of about $3\ \mu_B$ for the smallest sizes ($T \simeq 100 \text{ K}$). A similar behavior is also observed for Co and Ni clusters. For Ni_N different measurements [20, 21] still give some quantitatively different results for $\overline{\mu}_N$. The temperature dependence of $\overline{\mu}_N$ derived from experiment depends significantly on the TM considered. Notice also that the determination and control of the cluster temperature is a delicate problem. In Ni_N the experimental magnetization curves are qualitatively similar to the bulk, except of course for the broadening of the transition close to the Curie temperature T_C due to the finite cluster size [20]. Experiments on Co_N show that the magnetization per atom is about 0.1–$0.5\ \mu_B$ larger than the bulk one for cluster sizes of $50 \leq N \leq 600$ and for temperatures such that $100 \text{ K} \leq T \leq 1000 \text{ K}$ [20]. In addition, for Co clusters the magnetization is found to increase slightly with T for $100 \text{ K} \leq T \leq 500 \text{ K}$. These temperatures are still significantly lower than the bulk Curie temperature $T_C(\text{Co}) = 1388 \text{ K}$. In Fe clusters the finite-temperature behavior is qualitatively different from that of Ni or Co clusters. For example, for $250 \leq N \leq 600$ one observes a rapid, almost linear decrease of the magnetization with increasing T ($T \leq 500$–600 K). For $T \geq 300 \text{ K}$ the cluster magnetization per atom is smaller than in the bulk and for $T = 600 \text{ K}$ smaller by a factor 0.3 ($T_C(\text{Fe-bulk}) = 1043 \text{ K}$). As the cluster size increases ($250 \leq N \leq 600$) the difference between the cluster and the bulk magnetization gets even larger [20]. This trend should change at larger sizes.

In addition to the values of the intrinsic magnetic moments of isolated clusters, the Stern–Gerlach beam experiments yielded remarkable results for the magnetic behavior of a cluster ensemble [2, 3]. The asymmetric magnetic deflection and the magnetic-field dependence of the average magnetization of the cluster ensemble, which shows strong deviations from Langevin behavior, motivated several theoretical studies [14, 15, 24, 25]. Interestingly, different experiments on Fe_N clusters suspended in liquids also exhibited a non-Langevin behavior of the magnetic-field dependence of the average cluster magnetization [26]. Assuming superparamagnetic clusters and corresponding magnetic relaxation, the experimental findings could be analyzed consistent with electronic calculations of $\overline{\mu}_N$ [3, 14, 15, 24, 25]. However, a microscopic understanding of the remarkable spin relaxation process in an isolated cluster needs further study. The experiments on Fe_N in a magnetic field [19] also revealed the breakdown of the superparamagnetic regime in the case of supersonically cooled clusters. This was explained as due to a resonant coupling between cluster rotations and Zeeman splittings [24, 25].

Magnetic properties have also been investigated for clusters deposited on a crystal surface and for an ensemble of clusters embedded in a matrix or in a thin film [27–29]. In these cases the hybridizations, distortions and bond-length changes induced by cluster–matrix interactions may result in magnetic properties that are considerably different from those of free clusters. For example, the average magnetization of Fe_N or Co_N clusters embedded in a nonmagnetic or antiferromagnetic matrix like Cu, V or Cr is found to be smaller than the corresponding bulk magnetization. Magnetic clusters deposited on nonmagnetic surfaces show an interesting competition between the reduction of local coordination number which favors local moment formation and the sd cluster–substrate hybridization which weakens magnetism. The study of supported clusters provides complementary information that emphasizes the close relations between the magnetic properties of transition-metal clusters, nanostructured granular systems and thin films.

6.2   THEORY

The determination of the magnetic properties of transition-metal clusters requires a correct treatment of the interplay between the electronic kinetic energy and the interactions between the d electrons, which are mainly responsible for magnetism. From a microscopic point of view, the size-dependent magnetic properties are the result of a delicate balance between the effect of hybridizations, which favor equal filling of spin states, and the effect of Coulomb interactions, which according to Hund's rules favor the formation of local magnetic moments. Hybridizations and the resulting electron delocalization tend to reduce the ground-state kinetic energy E_K, but at the same time they also involve local charge fluctuations which increase the Coulomb-interaction energy E_C. The interplay between E_K and E_C introduces correlations in the electronic motion, which are fundamental in determining the magnetic properties. In general one expects that E_C should become more important as the dimensions of the system are reduced, since E_K decreases with decreasing coordination number z. The theory described in this section is based upon a realistic many-body Hamiltonian H for the valence s, p and d electrons which in the usual second quantization notation is given by

$$H = \sum_{i\alpha\sigma} \varepsilon_{i\alpha}^0 \hat{n}_{i\alpha\sigma} + \sum_{\substack{i\neq j \\ \alpha\beta\sigma}} t_{ij}^{\alpha\beta} \hat{c}_{i\alpha\sigma}^\dagger \hat{c}_{j\beta\sigma} + \frac{1}{2} \sum_{\substack{i\alpha\beta \\ \sigma\sigma'}}' U_{\alpha\beta}^{\sigma\sigma'} \hat{n}_{i\alpha\sigma} \hat{n}_{i\beta\sigma'}. \tag{1}$$

Here, $\varepsilon_{i\alpha}^0$ denotes the energy of the orbital α ($\alpha \equiv$ s, p, d) at atomic site i and $\hat{n}_{i\alpha\sigma}$ are the corresponding occupation-number operators. The $t_{ij}^{\alpha\beta}$ are the hopping integrals describing electronic transitions between atoms i and j. The third term in Eq. (1) approximates the electron–electron interactions [30, 31]. The prime in the sum indicates that the terms with $\alpha = \beta$ and $\sigma = \sigma'$ are to be excluded. $U_{\alpha\beta}^{\sigma\sigma'}$ is the effective on-site interaction between electrons with spin σ and σ' in the orbitals α and β of atom i. In Eq. (1) only the Coulomb and exchange intra-atomic integrals $U_{\alpha\beta} = \langle \alpha\beta | 1/r_{12} | \alpha\beta \rangle$ and $J_{\alpha\beta} = \langle \alpha\beta | 1/r_{12} | \beta\alpha \rangle$ are taken into account ($U_{\alpha\beta}^{\uparrow\downarrow} = U_{\alpha\beta}^{\downarrow\uparrow} = U_{\alpha\beta}$ and $U_{\alpha\beta}^{\uparrow\uparrow} = U_{\alpha\beta}^{\downarrow\downarrow} = U_{\alpha\beta} - J_{\alpha\beta}$). In practice it is usually a good approximation to replace these integrals by orbital-independent constants U and J. The parameters $t_{ij}^{\alpha\beta}$, U and J control the interplay between kinetic and Coulomb energies or equivalently the degree of itinerant as opposed to local moment behavior. Notice that Eq. (1) does not preserve spin rotation invariance, except for the case of only one orbital per atom (i.e. the single-band Hubbard model). This is of no major importance as long as symmetry-breaking Hartree-type approximations are used in the actual calculations. Spin rotation invariance could be restored by including the transversal exchange terms $H_{xy} = -\sum_{i,\alpha<\beta} J_{\alpha\beta}(S_{i\alpha}^+ S_{i\beta}^- + S_{i\alpha}^+ S_{i\beta}^-)$ [31].

For the sake of clarity we shall first discuss in Section 6.2.1 ground-state properties in the framework of the unrestricted Hartree–Fock approximation. In Section 6.2.2 the theory is extended to finite temperatures by using a functional integral formalism including spin fluctuations. Finally, in Section 6.2.3 we analyze the problem of electron correlations by exact diagonalization of the simpler single-band Hubbard model.

6.2.1 Ground-State Properties

Within a self-consistent mean-field approximation Eq. (1) can be rewritten as [10, 32]

$$H = \sum_{i\alpha\sigma} \varepsilon_{i\alpha\sigma} \hat{n}_{i\alpha\sigma} + \sum_{\substack{i\neq j \\ \alpha\beta\sigma}} t_{ij}^{\alpha\beta} \hat{c}_{i\alpha\sigma}^\dagger \hat{c}_{j\beta\sigma}, \tag{2}$$

with

$$\varepsilon_{i\alpha\sigma} = \varepsilon_{i\alpha}^0 + \sum_\beta (U_{\alpha\beta} - J_{\alpha\beta}/2)\Delta\nu_{i\beta} - \frac{\sigma}{2}\sum_\beta J_{\alpha\beta}\mu_{i\beta}. \tag{3}$$

$\Delta\nu_{i\beta} = \nu_{i\beta} - \nu_{i\beta}^0$ is the charge transfer on the orbital β at atom i and $\mu_{i\beta}$ the magnetic moment. The direct Coulomb integrals are denoted by $U_{\alpha\beta}$ and the exchange integrals by $J_{\alpha\beta} = U_{\alpha\beta}^{\uparrow\downarrow} - U_{\alpha\beta}^{\uparrow\uparrow}$. The local Green's functions $G_{i\alpha\sigma}(\varepsilon) = \langle i\alpha\sigma|G(\varepsilon)|i\alpha\sigma\rangle$ are determined from $G(\varepsilon) = (\varepsilon - H)^{-1}$ [33] and then the magnetic moments $\mu_{i\alpha}$ and average occupations $\nu_{i\alpha}$ are calculated self-consistently by integrating the local densities of states $\rho_{i\alpha\sigma}(\varepsilon) = -(1/\pi)\text{Im}\{G_{i\alpha\sigma}(\varepsilon)\}$.

The number of electrons with spin σ at the orbital α of atom i is given by

$$\nu_{i\alpha\sigma} = \langle \hat{n}_{i\alpha\sigma} \rangle = \int_{-\infty}^{\varepsilon_F} d\varepsilon \rho_{i\alpha\sigma}(\varepsilon). \tag{4}$$

Here, ε_F denotes the Fermi energy. The local magnetic moments are obtained from

$$\mu_i = \sum_\alpha (\nu_{i\alpha\uparrow} - \nu_{i\alpha\downarrow}). \tag{5}$$

Notice that self-consistency is required for Eqs (2)–(5). The magnetic order is inferred from the spatial distribution of the moments μ_i. In case of multiple self-consistent solutions for the magnetic order or when different atomic structures are involved, the corresponding free energies $F = E - TS$ are compared. Then, the atomic and magnetic structure with lowest F is taken. At $T = 0$ the energy is given by

$$E = \sum_{i\alpha\sigma} \int_{-\infty}^{\varepsilon_F} \varepsilon \rho_{i\alpha\sigma}(\varepsilon) \, d\varepsilon - E_{dc} - E_R, \tag{6}$$

where $\rho_{i\alpha\sigma}(\varepsilon) = -(1/\pi)\text{Im}\{G_{i\alpha\sigma}(\varepsilon)\}$ is the local density of states and $E_{dc} = \sum_{i\alpha\sigma}(\varepsilon_{i\alpha\sigma} - \varepsilon_{i\alpha}^0)\nu_{i\alpha\sigma}$ corrects for double counting. The repulsive energy E_R due interatomic interactions can be approximated by the Born–Mayer-type potential

$$E_R = \frac{1}{2} \sum_{i\neq j} A e^{-p\left(\frac{R_{ij}}{d_b}-1\right)}, \tag{7}$$

with parameters A and p fitted to bulk data, d_b being the bulk nearest-neighbor (NN) distance. Minimizing E with respect to the interatomic distances R_{ij}, and taking into account the resulting order of the magnetic moments, one obtains the atomic structure and spin configuration of the cluster [10, 27]. Correlations between the magnetic moments could also be calculated by using the Bethe–Peierls approximation. Alternatively the recently developed dynamical mean-field theory could also be applied to calculate μ_i and magnetic order [34].

The unrestricted Hartree–Fock theory tends to overestimate the formation of the magnetic moments and the stability of the magnetic order. However, the main interplay between the kinetic energy, the resulting size-dependent d bandwidth $W_d(N)$, and the exchange interactions J_{dd} is expected to be correctly described. This is essential for a systematic understanding of transition-metal magnetism. Let us recall that $W_d \propto t_{ij} \propto R_{ij}^{-5}$. One observes magnetism if $J > J_c$, where J_c is a critical value that depends on the size and structure of the cluster. In clusters one may obtain a magnetization for values of J/W for which the bulk metal is not magnetic [11]. Of course, s electrons and sd hybridizations must be included for a more complete treatment. This extension is straightforward [12].

In order to account for magnetic anisotropy, i.e. the dependence of the energy, spin and orbital magnetic moments on the orientation of the magnetization with respect to the cluster structure, one must add to the Hamiltonian given by Eq. (1) the spin–orbit interaction term

$$H_{SO} = -\xi \sum_{\substack{i,\alpha\beta \\ \sigma\sigma'}} (\vec{L}_i \cdot \vec{S}_i)_{\alpha\sigma,\beta\sigma'} \hat{c}_{i\alpha\sigma}^\dagger \hat{c}_{i\beta\sigma'}, \tag{8}$$

where $\vec{S}_i$ is the spin operator $\vec{L}_i$ the orbital moment operator, and ξ the spin–orbit coupling constant. The energy E_δ now depends on the direction of the magnetization δ ($\delta = x$, y or z) and is given as before by Eq. (6), where the density of states $\rho_{i\alpha\sigma}(\varepsilon)$ is now determined from $H + H_{SO}$. The resulting spin–orbit energy $E_\delta^{SO} = E_\delta(\xi) - E(\xi = 0)$ depends sensitively on the atomic structure of the cluster and is expected to be larger at the cluster surface than in the bulk or at planar extended surfaces. The magnetic anisotropy energy (MAE) ΔE is given by the change in the electronic energy E_δ occurring when the orientation of the magnetization $\langle \vec{S} \rangle$ changes for a fixed position of the atoms. For example, $\Delta E = E_x - E_z$ measures the relative stability of the magnetization direction along the x and z axes.

For many problems treating magnetic anisotropy effects and the magnetic relaxation of a rotating cluster in a magnetic field the phenomenological Hamiltonian

$$H' = H_{ex} + H_z + H_{anis} \tag{9}$$

has been used. In Eq. (9), the exchange interaction is given by

$$H_{ex} = -J \sum_{i,j} \vec{S}_i \cdot \vec{S}_j, \tag{10}$$

the Zeeman term due to an external magnetic field $\vec{h}$ by

$$H_z = -\sum_i \vec{\mu}_i \cdot \vec{h}, \tag{11}$$

and the magnetic anisotropy term with constant K_2 by

$$H_{\text{anis}} = -K_2 \sum_i \left(\frac{\vec{\mu}_i}{\mu_i} \cdot \vec{c} \right)^2. \tag{12}$$

Due to H_{anis} the magnetic moment $\vec{\mu}_i$ of atom i tends to be tied to the easy axis $\vec{c}$. H_{anis} describes the situation of uniaxial anisotropy. In terms of the electronic theory $K_2 = |\Delta E|$ is given by the anisotropy energy between the easy and hard magnetization directions. Notice that the interplay of the parameters J, h and K_2 controls the behavior of rotating magnetic clusters in an external magnetic field, the average magnetization of a cluster ensemble and thus the deflection in Stern–Gerlach experiments [24].

6.2.2 Finite-Temperature Properties: Spin Fluctuations

The magnetic properties of clusters at finite temperatures are of considerable interest. In view of possible applications in magnetic recording media, it is not enough to identify the clusters showing large average magnetizations per atom at $T = 0$, but it is also crucial to get information on the cluster magnetization as a function of T and on the size and structural dependence of the 'Curie' temperature $T_C(N)$ above which ferromagnetic order is destroyed by thermal fluctuations. Recently, the temperature dependence of the magnetization of Fe_N, Co_N and Ni_N clusters has been derived from Stern–Gerlach deflection measurements [19–21]. Particularly in the case of Fe_N the behavior of the magnetization seems quite anomalous. One observes strong deviations from the predictions of the Heisenberg model and rather large high-temperature magnetization values ($T > T_C$). From a theoretical point of view, the determination of finite-temperature properties of itinerant ferromagnets poses a serious challenge. In the following we outline the extension of the self-consistent tight-binding theory presented in the previous section to finite temperatures [35] by using a functional-integral formalism [36–38]. This is an itinerant electron approach that takes into account the environment dependence of the local magnetic moments by including at the same time the temperature-induced fluctuations of the spin degrees of freedom.

In order to calculate the partition function of the d-electron Hamiltonian, we rewrite the many-body interaction H_I, the third term in Eq. (1), as

$$H_I = \frac{1}{2} \sum_i \left[(U - J/2)\hat{N}_i^2 - 2J\hat{S}_{iz}^2 - (U - J)\hat{N}_i \right], \tag{13}$$

where $\hat{N}_i = \sum_{\alpha\sigma} \hat{n}_{i\alpha\sigma}$ is the electron number operator at atom i and $\hat{S}_{iz} = (1/2)\sum_\alpha (\hat{n}_{i\alpha\uparrow} - \hat{n}_{i\alpha\downarrow})$, the z component of the spin operator. U and J stand for the direct and exchange Coulomb integrals which as before are taken to be orbital-independent. The quadratic terms in Eq. (13) are linearized by means of a two-field Hubbard–Stratonovich transformation within the static approximation. A charge field η_i and an exchange field ξ_i are thus introduced at each cluster site i, and represent the local finite-temperature fluctuations of the d-electron energy levels and exchange splittings, respectively [35]. The saddle-point approximation for the charge fields yields a set of self-consistent equations for $\overline{\eta}_i = \overline{\eta}_i(\xi_1, \ldots, \xi_N)$. The d-electron energy levels of atom i are then given by

$$\varepsilon_{i\sigma} = \varepsilon_i^0 + (U - J/2)\Delta\nu_i - \frac{\sigma}{2}J\xi_i. \tag{14}$$

Comparison with Eq. (3) shows that the self-consistent local magnetic moment μ_i at $T = 0$ is now replaced by the fluctuating field ξ_i. The magnetic properties at $T > 0$ are obtained as an average over all possible exchange-field configurations $\vec{\xi} = (\xi_1, \ldots, \xi_N)$. For example, the local magnetization $\mu_i(T)$ at the atom i is given by the average of the local magnetic moment $2\langle \hat{S}_{iz} \rangle = \sum_\alpha (\langle \hat{n}_{i\alpha\uparrow} \rangle - \langle \hat{n}_{i\alpha\downarrow} \rangle)$ which fluctuates at $T > 0$ and depends on $\vec{\xi}$:

$$\mu_i(T) = \frac{1}{Z} \int 2\langle \hat{S}_{iz} \rangle e^{-\beta F_i(\xi)} d\xi = \frac{1}{Z} \int \xi e^{-\beta F_i(\xi)} d\xi. \tag{15}$$

Here, $F_i(\xi) = -k_B T \ln[\int \prod_{j \neq i} d\xi_j e^{-\beta F(\xi_1 \ldots \xi \ldots \xi_N)}]$ is the free energy associated with an exchange field ξ acting at atom i and $Z = \int e^{-\beta F_i(\xi)} d\xi$ is the partition function. $P_i(\xi) = e^{-\beta F_i(\xi)}/Z$ represents the probability for the value ξ of the exchange field at atom i. Notice that the last equality in (15) justifies the intuitive association between the fluctuations of the local moments $2\langle \hat{S}_{iz} \rangle$ and of the exchange field ξ_i at atom i. The cluster magnetization results from the average of the local magnetizations $\mu_i(T)$ [35]. This many-body treatment of magnetism was originally used by Hubbard and Hasegawa for studying bulk 3d TMs at finite temperatures [36–38].

A first insight regarding the magnetic behavior of Fe_N and Ni_N clusters at $T > 0$ can be obtained by considering the low-temperature limit of $F_i(\xi)$. The local free-energy difference

$$\Delta F_i(\xi) = F_i(\xi) - F_i(\xi_0), \tag{16}$$

with $\xi_0 = \mu_i^0$ being the local magnetic moment of atom i at $T = 0$, often has two minima at $\xi^+ = \xi_0$ and $\xi^- = -\xi_0$. At finite temperatures the ferromagnetic order between the magnetic moments is reduced or destroyed by spin fluctuations involving transitions between these two minima. For small Fe or Ni clusters mainly directional fluctuations of $\vec{\mu}_i$ are expected to dominate, since the reduction of the local coordination numbers should enhance local-moment behavior. As the cluster size increases fluctuations of the amplitude of μ_i or ξ_i also become important, indicating a crossover to a more itinerant behavior. The spin-fluctuation free-energy $\Delta F_i(\xi)$ depends strongly on the atomic site i, the cluster structure and the interatomic distances. For decreasing interatomic distance magnetism becomes less stable and $\Delta F_i(\xi^-)$ decreases.

As a first approximation we may estimate the cluster 'Curie' temperature $T_C(N)$ as the energy required to flip a local magnetic moment $\langle \hat{S}_{iz} \rangle$ or an exchange field ξ_i. Thus $T_C(N) \propto \langle \Delta F_i \rangle$, where $\langle \ldots \rangle$ indicates a cluster average. However, note that this is a crude approximation. Equation (16) for $\Delta F_i(\xi)$ ignores the fact that close to T_C all the exchange fields at different atomic sites fluctuate in some correlated way. The low-temperature limit, where all but one field ξ_i are kept equal to the $T = 0$ value, certainly overestimates $\Delta F_i(\xi)$ at $T > 0$ and thus T_C. Fluctuations of different exchange fields modify the magnetic environment at which individual spin fluctuations occur and should therefore be taken into account. The Curie temperature is in fact the temperature at which it costs no free energy to flip spins [36].

The finite-temperature fluctuations of the local moments $\langle \hat{S}_{iz} \rangle$ at different atomic sites i are not independent of each other. Short-range magnetic correlations are present even above the temperature T_C at which the average magnetization vanishes. They are usually referred to as short-range magnetic order (SRMO) [39–41]. In the framework of the previous functional integral theory, SRMO results in correlations between the exchange fields at neighboring sites. For example, in the ferromagnetic case $\langle \xi_i \xi_j \rangle > \langle \xi_i \rangle \langle \xi_j \rangle$, for i and j being nearest neighbors. Bethe–Peierls-type approximations are often used to include SRMO in bulk and surface calculations [41] (see also Ref. [34]). Short-range magnetic order is important for the magnetic behavior and is known to exist in macroscopic TMs. In small clusters, it should play a predominant role, particularly when the cluster size is comparable to the size of the SRMO domains. In this case, the entropy increase leading to a reduction of the ferromagnetic order within the cluster necessarily requires breaking the energetically favorable short-range magnetic correlations.

Taking into account SRMO is important for analyzing the deflections observed in Stern–Gerlach experiments, in particular for deriving from the deflections the temperature dependence of the cluster magnetization and the local magnetic moments. An approximate, model-independent approach has been recently proposed in order to take into account the effect of SRMO in clusters [42]. One may regard the transition-metal N-atom cluster at temperatures $T \geq T_C(N)$ as consisting of N/ν_{sr} ferromagnetic domains each having nearly the same number ν_{sr} of atoms. The ferromagnetic order is preserved within each domain, so the domains have a total magnetization $\nu_{sr}\mu(T = 0)$, but different domains are randomly oriented with respect to each other. Then, the average of the magnetization $\sqrt{\langle \mu_N^2 \rangle}$ is approximately

$$\overline{\mu}_N(T > T_C) \simeq \overline{\mu}_N(T = 0)\sqrt{\nu_{sr}/N}. \tag{17}$$

Note that Eq. (17) is derived assuming that $\nu_{sr} \ll N$. The degree of short-range order above $T_C(N)$ is characterized by ν_{sr}. In particular $\nu_{sr} = 1$ corresponds to the so-called disordered-local-moment picture, i.e. the absence of SRMO. The value of ν_{sr} depends of course on the transition metal under consideration. Taking into account bulk results, one expects stronger short-range order in Ni clusters than in Fe clusters. Equation (17) allows us to infer the degree of SRMO by comparing the low- and high-temperature values of $\overline{\mu}_N(T)$ and is useful for interpreting the cluster magnetic behavior.

6.2.3 Exact Treatment of Electron Correlations

A deeper understanding of the electronic correlations underlying the magnetic properties of clusters is not only important from a fundamental point of view, but it is likely to be crucial in investigations of more delicate properties such as the photoemission spectra or the magnetic behavior at finite temperatures. Consequently, it is important to go beyond simple mean-field approximations by including charge and spin fluctuations explicitly. Moreover, since the magnetic behavior of itinerant 3d electrons is very sensitive to the details of the electronic structure, the cluster structure and the local environment of the atoms [10], it is of considerable interest to study the relation between cluster structure and magnetism as well as the stability of cluster ferromagnetism with respect to structural changes and electronic excitations.

The determination of electron correlation effects, cluster geometry and their interplay is a difficult problem. Most theoretical studies performed so far have attempted to deal with one of these aspects at a time [4–11, 43–46]. For example, in Refs [4–7, 10, 11] the electron interactions were treated in mean-field approximations (X α, local spin density, unrestricted Hartree–Fock) and only a few, mostly highly symmetric structures were considered. Optimizations of the cluster geometry were performed in Ref. [8] for

Fe$_N$ with $N \leq 5$ by using density-functional theory with nonlocal gradient corrections. Simulated-annealing calculations of cluster structures were done in Ref. [9] by using a plane-wave basis and the local spin density approximation. The localized character of the atomic-like 3d orbitals and the complicated dependence of the magnetic moments and magnetic order on cluster geometry have so far prevented the application of rigorous first-principles methods to the study of the problem.

Electron-correlations effects and cluster geometry could be determined in the framework of the spd-band model Hamiltonian given by Eq. (1). The Hartree–Fock approximation discussed in Section 6.2.1 should be improved systematically, for example, by treating the residual interactions by perturbation methods or by using Gutzwiller or Jastrow variational *Ansätze* [47, 48]. However, a general implementation of such calculations for TM clusters would be very demanding, particularly if the structure is arbitrary, i.e. lacking any symmetry. An alternative approach that has provided numerous significant results in the field of itinerant magnetism is to consider a simpler electronic model that should allow an exact or at least very accurate solution of the many-body problem and at the same time should contain enough complexity to be able to shead light on the physics of real systems (e.g. 3d TMs). A model with such characteristics is given by the well-known Hubbard Hamiltonian [30],

$$H = -t \sum_{\langle i,j \rangle, \sigma} \hat{c}_{i\sigma}^{\dagger} \hat{c}_{j\sigma} + U \sum_i \hat{n}_{i\uparrow} \hat{n}_{i\downarrow}, \qquad (18)$$

where $\hat{c}_{i\sigma}^{\dagger}(\hat{c}_{i\sigma})$ refers to the creation (annihilation) operator for an electron at site i with spin σ, and $\hat{n}_{i\sigma} = \hat{c}_{i\sigma}^{\dagger} \hat{c}_{i\sigma}$ to the corresponding number operator. At low energies the electronic behavior results from a delicate balance between the tendency to delocalize the valence electrons in order to reduce their kinetic energy, and the effect of the Coulomb repulsions associated with local charge fluctuations. The relative importance of these contributions depends strongly on the value of the total spin S and on the ratio between the Coulomb interaction strength U and the nearest-neighbor (NN) hopping t. Therefore, the determination of the structure and magnetic behavior requires an accurate treatment of electron correlations, particularly for $U \gg t$. Moreover, the ground state corresponding to different cluster structures may have very different magnetic properties [43–45]. For this reason, it is also necessary to optimize the geometry, if rigorous conclusions about the magnetic behavior of these clusters are to be achieved. The point of view is thus complementary to that of the previous sections, where the description of the electronic structure is realistic, but the effects of Coulomb interactions are treated in a mean-field approximation.

The Hubbard model for small clusters is solved by expanding the eigenfunctions $|\Psi_l\rangle$ into the complete basis given by the states $|\Phi_m\rangle$ which have definite occupation numbers $n_{i\sigma}^m$ for all orbitals $i\sigma$ ($\hat{n}_{i\sigma}|\Phi_m\rangle = n_{i\sigma}^m|\Phi_m\rangle \forall i\sigma, n_{i\sigma}^m = 0, 1$). The eigenstate $|\Psi_l\rangle$ is written as

$$|\Psi_l\rangle = \sum_m \alpha_{lm}|\Phi_m\rangle, \qquad (19)$$

where

$$|\Phi_m\rangle = \left[\prod_{i\sigma} (\hat{c}_{i\sigma}^{\dagger})^{n_{i\sigma}^m} \right] |vac\rangle. \qquad (20)$$

The values of $n_{i\sigma}^m$ satisfy the usual conservation of the number of electrons $\nu = \nu_\uparrow + \nu_\downarrow$ and of the z component of the total spin $S_z = (\nu_\uparrow - \nu_\downarrow)/2$, where $\nu_\sigma = \sum_i n_{i\sigma}^m$. Taking into account all possible electronic configurations may imply a considerable numerical effort. For example, for $N = 12, 13$ and 14 the size of the Hilbert space $\binom{N}{\nu_\uparrow}\binom{N}{\nu_\downarrow}$ at half-band filling is $853\,776$, $2\,944\,656$ and $11\,778\,624$, respectively. For not too large clusters, the expansion coefficients α_{lm} corresponding to the ground state ($m = 0$) and low-lying excited states can be determined by sparse-matrix diagonalization procedures (e.g. Lanczos iterative method [49]). In order to calculate the ground state $|\Psi_0\rangle$ one considers the subspace of minimal S_z, since this ensures that there are no *a priori* restrictions on the total spin S. Several ground-state ($|\Psi_0\rangle$) and excited-state ($|\Psi_l\rangle$, $l > 0$) properties of the cluster can be calculated in terms of the coefficients α_{lm} by simple operations on the basis states $|\Phi_m\rangle$. Within a controlled accuracy the results are thus exact in the framework of the Hubbard model.

In order to determine the optimal cluster geometry one should first note that in the Hubbard model (Eq. (18)) only a single s-like orbital per site and nearest-neighbor (NN) hoppings are taken into account. In this case the hopping integrals t_{ij} between sites i and j take only two possible values, namely $t_{ij} = -t$, if $R_{ij} = R_0$ and $t_{ij} = 0$, if $R_{ij} > R_0$, where R_{ij} refers to the interatomic distance and R_0 to the NN distance. Thus, only the topological aspect of the structure is relevant for the electronic properties. In other words, defining the cluster structure is equivalent to defining for each atom i those atoms j that are connected to i by a hopping element t. This implies a discretization of the configurational space, since the set of all possible non-equivalent cluster structures is a subset of the set of *graphs* [50] with N vertices [46]. Notice, however, that for the study of clusters we must restrict ourselves to graphs that can be represented as a true *structure* in space. A *graph* is acceptable as a cluster *structure* if a set of atomic coordinates $\vec{R}_i$ ($i = 1, \ldots, N$) exists, such that the interatomic distances R_{ij} satisfy the conditions $R_{ij} = R_0$ if the sites i and j are connected in the graph and $R_{ij} > R_0$ otherwise. For $N \leq 4$ all graphs are possible cluster structures, but for $N \geq 5$ it is not possible, in a 3-dimensional space, to have all atoms being NNs of each other. As shown in Ref. [46], the number of geometry configurations increases extremely rapidly with the number of sites N. For this reason, full geometry optimizations of this kind have only been done for $N \leq 8$ [51] ($N \leq 9$ for $U = 0$ [46]). The optimal cluster structure is determined as a function of U/t and the number of valence electrons ν by comparing the ground-state energies of *all* possible non-equivalent structures. Once the optimal structure is found, the spin correlation functions and the total spin S are calculated, from which the ground-state magnetic order is determined [51].

As important as identifying the ferromagnetic ground state with $S \geq 1$ of small clusters is the determination of the stability of the ferromagnetic state at $T > 0$ and the characterization of the microscopic mechanism responsible for the temperature dependence of S. The energy difference ΔE between the ferromagnetic ground state $E(S \geq 1)$ and the lowest-lying nonferromagnetic state E' with $S = 0$ or $S = 1/2$, namely

$$\Delta E = E' - E(S \geq 1), \tag{21}$$

gives a first estimation of the 'Curie' temperature T_C above which the ferromagnetic order is strongly reduced by thermal excitations. The excitation energy ΔE results mainly from two different contributions. The first one is a purely *electronic* excitation energy ΔE_{el} in which the cluster structure remains fixed to the optimal structure at $T = 0$. The other one is a purely *structural* excitation energy ΔE_{st}, in which the electrons remain in the ground state and the structure is changed until the ground state corresponding to this new structure shows no ferromagnetism. Comparing ΔE_{el} and ΔE_{st} one may conclude whether the electronic excitations, the structural changes or both are principally responsible for the decrease of the cluster magnetization with increasing temperature.

6.3 RESULTS

In the following we present results obtained using the theoretical methods described in the previous section and which are characteristic of the state of the art of various problems.

6.3.1 Magnetic Moments and their Order

In this section we discuss a few representative results for the local magnetic moments $\mu(i)$ of atoms i and for the average magnetic moment per atom $\overline{\mu}_N$ of TM clusters, which were obtained using the self-consistent tight-binding (SCTB) theory presented in Section 6.2.1 (Eqs (2–5)). The parameters required for the calculations are determined as follows. The interatomic hopping integrals $t_{ij}^{\alpha\beta}$ between the orbitals at different atoms are taken from bulk band-structure calculations. For d electrons they are usually given by the canonical two-center approximation in terms of the bulk d bandwidth, W_d, taking into account the dependence on the interatomic distance R_{ij}: $dd(\sigma, \pi, \delta) = (-6, 4, -1)(W_d/2.5)(R_0/R_{ij})^5$, where R_0 is the bulk NN distance, $W_d(\text{Fe}) = 6.0$ eV and $W_d(\text{Rh}) = 7.4$ eV. The direct Coulomb integrals U are estimated from spectroscopic data. Thus, $U_{dd}(\text{Fe}) = 6.0$ eV and $U_{dd}(\text{Rh}) = 7.8$ eV. This leads approximately to local charge neutrality so that the results of SCTB calculations are not very sensitive to the precise value of U [10]. The main parameter for the determination of magnetic properties is the d-electron exchange integral J_{dd}, which is usually fitted to yield the proper bulk moment μ_b at $T = 0$ ($J_{dd}(\text{Fe}) = 0.70$ eV). For nonmagnetic TM's such as Rh the exchange integral is obtained from local-spin-density (LSD) Stoner theory [52] by taking into account a reduction of the LSD approximation results [53] by 20% due to correlation effects. In the case of Rh this yields $J_{dd}(\text{Rh}) = 0.48$ eV [54]. Applying the latter procedure to Fe one obtains $J_{dd}(\text{Fe}) = 0.71$ eV, which is very close to the value derived from the bulk magnetization. If sp electrons are included explicitly in the self-consistent calculations, the corresponding hopping and Coulomb integrals are obtained in a similar way [12, 55].

In Figure 6.1 the local magnetic moments $\mu(i)$ and the average magnetic moment per atom $\overline{\mu}_{15}$ of a bcc-like Fe_{15} cluster are given as a function J/W, where J is the intraatomic d-electron exchange integral and W is the bulk d bandwidth [11]. Figure 6.1 shows how the cluster magnetic moments develop as the exchange interaction increases and allows us to compare cluster and bulk behavior. Notice that J/W varies if the interatomic distances R_{ij} change, since $W \propto (R_0/R_{ij})^5$, or if the spatial extension of the d-electron wave function changes (for example, as one moves from a 3d to a 4d or 5d element within the same group of the periodic table). Therefore, these results are also of interest to illustrate the dependence of cluster magnetism on the interatomic distances and to infer the trends for different TMs. Cluster magnetism sets in for $J > J_c$, where the critical value J_c depends on the size and structure of the cluster. For small bcc clusters J_c is smaller than in the bulk. Therefore, in the case of Fe $\mu(i)$ the average magnetic moment per atom $\overline{\mu}_N$ is larger than in the solid [10]. The same holds for larger bcc Fe clusters, as shown in Table 6.1. Another important consequence derived from these calculations is

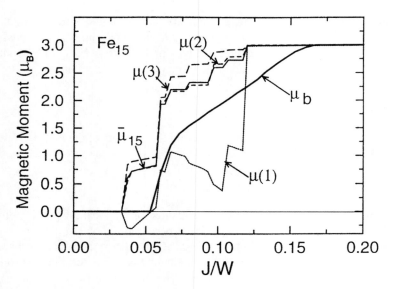

Figure 6.1 Local magnetic moments $\mu(i)$ and average magnetic moment per atom $\bar{\mu}_{15}$ (full curve) of Fe_{15} with bcc-like structure as a function of the ratio between the d-electron exchange integral J and the bulk d-bandwidth W, see Ref. [11]. $i = 1$ refers to the central atom, $i = 2$ to its 1st NN and $i = 3$ to the outer shell of the 2nd NN of the central atom. The corresponding bulk magnetization μ_b is also given

that clusters can be magnetic for values of J/W for which the bulk is still paramagnetic (see Figure 6.1). This phenomenon was experimentally observed for the first time in Rh_N clusters ($N \leq 60$) [18]. The results for Rh_N presented in Table 6.2 and first-principles calculations support this conclusion [23, 54, 56].

In Tables 6.1 and 6.2 we present typical results for the local magnetic moments μ_i, their order and the average magnetic moment per atom $\bar{\mu}_N$ in small Fe_N and Rh_N clusters [10, 11, 54]. For the smaller Fe clusters we obtain saturated magnetic moments which are in good agreement with those calculated by Lee, Callaway and Dhar [5] and observed in experiment [18–20]. As the cluster size increases, the d-band calculations for bcc Fe clusters yield a much more rapid decrease of $\bar{\mu}_N$ than the experimental results [19]. This is probably due to an underestimation of the effective J due to size-dependent screening effects. Comparing the local magnetic moments $\mu(i)$ at different atomic sites i, the magnetic order within the cluster can be inferred. Note that bcc Fe clusters mainly order ferromagnetically. Only for a small range of values of J/W (or NN distances) may the magnetic moment of the central atom point in the direction of the minority spins (see Figure 6.1). Except when the magnetic moments are saturated, $\mu(i)$ depends significantly on the local environment of atom i. The moments at the surface of the clusters are the largest. The results for bcc- and fcc-like Fe clusters demonstrate the sensitivity of the magnetic properties on the atomic structure. Strong dependence on structure and NN distances is also obtained for Rh_N clusters (see Table 6.2) [54, 56]. This kind of behavior is characteristic of itinerant TM magnetism.

The interdependence of cluster structure and magnetism is particularly delicate for transition metals like Mn, Cr or even γ-Fe, which in the solid show antiferromagnetic order. This can be seen for example from the results presented in Table 6.1 for fcc-like

Table 6.1 Size and structural dependence of the magnetic properties of Fe_N clusters. Results are given for the average magnetic moment per atom $\overline{\mu}_N$ (in units of μ_B) and for the local magnetic moments $\mu(i)$ of different cluster atoms i (see Ref. [10]). The considered icosahedral, bcc- and fcc-like cluster structures are composed of a central atom and the successive shells of its nearest neighbors (NN). $i = 1$ refers to the central atom, $i = 2$ to the atoms in the 1st NN shell of atom 1, $i = 3$ to the 2nd NN shell and so on. Positive (negative) signs correspond to moments that are parallel (antiparallel) to $\overline{\mu}_N$

Cluster	$\overline{\mu}_N$	$\mu(1)$	$\mu(2)$	$\mu(3)$	$\mu(4)$	$\mu(5)$
Fe_9 (bcc)	3.00	2.96	3.01			
Fe_{11} (bcc)	3.00	2.96	3.02	2.97		
Fe_{13} (bcc)	3.00	2.95	3.01	2.98		
Fe_{13} (fcc)	2.08	−1.79	2.40			
Fe_{13} (ico)	2.23	−2.20	2.60			
Fe_{15} (bcc)	2.73	1.28	2.88	2.76		
Fe_{19} (fcc)	1.95	−1.28	1.91	2.56		
Fe_{27} (bcc)	2.85	2.88	2.67	2.85	2.97	
Fe_{43} (fcc)	1.23	−1.37	−0.90	0.89	2.49	
Fe_{51} (bcc)	2.45	1.28	1.87	1.57	2.62	2.83
Bulk (bcc)	2.21	2.21				

Table 6.2 Size and structural dependence of the magnetic properties of Rh_N clusters. Results are given for the cohesive energy per atom E_{coh} (in eV), the equilibrium bond length R/R_0 (R_0 = bulk NN distance), the average magnetic moment per atom $\overline{\mu}_N$ (in units of μ_B), and the local magnetic moments $\mu(i)$ of different atoms i (see Ref. [54]). The considered structures are bcc-like (bcc), fcc-like (fcc), icosahedral-like (ico) and twisted double square (tw)

Cluster	E_{coh}	R/R_0	$\overline{\mu}_N$	$\mu(1)$	$\mu(2)$	$\mu(3)$	$\mu(4)$	$\mu(5)$
tw_9	2.38	0.95	0.66	−0.13	0.71	0.79		
bcc_9	2.29	0.88	1.34	0.34	1.44			
ico_{11}	2.43	0.98	0.73	−0.13	0.83			
bcc_{11}	2.36	0.91	0.73	−0.10	0.80	0.57		
bcc_{13}	2.41	0.93	0.57	0.15	0.53	0.76		
ico_{13}	2.38	0.93	0.92	0.09	0.97			
fcc_{13}	2.38	0.96	0.77	0.26	0.81			
fcc_{15}	2.44	0.96	0.80	0.78	0.99	0.92	−0.27	
bcc_{15}	2.36	0.95	1.33	0.05	1.24	1.58		
fcc_{19}	2.52	0.97	0.95	1.16	1.07	0.64		
ico_{19}	2.48	0.89	0.11	0.20	0.14	0.00	−0.07	
bcc_{19}	2.48	0.93	0.21	0.01	0.19	0.05	0.33	0.49
fcc_{43}	2.63	0.98	0.28	0.51	0.26	0.02	0.39	
ico_{55}	2.75	0.96	0.00	0.00	0.00	0.00	0.00	
fcc_{55}	2.68	0.98	0.44	−0.31	0.19	0.15	0.57	0.60
Bulk	3.07	1.00	0.00					

Fe clusters as well as from results obtained for Cr clusters [10]. Notice that fcc-like Fe_{13} and Fe_{19} show antiferromagnetic-like order and that their local magnetic moments μ_i are smaller than in bcc clusters. Moreover, for transition-metal clusters with a tendency towards antiferromagnetic order one expects magnetic frustration effects particularly if the structures are compact. This often leads to more complex magnetic configurations like noncollinear spin arrangements, as was shown in the framework of the single-band Hubbard model [57]. In this context it is also very interesting to study mixed clusters involving ferromagnetic and antiferromagnetic TMs like $(Cr_xFe_{1-x})_N$ [58] as well as Fe_N clusters embedded in a Cr matrix or deposited on a Cr surface.

In Table 6.3 and Figure 6.2 we present results for the local magnetic moments μ_i of Fe_N clusters embedded in a Cr matrix [55]. These materials are relevant from a technological point of view, since their magnetic behavior is very sensitive to the structural and chemical environment of the atoms. Moreover, the competition between the antiferromagnetic order of the matrix and the tendency of Fe_N clusters towards ferromagnetic order leads to particularly interesting magnetic interactions between the cluster and the matrix. The main results may be summarized as follows. (i) As long as the size of the embedded cluster is very small ($N \leq 4$), the magnetic order *within* the Fe_N cluster is antiferromagnetic. For $N \leq 4$, the size of the local magnetic moments $\mu(i)$ is strongly reduced. In other words, the Fe clusters adopt the magnetic order given by the spin-density wave of the matrix. This is in contrast to the behavior observed for free clusters which show ferromagnetic order and $\mu(i) > \mu_b$. (ii) For $N \geq 6$ one observes the expected transition from antiferromagnetic to ferromagnetic order within Fe_N. This is followed by a considerable increase of the local

Table 6.3 Local magnetic moments $\mu(i)$ (in units of μ_B) for bcc Fe_N clusters embedded in Cr, see Ref. [55]. i refers to increasing distance from the Fe atom, $i = 1$. The second row for each N indicates whether atom i is an Fe or a Cr. The structures of the Fe_N clusters are compact substructures of the bcc lattice: distorted tetrahedron ($N = 4$), square bipyramid ($N = 6$) and a central atom with the successive shells of nearest neighbors ($N \geq 9$). The direction of the magnetic moments of Cr atoms close to the Fe/Cr interface is always the same as in pure antiferromagnetic Cr

i / N	1	2	3	4	5	6	7	8	9	10	11	12
1	0.26	−0.61	0.61	0.60								
	Fe	Cr	Cr	Cr								
2	0.26	−0.26	0.61	−0.61	−0.61	0.61	−0.61	0.60	0.60	−0.60		
	Fe	Fe	Cr	Cr	Cr	Cr	Cr	Cr	Cr	Cr		
4	0.26	−0.26	0.62	−0.62	−0.61	0.61	0.60	−0.60	−0.61	0.61	−0.60	0.60
	Fe	Fe	Cr	Cr	Cr	Cr	Cr	Cr	Cr	Cr	Cr	Cr
6	1.20	0.78	−0.60	0.54	−0.58	−0.60	0.58	−0.59	0.58	0.60		
	Fe	Fe	Cr	Cr	Cr	Cr	Cr	Cr	Cr	Cr		
9	2.24	0.61	−0.65	−0.62	0.59							
	Fe	Fe	Cr	Cr	Cr							
15	2.21	1.42	1.59	0.39	−0.60	0.49						
	Fe	Fe	Fe	Cr	Cr	Cr						
27	2.23	2.18	1.57	0.99	−0.75	0.53	0.59					
	Fe	Fe	Fe	Fe	Cr	Cr	Cr					
51	2.42	2.14	2.31	1.97	1.29	−0.89	−0.88	0.50				
	Fe	Fe	Fe	Fe	Fe	Cr	Cr	Cr				

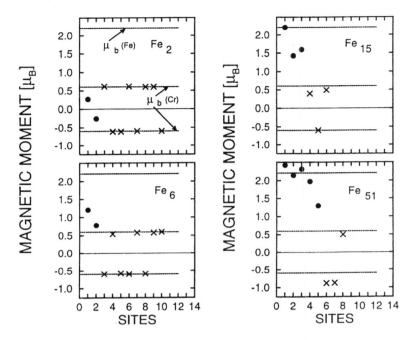

Figure 6.2 Spatial distribution of the local magnetic moments $\mu(i)$ for Fe_2, Fe_6, Fe_{15} and Fe_{51} clusters embedded in an antiferromagnetic Cr matrix. The non-equivalent atomic sites i are ordered according to increasing distance from the central Fe atom of the cluster. Dots correspond to Fe atoms and crosses to Cr atoms close to the cluster–matrix interface. Notice the decrease of the $\mu(i)$ of the Fe atoms close to the Fe–Cr interface

magnetic moments at the Fe atoms, although close to the interface with the matrix the $\mu(i)$ are still smaller than μ_b. This is clearly illustrated by the spatial distribution of $\mu(i)$ given in Figure 6.2. (iii) At the interface there is a strong tendency towards antiferromagnetic coupling between Fe and Cr moments. The magnetic moments $\mu(i)$ of the Fe atoms are very sensitive to the local atomic environment. For larger clusters ($N \geq 9$), the magnetic moment at the center of the cluster is quite close to the Fe-bulk value. However, the average magnetization per atom is always smaller than μ_b (Fe) due to the contributions of interface Fe atoms. The trend is thus opposite to that of free clusters or surfaces. (iv) Fe clusters embedded in a Cr matrix have many properties in common with Fe/Cr multilayers. The antiferromagnetic order in the Cr even close to the interfaces, the antiferromagnetic coupling between Fe and Cr moments at the interface, the strong reduction of the local magnetic moments at Fe atoms close to the interface, the possibility of a slight enhancement of the Fe moment beyond the bulk value at the center of a cluster or film, the strong changes of $\mu(i)$ associated to frustrations, are all characteristics shared by Fe–Cr systems both in the form of embedded clusters and of multilayer structures [59, 60]. One concludes that the immediate local environment of the atoms gives the dominant contribution to the magnetic behavior, even though, for quantitative predictions, the details of the electronic structure and therefore the geometrical structure at a larger length scale are also important.

It is worth recalling that all the results discussed so far were obtained using a mean-field Hartree–Fock-type theory which neglects quantum fluctuations and which tends to overestimate the stability of ferromagnetism. An explicit treatment of electronic correlations and fluctuations seems therefore necessary in order to obtain more definitive quantitative results. Improvements of mean-field theory are expected to affect, for example, the onset of magnetism in 4d TM clusters where the energy of low- and high-spin states are close, the many-body excitations involving states of different spin, and the relative stability of cluster structures showing different magnetic behavior. Electron correlation effects are discussed again in Section 6.3.4 by applying exact diagonalization methods to the single-band Hubbard model.

6.3.2 Magnetic Anisotropy Effects

The magnetic anisotropy is one of the main characteristics of a magnetic material. It determines the low-temperature orientation of the magnetization with respect to the atomic structure of the system and the stability of the magnetization direction in the case of single-domain particles. These are properties of crucial importance in technological applications (e.g. magnetic recording or memory devices) where the magnetization must be pinned to a given direction in space. The magnetic anisotropy of small 3d transition-metal (TM) clusters deserves special attention, not only from a purely theoretical point of view but also because of its implications for cluster-beam Stern–Gerlach experiments [2, 3]. In this section we present results on several magneto-anisotropical properties of transition-metal clusters. First, the spin relaxation process in Stern–Gerlach deflection experiments is analyzed using the model Hamiltonian H' given by Eqs (9)–(12). Then, the magnetic anisotropy energy is studied in the framework of the self-consistent electronic theory presented in Section 6.2.1.

Concerning the Stern–Gerlach deflection experiments, one should first note that the clusters in the beam that cross the inhomogeneous magnetic field are rotating rapidly due to the cluster formation process. As a result of magnetic anisotropy, the average magnetization $\vec{M}$ rotates with the cluster, since $\vec{M}$ is tied to the easy axis $\vec{c}$ (see Eq. (12)

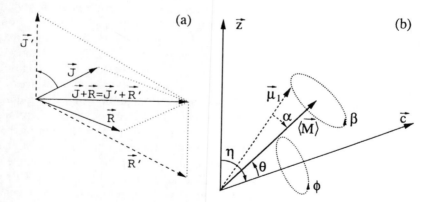

Figure 6.3 (a) Sketch of the change of electronical angular momentum $\vec{J}$ and cluster rotational moment $\vec{R}$ due to alignment of $\vec{\mu} \propto \vec{J}$ in an external magnetic field $\vec{h} \parallel \vec{J}'$. $\vec{J} \to \vec{J}'$ and $\vec{R} \to \vec{R}'$ due to the alignment of the magnetization $\langle \vec{M} \rangle$. The total angular momentum $\vec{J} + \vec{R}$ is conserved. (b) Illustration of the orientation of the atomic magnetic moments $\vec{\mu}_1$, the average magnetization $\langle \vec{M} \rangle$ and the uniaxial c axis of the cluster. The external magnetic field is assumed to be parallel to the z axis

and Figure 6.3). Hence, in an external magnetic field $\vec{h}$ one expects competing effects for aligning $\vec{M}$ in the direction of $\vec{h}$.

The dependence of the Stern–Gerlach deflection profile on the magnetic field is interesting. The anomalous asymmetric deflection observed for magnetic clusters in a collision-free beam [2, 3] implies internal transfer between spin and orbital angular momenta mediated by spin–orbit coupling, since the total angular momentum must be conserved.

The interplay of magnetic anisotropy and coupling to an external magnetic field is shown in Figures 6.4 and 6.5. Depending on the strength of the coupling of the cluster

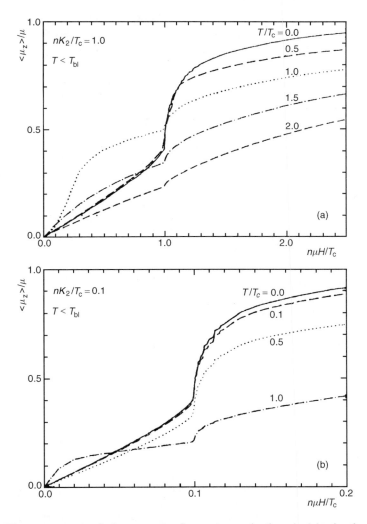

Figure 6.4 The component of the average cluster magnetization $\langle \mu_z \rangle / \mu$ in the direction of the magnetic field H for $T < T_{bl}$ and different anisotropy constant K_2: (a) $nK_2/T_c = 1.0$ and (b) $nK_2/T_c = 0.1$. The blocking temperature is denoted by T_{bl}. The different curves refer to different cluster temperatures T/T_c. μ denotes the magnetic moment per cluster atom. nK_2, $n\mu H$ and T are given in units of the Curie temperature T_c of the Heisenberg model (n = number of cluster atoms)

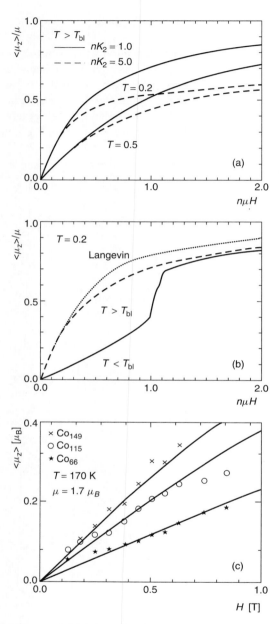

Figure 6.5 (a) The component of the average cluster magnetization $\langle \mu_z \rangle$ in the direction of the magnetic field $\vec{H}$ for two values of the lattice anisotropy constant K_2 and for two cluster temperatures. We use $T \ll T_c$ and give T, $n\mu H$ and nK_2 in units of T_c. (b) Comparison of $\langle \mu_z \rangle / \mu$ for two limiting cases $T < T_{bl}$ and $T > T_{bl}$ and for $nK_2 = 1.0$ and $T = 0.2$ in units of T_c. The dotted curve refers to the Langevin function obtained for $T > T_{bl}$ and $K_2 = 0$. (c) $\langle \mu_z \rangle$ against applied magnetic field H in tesla. Results of our theoretical model are compared with experimental results for Co$_N$ clusters [3]. For the cluster temperature T we use 170 K for all three cluster sizes. The magnetic moment per atom is assumed to be the same as the bulk value $\mu = 1.7\mu_B$. The anisotropy constant K_2 was found to have very little effect in this field range of H, since $T > T_{bl}$

magnetization $\vec{M}$ to the easy axis, the external magnetic field, cluster rotation and temperature, the average projection of the magnetization of the cluster ensemble $\langle \mu_z \rangle$ in the direction of an external magnetic field may be smaller than the one predicted by the Langevin function. One may introduce as in Néel's theory the blocking temperature $k_B T_{bl} = N K_2 / \ln(\tau_m / \tau_0)$, where τ_0 is the gyromagnetic precession time ($\tau_0 \simeq 10^{-9}$ s) and τ_m is the measuring time. Then, for temperatures $T < T_{bl}$, strong magnetic anisotropy effects occur. These are much less pronounced for $T > T_{bl}$ where the clusters behave like superparamagnetic particles as long as $T < T_C$. In Figure 6.4 results are shown for the dependence of $\langle \mu_z \rangle$ of a cluster ensemble on magnetic anisotropy constant K_2 and temperature T. As is physically obvious the deviation of $\langle \mu_z \rangle$ from Langevin behavior gets stronger as K_2 increases. At $n\mu H \sim T_c$ the curve changes to the Langevin curve $\langle \mu_z \rangle$, since then K_2 gets smaller than μH and the latter energy starts to dominate. As temperature T increases the jump in $\langle \mu_z \rangle$ gets smaller, since $K_2 < K_B T$. In Figure 6.5 we show results for $\langle \mu_z \rangle$ of a cluster ensemble for $T > T_{bl}$ and $T < T_{bl}$. For $\mu H > K_2$ the cluster magnetization is uncoupled from the easy axis $\vec{c}$ and then $\langle \mu_z \rangle$ should approach the Langevin behavior. The saturation of $\langle \mu_z \rangle$ as H increases occurs faster for $T \rightarrow 0$ in the case of $T > T_{bl}$ and $K_2 \gg \mu H$. For $T < T_{bl}$ and $T \ll T_c$ note the strong reduction of $\langle \mu_z \rangle$ compared to Langevin behavior. The comparison with experiment shown in Figure 6.5(c) indicates that the Co$_N$ clusters were studied at temperatures $T > T_{bl}$ [61]. This analysis demonstrates that it is important for the interpretation of Stern–Gerlach deflection experiments on clusters to take into account magnetic anisotropy. For rare-earth clusters one expects even larger effects due to the stronger spin–orbit coupling.

The magnetic anisotropy of the clusters can be characterized by a lattice anisotropy field $\vec{H}_a$ describing the alignment of the magnetization along the easy axis $\vec{c}$, see Figure 6.3(b). In the reference frame of the external field $\vec{H}$, the lattice anisotropy $\vec{H}_a$ becomes an oscillating field for rotating clusters acting in addition to $\vec{H}$. As in the case of nuclear magnetic resonance (NMR) experiments, $\vec{H}_a$ acts like a radio-frequency (r.f.) field and will cause a diminished effective cluster magnetization. This effect is strongest for resonance when $\gamma H \sim p\omega_{rot}$ [62]. Here, ω_{rot} is the rotational frequency of the cluster and the factor p depends on the symmetry ($p = 2$ for uniaxial, $p = 4$ for cubic magnetic symmetry, etc.). Using the Bloch equations, $\langle \mu_z \rangle$ has been calculated by Jensen et al. as a function of $\vec{H}_a$ and ω_{rot}. In Figure 6.6 results are shown for $\langle \mu_z(H) \rangle$. These explain fairly well the experimental findings by de Heer et al. [2, 63].

Note that a related behavior is expected for the magnetic susceptibility of liquid-suspended clusters in an oscillating magnetic field [26]. Due to the interplay of exchange coupling, dipolar interactions, magnetic anisotropy and external magnetic field one also expects interesting behavior for an ensemble of magnetic clusters embedded in a metallic matrix [64]. This interplay is reflected, for example, in the magnetoresistance of thin films into which magnetic clusters are embedded.

In Figure 6.7 typical results are shown for the magnetoresistance $\Delta\rho(H) = [\rho(0) - \rho(H)]/\rho(0)$. $\Delta\rho(H)$ starts to decrease drastically as $\vec{H}$ begins to align the magnetic clusters. This occurs for $\mu H > K_2$. Then the external magnetic field overcomes the anisotropy fields characterized by K_2 and which were responsible for a random orientation of the local easy axes of the cluster ensemble. Note that the clusters represent a superparamagnetic system with vanishing magnetization at $H = 0$. Then, for $H > H_c$, the clusters get ferromagnetically aligned in the direction of $\vec{H}$ and thus $\Delta\rho(H) \rightarrow 0$. Note also, for a magnetic cluster deposited on the surface of a crystal, that the magnetic dipole interaction

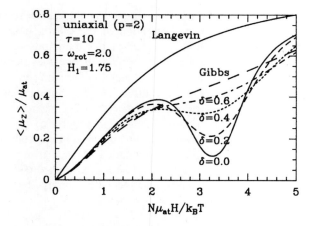

Figure 6.6 Magnetic field dependence of $\langle \mu_z(H) \rangle$ for different distributions $f(\omega_{rot})$ of the cluster rotation frequency ω_{rot}. σ is the width of the normal distribution around the mean value ω_{rot} in units of ω_{rot}. In addition $\langle \mu_z(H) \rangle$ assuming Gibbs distribution with an average equal to ω_{rot} is shown. N is number of cluster atoms. τ in units of $2\pi/\omega_{rot} = (2\pi N \mu_{at}/\gamma k_B T_{vib})$ and H_1 in units of $(k_B T_{vib}/N\mu_{at})$ refer to the relaxation time and the rotating magnetic field, respectively. T_{vib} is the vibrational cluster temperature

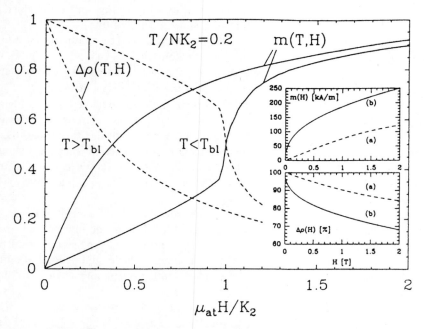

Figure 6.7 Magnetization $m(T, H)$ and the corresponding magnetoresistance $\Delta\rho(T, H)$ as a function of the external magnetic field H of an ensemble of clusters each having N atoms in a matrix with random orientation of the anisotropy easy axes. The cluster temperature is chosen to be $T = 0.2NK_2$, where K_2 is the anisotropy constant. The magnetization is calculated for statistical equilibrium ($T \gg T_{bl}$) and for strong blocking ($T \ll T_{bl}$), where T_{bl} is the blocking temperature. A linear dependence between $m(T, H)$ and $\Delta\rho(T, H)$ is assumed. The inset indicates measurements of $m(H)$ and $\Delta\rho(T, H)$ for layered perovskites for a sample (a) as deposited and (b) annealed

between the clusters comes into play. Depending on the distance between the clusters and on temperature, one may arrange a magnetization that is perpendicular or parallel to the surface plane, for example.

In Figure 6.8 results are given for the magnetic shell structure of clusters. This results from the dependence of the magnetic moments μ_i on the atomic environment. We use

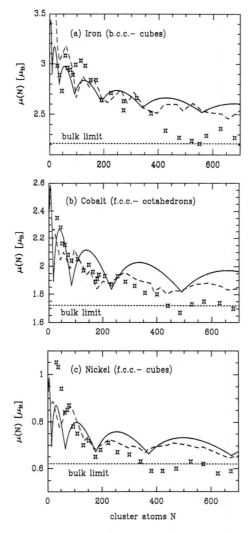

Figure 6.8 Results for the shell structure of the magnetic moment μ_N of (a) Fe, (b) Co and (c) Ni clusters as a function of the number N of cluster atoms. A bcc-cube for Fe clusters, an fcc-cube for Ni clusters and an fcc-octahedron for Co clusters is assumed. The solid curves are calculated assuming a statistical shell-by-shell growth of the cluster. Only cluster structures with completely filled atomic shells are considered. For Fe the values $\mu_{at} = 4.0\mu_B$, $\mu_{surf} = 3.0\mu_B$ and $\mu_{bulk} = 2.21\mu_B$; for Co $\mu_{at} = 3.0\mu_B$, $\mu_{surf} = 1.9\mu_B$ and $\mu_{bulk} = 1.72\mu_B$; and for Ni $\mu_{at} = 1.2\mu_B$, $\mu_{surf} = 0.7\mu_B$ and $\mu_{bulk} = 0.62\mu_B$ are used. The dashed curves are calculated with magnetic moments $\mu(z_i)$ which depend on the number of nearest neighbors q_i of lattice site i, see Ref. [24]. The stars refer to measurements of transition metal clusters [19, 20]

two models. First, we assume shell-by-shell growth and calculate the moments of the topmost atomic shell by

$$\bar{\mu}_o = x_o(1 - x_o)\mu_{at} + x_o^2\mu_s,$$

where x_o is the concentration of occupied sites in this shell, and the moments in the next shell by

$$\bar{\mu}_1 = (1 - x_o)\mu_s + x_o\mu_b.$$

Furthermore, all $\mu_i = \mu_b$ for the inner shells. Secondly, we use $\mu_i = \mu_1$, if one has for the coordination number $q_i \leq q_c$, and $\mu_i = \mu_2$, if $q_i > q_c$. There are discrepancies between the theoretical results obtained by using these models and the experimental ones. This needs further analysis, since it might advance our understanding of the formation of moments in clusters. For details see Ref. [24].

The magnetic anisotropy and the resulting coupling of the magnetization to rotations and internal vibrations provide a natural mechanism for describing the phenomenon of spin relaxation within *isolated* clusters in an external magnetic field and which was first observed in Fe_N by de Heer *et al.* [2, 3]. Furthermore, the magnetic anisotropy energy (MAE) — the energy involved in rotating the magnetization from a low-energy direction (easy axis) to a high-energy direction (hard axis) — is one of the key parameters characterizing the dynamics of rotating clusters in a Stern–Gerlach magnet. It determines the blocking temperature above which a superparamagnetic behavior holds and may lead to a resonance-like coupling between the rotational frequency and the Zeeman splittings [2, 3, 14, 15, 24, 25, 63]. From these investigations as well as from Mössbauer studies on supported Fe nanoparticles [65] it was inferred that the MAE of TM clusters is considerably larger than in the corresponding crystals. It is therefore very interesting to determine the MAE using the self-consistent electronic theory of Section 6.2.1. In particular, this will demonstrate how the MAE depends on cluster size, structure, bond length and d-band filling.

In Figure 6.9 results are given for the MAE ΔE, the orbital angular momentum $\langle L_\delta \rangle$ along the magnetization direction δ ($\delta = x, y, z$) and for the average spin projection $\langle S_z \rangle$ of an Fe_4 cluster. The results are representative of a large number of studied cluster sizes and structures [13]. The two-center hopping integrals and the intra-atomic Coulomb integrals used in the calculations are the same as in Ref. [10]. The value of the spin–orbit coupling constant corresponding to Fe is $\xi = 0.05$ eV [13]. The results are given as a function of the bond length d in order to analyze the role of cluster relaxation and to infer the possible coupling of the magnetization direction to vibrations and distortions. Only $\langle S_z \rangle$ is shown, since the magnitude of the spin magnetization $|\langle \vec{S} \rangle|$ depends very weakly on the direction of the magnetization (typically, $|\langle S_z \rangle - \langle S_x \rangle| \sim 10^{-3} - 10^{-4}$). Comparing Figures 6.9(a) and 6.9(c) it is clear that the variations of the MAE are related to the variations of $\langle S_z \rangle$ and to the resulting changes in the electronic spectrum. For large values of d/d_B the spin magnetic moments are saturated, i.e. $\langle S_z \rangle \simeq (10 - \gamma_d)/2 = 3/2$. When d/d_B decreases, discrete changes in the spin polarization, $\Delta \langle S_z \rangle$, occur and nonsaturated spin magnetizations are obtained (see Section 6.2.1). For constant values of $\langle S_z \rangle$, the MAE and orbital momentum $\langle L_\delta \rangle$ vary continuously, since the electronic spectrum and the local magnetic moments are continuous functions of d/d_B. However, this is not the case when $\langle S_z \rangle$ changes, since then a strong and discontinuous redistribution of the spin-polarized electronic density takes place. Important changes in the energy-level structure around the Fermi energy ε_F occur and these modify the details of the spin-orbit (SO) mixing. The

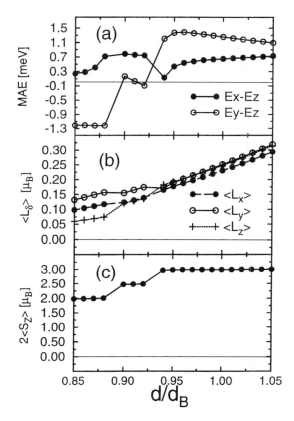

Figure 6.9 (a) Magnetic anisotropy energy (MAE), (b) orbital magnetic moment $\langle L_\delta \rangle$ and (c) spin magnetic moment $\langle S_z \rangle$ of Fe_4 with rhombohedral structure as a function of the bond length d (d_B = bulk NN distance). The magnetization direction $\delta = x$ is along the middle bond, the direction $\delta = y$ is perpendicular to x and within the plane of the cluster, and $\delta = z$ is perpendicular to the cluster plane spanned by x and y. (see Ref. [13])

resulting changes in the electronic energy depend on the explicit form of H_{SO} and therefore on the direction of the magnetization (see Figure 6.9). Consequently, very significant and discontinuous variations of the MAE are observed which may even lead to a change of sign of the MAE as the spin moment $\langle S_z \rangle$ decreases. Notice that the rhombohedral Fe_4 cluster presents a remarkable in-plane anisotropy $(E_x - E_y) \simeq 0.4$ eV. A similar situation is found for other clusters and band-fillings [13]. This indicates that uniaxial anisotropy models are not directly applicable to clusters that have strongly reduced symmetry.

In Table 6.4 results are given for the MAE of Fe clusters with different structures and interatomic distances [13]. The following general trends may be summarized. (i) The MAE is much larger in small clusters than in the corresponding crystalline solids. In fact, the anisotropy energy ΔE is often even larger than in thin films. For instance, values of $\Delta E \sim 4$–5 meV are typical for clusters (see Table 6.4). These conclusions are in agreement with experiments on free clusters [15, 24, 63] and supported Fe nanoparticles [65]. (ii) ΔE depends much more sensitively than the magnetic moments on the geometrical structure of the cluster. Indeed, changes of sign in ΔE are found as a function of the

Table 6.4 Size and structural dependence of the magnetic anisotropy energy (MAE) of Fe_N clusters. The off-plane MAE $\Delta E = E_x - E_z$ and the in-plane MAE $\Delta E = E_x - E_y$ (results in brackets) are given in meV for different values of the interatomic bond-length d (d_B = bulk NN distance). Different structures are considered for each cluster size: (a) triangle, (b) chain, (c) square, (d) rhombus, (e) trust, (f) square pyramid, (g) triangle, (h) square bipyramid, (i) hexagon and (j) pentagonal bipyramid (see Ref. [13])

N	Structure	$\Delta E(d/d_B = 1.05)$	$\Delta E(d/d_B = 1.00)$	$\Delta E(d/d_B = 0.90)$
3	(a)	5.13 (1.43)	5.30 (−1.01)	−0.11 (−0.10)
	(b)	2.01	1.69	1.25
4	(c)	5.02 (−0.48)	5.21 (−0.59)	−1.03 (−0.11)
	(d)	0.30 (−0.12)	0.27 (−0.12)	0.33 (0.31)
5	(e)	0.29 (−1.01)	0.28 (−0.82)	−0.10 (−0.75)
	(f)	1.09 (−0.01)	0.88 (−0.03)	−0.07 (0.04)
6	(g)	−1.25 (−0.76)	−1.12 (−0.62)	0.33 (0.29)
	(h)	4.66 (0.04)	4.82 (−0.02)	−0.07 (−0.23)
7	(i)	1.78 (0.40)	1.91 (0.09)	2.22 (0.30)
	(j)	4.32 (−0.03)	4.52 (−0.02)	−0.39 (0.00)

interatomic distance d even in situations where the magnetic moments are saturated and consequently do not depend on the cluster structure. (iii) The in-plane MAEs are considerably important in general. In some cases they are even larger than the usually considered off-plane anisotropy. The in-plane MAE is of course largest for low-symmetry structures and decreases, though not monotonically, as the angle between non-equivalent x and y directions decreases. Experiments support these conclusions.

Further studies of the MAE of transition-metal clusters are worthwhile and necessary. In particular, they should help to improve our understanding of the magneto-anisotropic behavior of clusters embedded in a matrix and of various nanostructures such as thin wires and inhomogeneously growing films.

6.3.3 Magnetic Properties at Finite Temperatures: Spin-Fluctuation Effects

The study of spin fluctuations in clusters and the resulting temperature dependence of the magnetic properties is a subject of considerable current interest. In this section we discuss results derived from the spin-fluctuation theory described in Section 6.2.2 which takes into account both the fluctuations of the magnetic moments and the itinerant character of the d electron states.

In Table 6.5 results are given for the average cluster magnetization per atom (in μ_B) of Fe_N and Ni_N clusters at temperatures $T \geq T_C(N)$ as calculated using Eq. (17). The value of the magnetic moments at $T = 0$ is taken from experiment [19, 20]. The clusters are assumed to consist of ferromagnetic domains. The degree of SRMO within the cluster at $T > T_C$ is characterized by the average number of atoms ν_{sr} in a ferromagnetically ordered domain. In particular, the disordered-local-moment picture (no SRMO) corresponds to $\nu_{sr} = 1$. The results shown in Table 6.5 demonstrate the importance of short-range magnetic order in TM clusters. Indeed, very good agreement with experiment is found for Fe_N with $\nu_{sr} = 15$ and for Ni_N with $\nu_{sr} = 19$–43 [42]. In contrast, the results obtained neglecting SRMO and hence using $\nu_{sr} = 1$ compare very poorly with experiment. This rules out the disorder-local-moment picture for TM clusters at $T > T_c$, as is already

Table 6.5 Average magnetization per atom $\overline{\mu}_N(T > T_C)$ of Fe_N and Ni_N clusters with short-range magnetic order (SRMO) at high temperatures (in units of μ_B) as obtained from Eq. (17). ν_{sr} refers to the number of atoms within a ferromagnetically ordered domain due to SRMO. We use $\nu_{sr} = 15$ for Fe_N and $\nu_{sr} = 19$–43 for Ni_N, see Ref. [42]. Results are also given for $\nu_{sr} = 1$, which corresponds to neglecting SRMO. In the case of Fe_N the ranges of theoretical values given for $\overline{\mu}_N$ result from the range of cluster sizes N. The experimental results are estimated from Ref. [19]

	N	$\overline{\mu}_N(\nu_{sr} = 1)$	$\overline{\mu}_N(\nu_{sr})$	$\overline{\mu}_N$(Experiment)
Fe_N	50–60	0.47–0.38	1.61–1.48	1.6±0.2
	82–92	0.33–0.31	1.26–1.20	1.2±0.2
	120–140	0.27–0.25	1.05–0.97	0.9±0.1
	250–290	0.16–0.15	0.63–0.59	0.4±0.05
	500–600	0.10–0.09	0.38–0.36	0.4±0.05
Ni_N	140–160	0.06–0.05	0.39–0.24	0.36±0.16
	200–240	0.05–0.04	0.33–0.20	0.24±0.16
	550–600	0.03–0.02	0.17–0.11	0.11±0.08

known to be the case for the bulk and near the surfaces of macroscopic TMs [39–41]. The degree of SRMO derived for the clusters seems to be somewhat larger than what is generally accepted in the corresponding solids. This could be related to the increase of the local magnetic moments and exchange-splittings as calculated for $T = 0$ (see Section 6.3.1). Thus, SRMO is expected to play a significant role for the finite-temperature behavior of magnetic TM clusters.

In order to discuss the temperature-induced spin-fluctuations in small Fe_N and Ni_N clusters we consider the low-temperature limit of the local free-energy $F_i(\xi)$ (Eq. (16)). For these clusters one always obtains $\Delta F_i(\xi) = F_i(\xi) - F_i(\xi = \mu_i^0) > 0$, which indicates, as expected, that the ferromagnetic order is stable at low temperatures. Comparing the behavior of $\Delta F_i(\xi)$ for different atoms i within the ferromagnetic cluster provides useful information concerning the stability of the local magnetizations and its environmental dependence.

In Figure 6.10 results are given for $\Delta F_i(\xi)$ at the central atom and at one of the surface atoms of an Fe_9 cluster with assumed bcc-like structure. The parameters used for the calculations are the same as in Ref. [10], namely bulk d bandwidth $W = 6.0$ eV, direct Coulomb integral $U = 6.0$ eV and exchange integral $J = 0.70$ eV. Two values are considered for the NN distance: $d/d_B = 1.0$ and $d/d_B = 0.92$, where d_B refers to the NN distance in bulk Fe. For the surface atoms, which have the largest local magnetic moments μ_i^0 at $T = 0$, $F_i(\xi)$ shows two minima located at the molecular fields $\xi^+ = \mu_i^0$ and $\xi^- \simeq -\mu_i^0$ [35]. This double-minimum structure indicates that the flips of the surface magnetic moments keeping their amplitude approximately constant are the dominant magnetic excitations. In contrast, for the central atom which has a much smaller local magnetic moment at $T = 0$, one finds that $F_i(\xi)$ has one single minimum. This indicates that here the amplitude fluctuations of the local moments dominate. At the surface of the cluster only small fluctuations of ξ are possible with an excitation energy $\Delta F_i(\xi)$ smaller than the energy $\Delta F_i(\xi^-) = F_i(\xi^-) - F_i(\xi^+)$ required to flip a local magnetic moment. The probability $P_i(\xi) \propto \exp\{-\beta \Delta F_i(\xi)\}$ has two sharp maxima at $\xi \simeq \xi^+$ and $\xi \simeq \xi^-$ with $P_i(\xi^-) \gg P_i(\xi = 0)$. The fact that very small clusters and particularly cluster

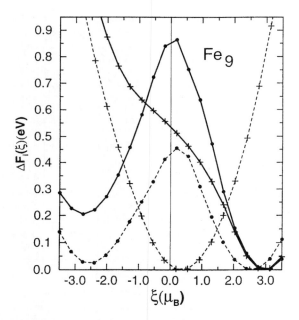

Figure 6.10 Local spin-fluctuation energy $\Delta F_i(\xi) = F_i(\xi) - F_i(T = 0)$ as a function of the exchange field ξ at different atomic sites i of Fe_9 with bcc-like structure. Crosses refer to the central atom and dots to one of the eight surface atoms (NN of the central one). Results are given for $d/d_B = 1.0$ (full curves) and $d/d_B = 0.92$ (dashed curves), where d refers to the NN distance (d_B = bulk NN distance)

atoms having small local coordination numbers z_i show such a Heisenberg- or Ising-like behavior is not surprising, since the kinetic energy loss caused by flipping a local magnetic moment ($\xi \simeq \mu_i \to \xi \simeq -\mu_i$) is smaller when z_i is smaller ($E_K \propto \sqrt{z_i}$). At the same time the exchange energy $\Delta E_x = (J/4) \sum_i \mu_i^2$, being a local property, is much less affected by the change of sign of ξ. Even Ni, which in the solid state has a single minimum in $F_i(\xi)$ and which is therefore dominated by amplitude fluctuations of the local magnetic moments [38], tends to show a double-minimum structure in $F_i(\xi)$ for sufficiently small N. These results indicate the existence of a transition from Heisenberg-like to itinerant-like behavior with increasing cluster size [35]. Physically this can be interpreted as a consequence of the competition between the Coulomb interaction energy, which is relatively more important in small clusters, and the kinetic d-band energy, which is more important in the bulk.

It is also interesting to determine how $\Delta F_i(\xi)$ depends on the interatomic distances in order to infer effects due to structural distortions on the cluster magnetization curves. If the free energy is optimized by changing the NN distances and by keeping the cluster symmetry unchanged (uniform relaxation) usually a bond-length contraction is obtained ($d < d_B$) [10]. Therefore, we also show in Figure 6.10 results for Fe_9 using $d/d_B = 0.92$ [10]. In this case one finds no qualitative, but strong quantitative, changes in $F_i(\xi)$ as a function of ξ. Besides the shift of the position of the minimum at ξ^+ reflecting the reduction of the local magnetic moments μ_i at $T = 0$, one observes a remarkable reduction (by a factor of about 10) of the free energy $\Delta F_i(\xi^-) = F_i(\xi^-) - F_i(\xi^+)$ required to flip a local magnetic moment (or to change the exchange field from ξ^+ to ξ^-) at the surface atoms. A similar large reduction of the Curie temperature $T_C(N)$ is expected to occur upon relaxation, since in first approximation $T_C(N) \propto \Delta F_i$. Once more, the results clearly show the strong sensitivity of the magnetic properties of 3d TM clusters to changes in the local environment. Recent calculations including correlations effects exactly within the single-band Hubbard model [51] have also revealed the importance of structural changes and structural fluctuations for the temperature dependence of the magnetic properties of clusters. As in the unrelaxed case, $F_i(\xi)$ depends strongly on the position of the atom i within the cluster. Comparing $\Delta F_i(\xi^-)$ at different atomic sites i one observes that $\Delta F_i(\xi^-)$ does not scale simply with the local coordination number z_i. This implies that the effective exchange coupling between local magnetic moments depends on the local environment. A similar behavior is found near the surface of macroscopic TMs [41].

For clusters one would like to characterize the stability of magnetism at finite temperatures by a size-dependent 'Curie' temperature $T_C(N)$. However, the critical behavior of finite systems is different from that of the bulk [66]. Strong deviations from bulk-like behavior are expected to occur when the correlation length $\lambda(T) \sim (T - T_C)^{-\gamma}$ becomes of the order of the cluster radius R. This is the case at temperatures $T \leq T^*$, where $(T^* - T_C) \sim R^{-1/\gamma}$. The divergence at T_C in the specific heat $C_V(T)$ and magnetic susceptibility $\chi(T)$ disappear, since the long-wavelength magnetic fluctuations are cutoff by the finite size of the cluster (i.e. $k \geq k_{min} \sim 1/R$). Instead, these properties have a peak at $T = T_C$ with a size-dependent width. Notice moreover that it might be difficult to define a unique critical temperature $T_C(N)$, since the position of the peak in $C_V(T)$ and $\chi(T)$ can be different [66]. The understanding of the size dependence of the 'Curie' temperature $T_C(N)$ of TM clusters remains a problem. From the experimental results on $\overline{\mu}_N(T)$ of free TM clusters [19–21] it is difficult to derive quantitative results on the size dependence of $T_C(N)$. To our knowledge, measurements of $C_V(T)$ or $\chi(T)$ are not available at present. In the case of Co_N the experimental results indicate that $T_C(N) > 1000$ K for $50 \leq N \leq 600$ [20] (T_C(Co-bulk) $= 1388$ K]. In Ni_N, with $140 \leq N \leq 600$, $T_C(N)$ seems to be of the same order of magnitude as T_C(Ni-bulk) $= 627$ K, though maybe somewhat smaller for $200 \leq N \leq 600$. Large experimental uncertainties for $T \geq 400$ K make estimates of T_c very difficult [20]. Fe clusters show the most interesting behavior. In this case the temperature and size dependence of the magnetization indicates that $T_C(N)$ is much smaller than T_C(Fe-bulk) $= 1043$ K and that T_c decreases with increasing N for $250 \leq N \leq 600$. It has been suggested that this anomalous behavior could be related to a temperature-induced structural change [20]. Note that a structural change is observed for thin Fe films.

6.3.4 Effects of Electron Correlations and Structure on Cluster Magnetism

In the following we review the main results of a recent study of electron correlations and cluster structure which was performed by using the Hubbard Hamiltonian (Eq. (18)) and by treating the electronic problem and the geometry optimization exactly and on the same level [51]. The results for the most stable structures and the corresponding total spin S as a function of the Coulomb repulsion strength U/t and the number of electrons ν are summarized in the form of 'phase' or structural diagrams. Physically, the variations of U/t can be viewed as corresponding to changes in the interatomic distance ($t_{ij} \propto R_{ij}^{-5}$ for TMs) or to changes in the spatial extension of the atomic-like valence wave function as occur for different elements within the same group. The value of v can be related either to simple-metal clusters in different ionic states ($v = N, N \pm 1$) or more indirectly to different band-fillings v/N as in a TM d series. An *ab initio* determination of the effective interaction parameters (U/t or Heisenberg's $J = t^2/U$) is in principle possible for some specific systems [67]. However, to examine the physical behavior as a function of U/t and v/N has a more universal character.

As shown in Figures 6.11 and 6.12, for low concentrations of electrons or holes [68] ($v/N \leq 0.4$–0.6 and $2 - v/N \leq 0.3$–0.6) the optimal cluster structure is independent of U/t. This means that the structure that yields the minimal kinetic energy (uncorrelated limit) remains the most stable one irrespective of the strength of the Coulomb interactions [69]. Furthermore, no magnetic transitions are observed: the ground state is always a singlet or a doublet. This indicates that for low carrier concentrations the Coulomb interactions are very efficiently suppressed by correlations, so that the magnetic and geometric structure of the clusters are dominated by the kinetic term. These trends are physically plausible, since the effects of electron–electron interactions should be roughly proportional to $(v/N)^2$, and can be justified by general analytical results [51]. Nonetheless, the fact that this holds for finite values of v/N and for $U/t \to +\infty$ does not seem obvious *a priori*.

For small v compact structures are obtained which have maximal average coordination number $\bar{z}$ ($t > 0$). For example, for $v = 2$, the ground-state structures are the triangle, tetrahedron, triangular bipyramid, capped bipyramid and pentagonal bipyramid for $N =$

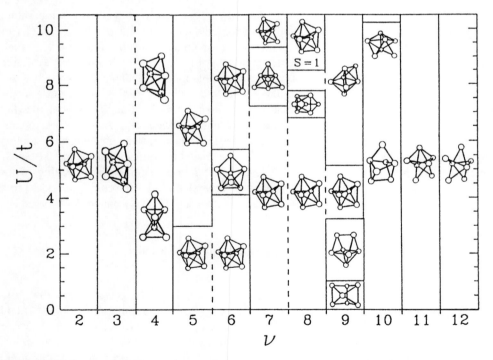

Figure 6.11 Phase diagram of clusters having $N = 7$ atoms, see Ref. [51]. The ground-state structures obtained for a Coulomb repulsion U, hopping integral t and total number of electrons v are illustrated. The ground-state spin S is minimal (i.e. $S = 0$ or $S = 1/2$) unless indicated. See also Figure 6.12

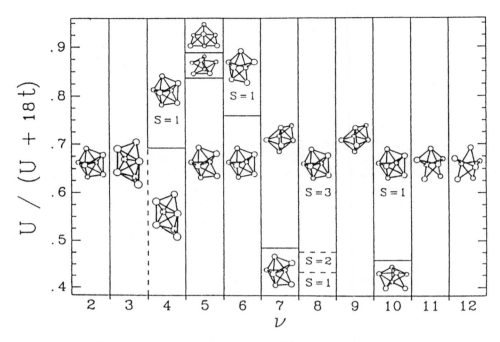

Figure 6.12   Phase diagram of clusters treated using the Hubbard Hamiltonian, see Ref. [51], having $N = 7$ atoms and $U/t > 10$. As in the previous figure, the ground-state spin S is indicated only if $S \geq 1$

3–7, respectively. These are all substructures of the icosahedron and have the largest possible number of *triangular* loops. In contrast, for large ν (small $\nu_h = 2N - \nu$) rather open ground-state structures are found (see Figures 6.11 and 6.12). In particular for $\nu_h = 2$ we obtain bipartite structures that have the largest possible number of *square* loops ($N \geq 4$). As indicated above, this can be qualitatively understood in terms of the single-particle spectrum. In the first case (small ν) the largest stability corresponds to the largest bandwidth for bonding states having a negative energy $\varepsilon_b \leq -\bar{z}t$. This is achieved by the most compact structure. In the second case (small ν_h) the largest stability is obtained for the largest bandwidth for antibonding (positive-energy) states, for the most compact *bipartite* structure.

A much more interesting interplay between electronic correlations, magnetism and cluster structure is observed around half-band filling (i.e. $|\nu/N - 1| \leq 0.2$–0.4). Here, several structural transitions are found as a function of U/t. Starting from the ground-state structures for $U = 0$ [46], we typically observe that as U is increased first one or more of the weakest bonds are broken to achieve stable structures. This change to more open structures occurs for $U/t \simeq 1$–4 and most frequently for $\nu < N$. In this case the $U = 0$ structures are more compact, while for $\nu > N$ the structures are already rather open for $U \to 0$ (see, for example, Figure 6.11). The trend to break bonds as U/t increases seems physically plausible, since the Coulomb interactions usually tend to weaken the kinetic energy contributions. However, if U is further increased ($U/t > 5$–6) it is energetically more advantageous to create additional new bonds. This can be interpreted as a consequence of the electronic correlations. Higher coordination gives the strongly correlated electrons more possibilities for mutually avoiding motions that lower the kinetic energy without excessively increasing the Coulomb energy due to local charge-fluctuations. Moreover, these compact structures are more symmetric. Therefore, the electron-density distribution is more uniform, which also contributes to a lowering of the static Coulomb-repulsion energy.

The structural changes at larger U/t are often accompanied by strong changes in the magnetic behavior. One may say that such structural changes are driven by magnetism. For half-band filling ($\nu = N$) the optimal structures have the minimal total spin S and for large U/t strong antiferromagnetic correlations. The optimal antiferromagnetic structures are *nonbipartite* (for example, the rhombus is more stable than the square for $N = \nu = 4$). The bonds that are 'frustrated' in a static picture of antiferromagnetism yield an appreciable energy-lowering when quantum fluctuations are taken into account. Therefore, clusters with one electron per site and large U/t can be regarded as frustrated quantum antiferromagnets.

For all cluster sizes studied ($N \leq 8$) the most stable structures for $\nu = N + 1$ show ferromagnetism for large enough U/t (typically $U/t > 4$–14). Increasing U/t further leads to a fully polarized ferromagnetic state ($S = (N - 1)/2$) that is in agreement with Nagaoka's theorem [70]. The value of the Coulomb interaction strength U_c/t, above which saturated ferromagnetism dominates, increases with increasing N. This is a consequence of the fact that the antiferromagnetic correlations tend to dominate as we approach half-band-filling ($\nu/N = 1 + 1/N$). U_c/t is approximately proportional to N and diverges in the thermodynamic limit. For the smaller clusters (viz. $N = 3$, 4 and 6) the $\nu = N + 1$ band-filling is the only case where the optimal structures are ferromagnetic. For $N = 5$ unsaturated ferromagnetism ($S = 1$) is also found for $\nu = 4$, though at rather large values of U ($U/t > 30$). However, for larger clusters, ferromagnetism extends more and more throughout the (U/t)–(ν/N) phase diagram. Indeed, for $N = 7$ ferromagnetism is found for $\nu = 4, 6, 8$ and 10, as shown in Figure 6.12. Clusters with $N = 8$ can be ferromagnetic for $\nu = 9$–12 [51]. The tendency towards ferromagnetism is much stronger above half-band-filling than for $\nu \leq N$. This consequence of the asymmetry of the single-particle spectra of compact structures is qualitatively in agreement with experiments on 3d TM clusters. In fact one observes, as already discussed in previous sections, that the magnetization per atom in V and Cr clusters is very small if not zero ($\overline{\mu}_N < 0.6$–$0.8\mu_B$) [17], while Fe, Co, Ni and even Rh clusters show large magnetizations [3, 19, 21]. In any case, it is worth noting that ground-state ferromagnetism is much less frequent than what one would expect from mean-field Hartree–Fock arguments (Stoner criterion).

Recently, the calculations have been extended to larger sizes in the range $N = 12$–14 by considering a few representative cluster geometries, like icosahedral, hcp, fcc and bcc structures [71]. These exact diagonalizations for larger clusters, which are very demanding in view of the size of Hilbert space, confirm the trends derived from the exhaustive geometry optimizations for $N \leq 8$ [51]. For example, for $N = 13$ sites one obtains that the icosahedron is the most stable geometry among the considered ones for $\nu \leq 9$, while for $\nu \geq 21$ the bipartite bcc structure dominates. This holds irrespective of the value of U/t. In contrast, a variety of structural and magnetic transitions is found around half-band-filling. For example, for small U/t the fcc clusters yield the lowest energy for $11 \leq \nu \leq 15$ and the hcp clusters for $16 \leq \nu \leq 19$. As U/t is increased ground-state ferromagnetism develops and the cubic structures are displaced by the icosahedron which becomes the most stable of the considered geometries. As before, the tendency towards ferromagnetism is much larger above half-band-filling. Saturated ferromagnetism in particular is quite common for sufficiently large U/t and $v/N > 1$, but is never found for $v \leq N$. Note this is in contrast to the ferromagnetic ground-state predicted in unrestricted Hartree–Fock calculations [57] and illustrates the drastic effects of electron correlations for large U/t.

It is interesting to compare exact and Hartree–Fock (HF) results of the Hubbard model in order to infer the possible limitations of mean-field calculations for more realistic models. The following main conclusions can be drawn:

(i) The HF approximation yields ferromagnetism at much smaller values of U/t than the exact results. For example, in compact structures below half-band-filling the ferromagnetic solution always dominates for large U/t, whereas the exact ground-state spin is in most cases minimal (i.e. $S = 0$ or $1/2$) even for $U/t \rightarrow +\infty$.

(ii) The magnetic behavior obtained using HF generally coincides with the exact result above half-band-filling and for large U/t where ferromagnetism dominates. However, the energy differences between high- and low-spin states are severely overestimated in HF.

(iii) Since the optimal cluster structures corresponding to low- and high-spin states are very different, the HF calculations fail to predict the correct ground-state structure when the HF and the exact ground-state spins are different. However, HF is quite successful in yielding the most stable structure when ferromagnetism dominates in the exact result.

(iv) Subtle changes in the ground-state structures such as the rearrangement of surface bonds found for intermediate values of U/t are not reproduced by the HF calculations. The error in the HF ground-state energy precludes us from properly discerning the small energy differences involved.

(v) Allowing for noncollinear spin arrangements considerably improves the accuracy of the HF results for compact structures with a tendency towards antiferromagnetism.

These conclusions illustrate the importance of correlations in low-dimensional systems [43–45, 51]. Similar trends are expected to hold for the realistic spd-band model given by Eq. (1). However, it should also be noted that the Hubbard model with the restriction to one orbital per site tends to exaggerate the effects of quantum fluctuations. Extensions of the model by including either several bands or nonlocal interactions should tend to weaken strong fluctuation effects. Moreover, the validity of the Hartree–Fock approximation is considerably improved by using effective interaction parameters that are fitted to experimental bulk results like the bulk magnetic moment μ_b and thus include part of the electron correlations.

In Figure 6.13 the energies ΔE_{el} and ΔE_{st} resulting from only electronic or structural changes, respectively, are shown for $v = N + 1$ and $N \leq 8$. Two values of U/t are considered: $U/t = 16$ which yields saturated ferromagnetism for $N \leq 5$ and a nonsaturated ferromagnetic ground state for $6 \leq N \leq 8$, and $U/t = 64$ which is representative of the strongly correlated limit for which the ground-state spin is saturated for all $N \leq 8$. Results for ΔE for intermediate values of U/t lie between the curves shown and can be inferred from the figure. As in the case of the exchange-field fluctuation energy $\Delta F_i(\zeta)$ (Section 6.3.3), the energies $\Delta E_{el}(N)$ and $\Delta E_{st}(N)$ are very sensitive to changes in the interatomic distances, in particular when the reduction of U/t implies a transition from saturated to unsaturated ground-state ferromagnetism. Concerning the size dependence one finds that ΔE shows strong even–odd oscillations as a function of N, which are due to the size dependence of the final-state singlet or doublet energy $E(S = 0, 1/2)$. The

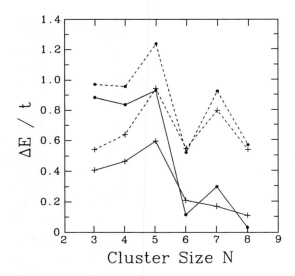

Figure 6.13 Size dependence of the spin excitation energy $\Delta E = E' - E(S)$, which is the energy difference between the ferromagnetic ground-state $E'(S)$, $(S \geq 1)$ having the optimal cluster structure and the lowest-lying nonferromagnetic state E' with $S = 0$ or $S = 1/2$. Dots correspond to a purely *electronic excitation* and crosses to a purely *structural change*. Results are obtained (see Ref. [51]) using the Hubbard model (Eq. (18)) and a Coulomb-repulsion energy $U/t = 16$ (full lines) or $U/t = 64$ (dashed lines). The number of electrons is $\nu = N + 1$. The ground-state spin S is maximal, except for $U/t = 16$ and $N \leq 6$, where $S = 3/2$ or 2

doublets, which correspond to an even number of sites ($\nu = N + 1$), are relatively more stable than the singlets (odd number of sites). This trend is opposite to the one observed at half-band-filling ($\nu = N$), where the singlets (ν even) are relatively more stable than the doublets (ν odd). Moreover, the amplitude of the even–odd oscillations *increases* with decreasing $J = 4t^2/U$ (i.e. increasing U/t). Therefore, they are not related to the coupling constant J, but to an interference effect present even in the limit of $U/t \rightarrow +\infty$. Besides these oscillations, we observe that ΔE decreases with increasing N. This is due to the fact that as $N \rightarrow \infty$ the band-filling approaches the half-filled case ($\nu/N = 1 + 1/N$) where antiferromagnetic correlations dominate. Therefore, for any finite U/t there is always a finite size $N_c \propto U/t$ such that the ground state is no longer ferromagnetic for $N > N_c$ (i.e. $\Delta E = 0$ for $N = N_c$).

A particularly interesting result shown in Figure 6.13 is that ΔE_{el} and ΔE_{st} have similar values and that ΔE_{st} can be smaller than ΔE_{el} [51]. This indicates that *structural changes* are at least as important as the *electronic excitations* for determining the temperature dependence of the magnetization of small clusters. A complete description of the finite-temperature magnetic properties of small clusters within the Hubbard model would require that the electronic and structural effects are treated simultaneously and at the same level. Experimentally, there seem to be indications that structural changes could be related to the drop of the magnetization observed in Fe_N for increasing T [19]. However, it still remains to be proven whether the conclusions obtained here from the calculation of ΔE_{el} and ΔE_{st} within the single-band Hubbard model and for $\nu = N + 1$ are also valid for more realistic d-band models and for other band-fillings. If this is so, structural fluctuations should play an important role in determining the temperature dependence of the magnetization of 3d TM clusters. In any case the already reported strong dependence of the spin-fluctuation energies on the cluster structure and NN distances (Section 6.3.3) lets us expect that the present trends derived within the Hubbard model are probably relevant for more realistic situations.

6.4 DISCUSSION

The different problems discussed reflect the remarkable activity in the field of itinerant cluster magnetism and suggest possible future studies. For example, we have seen that in free 3d transition-metal (TM) clusters the magnetic moments are larger than in the corresponding solids. This has been related to the reduction of the local coordination number and the associated reduction of the effective d bandwidth. In contrast, when these clusters are embedded in an antiferromagnetic or nonmagnetic matrix the magnetic moments are smaller than the bulk magnetization per atom. This was shown to be a consequence of cluster–matrix hybridizations. Therefore, a variety of interesting magnetic behavior is expected for magnetic clusters deposited on surfaces due to the interplay between reduction of local coordination number and cluster–substrate effects. The characterization of the magnetic behavior of clusters and atomic islands on surfaces as a function of size and structure is a subject of considerable experimental and theoretical interest. Submonolayer magnetic nanostructures involving clusters and finite atomic chains in low-symmetry arrangements are attracting increasing attention, particularly in view of developing magnetic materials with new magnetoanisotropic properties.

Electron correlations at finite temperature in finite-size systems, in particular magnetic clusters, are problems of fundamental importance with many open questions. Methodological improvements of both theoretical and experimental methods are necessary. For instance, a systematic theoretical study of photoemission spectra of magnetic TM clusters should allow the relation of the size dependence of the excitation spectra and the cluster magnetization at finite temperatures, thus adding valuable complementary information to the results of Stern–Gerlach deflection experiments. The determination of the size and structural dependence of electronic excitations will motivate the development of an improved treatment of many-body effects. Rigorous calculations using exact diagonalization methods and a realistic d-band model Hamiltonian could serve as a basis for testing approximate methods applicable to larger clusters. The analysis of the evolution of the many-body electronic structure as a function of size should help the understanding of the formation of magnetic moments, of magnetic anisotropy and the transition from localized to itinerant magnetism.

The enhancement of the local magnetic moments and the presence of significant short-range magnetic order at high temperatures ($T > T_C(N)$) are two characteristics of TM cluster magnetism that are expected to play a significant role in determining the finite temperature behavior. Presently, one needs more studies of the temperature dependence of cluster magnetic properties. In particular the size dependence of the cluster Curie temperature $T_C(N)$ must be understood more clearly. In view of the enhancement of the local magnetic moments μ_i and of the d-level exchange-splittings one could expect that $T_C(N)$ should be larger in small clusters than in the bulk. However, on the other hand it should become energetically easier to disorder the local magnetic moments in a cluster, since the local coordination numbers are smaller. If this dominates, then $T_C(N)$ should decrease with decreasing N. In addition, recent calculations suggest that changes or fluctuations of the cluster structure are likely to affect the temperature dependence of the magnetization. Such effects could be important for systems like Fe_N and Rh_N which already show remarkable structure-dependent properties at $T = 0$. In order to get reliable results for this problem, the electronic theory must take into account both the fluctuations of the magnetic moments and the itinerant character of the d electron states. Simple spin models, for example based on the Heisenberg or Ising model, should treat more explicitly the electronic nonlocal effects responsible for the size dependence of the interactions between the magnetic moments. The spin-fluctuation theory presented in Section 6.2.2 [35] addresses these two aspects of the problem. Monte Carlo simulations for small clusters could be used for determining the temperature dependence of the magnetic properties by taking into account spin fluctuations beyond the single-site approximation. For larger clusters simpler models could be derived using the effective exchange interactions obtained in the electronic calculations.

It is also very interesting to use clusters as test systems for studying magnetism and the relaxation dynamics in small systems not at equilibrium. In particular the dependence of the magnetic anisotropy on temperature and on non-equilibrium states deserves special attention. Magneto-optics is expected to become a useful tool for the analysis of cluster magnetism. For example, magnetic Mie-scattering could also contribute to the characterization of the magnetic properties.

In conclusion, cluster magnetism offers a challenging research area, fascinating both from the fundamental point of view and regarding technological applications. It is therefore expected that this field will continue to grow. Magnetic quantum-dot structures and spin-dependent tunneling will become an active field, for example.

6.5   ACKNOWLEDGMENTS

It is a pleasure to thank Prof. J. Dorantes-Dávila and Dr P. Jensen for numerous helpful discussions. The authors acknowledge support by the J. S. Guggenheim Memorial Foundation (GMP) and by the Deutsche Forschungsgemeinschaft (KHB).

6.6   REFERENCES

[1]   D. M. Cox, D. J. Trevor, R. L. Whetten, E. A. Rohlfing and A. Kaldor, *Phys. Rev. B*, **32**, 7290 (1985).
[2]   W. A. de Heer, P. Milani and A. Châtelain, *Phys. Rev. Lett.* **65**, 488 (1990).
[3]   J. P. Bucher, D. C. Douglass and L. A. Bloomfield, *Phys. Rev. Lett.* **66**, 3052 (1991).
[4]   D. R. Salahub and R. P. Messmer, *Surf. Sci.* **106**, 415 (1981); C. Y. Yang, K. H. Johnson, D. R. Salahub, J. Kaspar and R. P. Messmer, *Phys. Rev. B* **24**, 5673 (1981).
[5]   K. Lee, J. Callaway and S. Dhar, *Phys. Rev. B* **30**, 1724 (1985); K. Lee, J. Callaway, K. Wong, R. Tang and A. Ziegler, *Phys.Rev. B* **31**, 1796 (1985); K. Lee and J. Callaway, *Phys. Rev. B* **48**, 15358 (1993).
[6]   B. I. Dunlap, *Z. Phys. D* **19**, 255 (1991).
[7]   J. L. Chen, C. S. Wang, K. A. Jackson and M. R. Perderson, *Phys. Rev. B* **44**, 6558 (1991).
[8]   M. Castro and D. R. Salahub, *Phys. Rev. B* **49**, 11842 (1994).
[9]   P. Ballone and R. O. Jones, *Chem. Phys. Lett.* **233**, 632 (1995).
[10] G. M. Pastor, J. Dorantes-Dávila and K. H. Bennemann, *Physica B* **149**, 22 (1988); *Phys. Rev. B* **40**, 7642 (1989).
[11] J. Dorantes-Dávila, H. Dreyssé and G. M. Pastor, *Phys. Rev. B* **46**, 10432 (1992).
[12] A. Vega, J. Dorantes-Dávila, L. C. Balbás and G. M. Pastor, *Phys. Rev. B* **47**, 4742 (1993).
[13] G. M. Pastor, J. Dorantes-Dávila, S. Pick and H. Dreyssé, *Phys. Rev. Lett.* **75**, 326 (1995).

[14] S. N. Khanna and S. Linderoth, *Phys. Rev. Lett.* **67**, 742 (1991).

[15] P. J. Jensen, S. Mukherjee and K. H. Bennemann, *Z. Phys. D* **21**, 349 (1991).

[16] D. C. Douglass, A. J. Cox, J. P. Bucher and L. A. Bloomfield, *Phys. Rev. B* **47**, 12874 (1993).

[17] D. C. Douglass, J. P. Bucher and L. A. Bloomfield, *Phys. Rev. B* **45**, 6341 (1992).

[18] A. J. Cox, J. G. Louderback and L. A. Bloomfield; *Phys. Rev. Lett.* **71**, 923 (1993); A. J. Cox, J. G. Louderback, S. E. Apsel and L. A. Bloomfield, *Phys. Rev. B* **49**, 12295 (1994).

[19] I. M. L. Billas, J. A. Becker, A. Châtelain and W. A. de Heer, *Phys. Rev. Lett.* **71**, 4067 (1993).

[20] I. M. L. Billas, A. Châtelain and W. A. de Heer, *Science* **265**, 1682 (1994).

[21] S. E. Apsel, J. W. Emert, J. Deng and L. A. Bloomfield, *Phys. Rev. Lett.* **76**, 1441 (1996).

[22] R. Galicia, *Rev. Mex. Fis.* **32**, 51 (1985).

[23] B. V. Reddy, S. N. Khanna and B. I. Dunlap, *Phys. Rev. Lett.* **70**, 3323 (1993).

[24] P. J. Jensen and K. H. Bennemann, *Z. Phys. D* **26**, 246 (1993).

[25] A. Maiti and L. M. Falicov, *Phys. Rev. B* **48**, 13596 (1993).

[26] M. Hanson, C. Johansson and S. Mørup, *J. Cond. Mater.* **5**, 725 (1993); J. A. Becker *et al.*, *Proceedings of ISSPIC 7, Kobe, Japan* (1994).

[27] G. M. Pastor, in *Current Problems in Condensed Matter: Theory and Experiments*, ed. J. L. Morán-López (Plenum, New York, 1998).

[28] S. E. Weber, B. K. Rao and P. Jena, preprint (1997).

[29] B. J. Hickey M. A. Howson, S. O. Musa and N. Wiser, *Phys. Rev. B* **51**, 667 (1995); R. van Helmolt, J. Wecker, B. Holzapfel, L. Schultz and K. Samwer, *Phys. Rev. Lett.* **71**, 2331 (1993).

[30] J. Hubbard, *Proc. R. Soc. London* **A276**, 238 (1963); **A281**, 401 (1964); J. Kanamori, *Prog. Theo. Phys.* **30**, 275 (1963); M. C. Gutzwiller, *Phys. Rev. Lett.* **10**, 159 (1963).

[31] B. Mühlschlegel, *Z. Phys.* **208**, 94 (1968).

[32] R. H. Victora, L. M. Falicov and S. Ishida, *Phys. Rev. B* **30**, 3896 (1989).

[33] H. Haydock, *Solid State Physics* (Academic Press, London, 1980), Vol. **35**, pp. 215 ff.

[34] The dynamical mean-field theory is characterized by the Green's function $G_{i\sigma}(\omega) = [\omega - \varepsilon_{i\sigma} - \sum_j t_{ij} G_{j\sigma} - \sum_{i\sigma}(\omega)]^{-1}$, $\varepsilon_{i\sigma} = (n - \sigma\mu)U/2$, and self-energy $\sum_{i\sigma}(\omega) = U^2 \sum_{\omega'} G_{i\sigma}(\omega - \omega')\chi_{-\sigma}(\omega')$. Here, $\chi_\sigma(\omega')$ is the spin susceptibility and μ the magnetic moment. For details see P. Lombardo, M. Avignon, J. Schmalian and K. H. Bennemann, *Phys. Rev. B* **54**, 5317 (1996).

[35] J. Dorantes-Dávila and G. M. Pastor, to be published .

[36] J. Hubbard, *Phys. Rev. B* **19**, 2626 (1979); **20**, 4584 (1979); H. Hasegawa, *J. Phys. Soc. Jpn.* **49**, 178 (1980); **49**, 963 (1980).

[37] T. Moriya, ed., *Electron Correlation and Magnetism in Narrow-Band Systems*, Springer Series in Solid State Sciences **29**, Springer, Heidelberg, 1981.

[38] Y. Kakehashi, *J. Phys. Soc. Jpn.* **50**, 2251 (1981).

[39] V. Korenman and R. E. Prange, *Phys. Rev. Lett.* **53**, 186 (1984).

[40] E. M. Haines, R. Clauberg and R. Feder, *Phys. Rev. Lett.* **54**, 932 (1985).

[41] J. Dorantes-Dávila, G. M. Pastor and K. H. Bennemann, *Solid State Commun.* **59**, 159 (1986); **60**, 465 (1986).

[42] G. M. Pastor and J. Dorantes-Dávila, *Phys. Rev. B* **52**, 13799 (1995).

[43] L. M. Falicov and R. H. Victora, *Phys. Rev. B* **30**, 1695 (1984).

[44] Y. Ishii and S. Sugano, *J. Phys. Soc. Jpn.* **53**, 3895 (1984).

[45] J. Callaway, D. P. Chen and R. Tang, *Z. Phys D* **3**, 91 (1986); *Phys. Rev. B* **35**, 3705 (1987).

[46] Y. Wang, T. F. George, D. M. Lindsay and A. C. Beri, *J. Chem. Phys.* **86**, 3493 (1987).

[47] P. Fulde, *Electron Correlations in Atoms, Molecules and Solids*, Springer, Berlin, 1990.

[48] S. L. Reindl and G. M. Pastor, *Phys. Rev. B* **47**, 4680 (1993).

[49] C. Lanczos, *J. Res. Nat. Bur. Stand.* **45**, 255 (1950); B. N. Parlett, *The Symmetric Eigenvalue Problem*, Prentice-Hall, Engelwood Cliffs, NJ, 1980; J. K. Collum and R. A. Willoughby, *Lanczos Algorithms for Large Symmetric Eigenvalue Computations*, Vol. I, Birkhauser, Boston, 1985.

[50] A graph $G = (V, E)$ is defined as a set V of sites or vertices $i = 1, \ldots, N$ and a set E of connections, bonds or edges (i, j) with $i, j \in V$. On physical grounds (see Eq. (18)) we restrict ourselves to graphs that are simple (i.e. no more than one connection between two given sites), connected (i.e. for all $i, j \in V$ there is a sequence $\{i_1, i_2, \ldots, i_K\}$ with $i_1 = i$

and $i_K = j$ such that $(i_k, i_{k+1}) \in E$ for all $k < K$) and without on-site loops (i.e. if $(i, j) \in E$ then $i \neq j$).

[51] G. M. Pastor, R. Hirsch and B. Mühlschlegel, *Phys. Rev. Lett.* **72**, 3879 (1994); *Phys. Rev. B* **53**, 10382 (1996).

[52] N. E. Christensen, O. Gunnarsson, O. Jepsen and O. K. Andersen, *J. de Phys.* (Paris) **49** C8-17 (1988); O. K. Andersen, O. Jepsen and D. Glötzel, in *Highlights of Condensed Matter Theory*, ed. F. Bassani, F. Fumi and M. P. Tosi (North-Holland, Amsterdam, 1985) p. 59.

[53] G. Stollhoff, A. M. Oles and V. Heine, *Phys. Rev. B* **41**, 7028 (1990).

[54] P. Villaseñor-González, J. Dorantes-Dávila, H. Dreyssé and G. M. Pastor, *Phys. Rev. B* **55**, 15084 (1997).

[55] A. Vega, L. C. Balbás, J. Dorantes-Dávila and G. M. Pastor, *Phys. Rev. B* **50**, 3899 (1994); *Computational Materials Science* **2**, 463 (1994).

[56] Y. Jinlong, F. Toigo and W. Kelin, *Phys. Rev. B* **50**, 7915 (1994); Z.-Q. Li, J.-Z. Yu, K. Ohno and Y. Kawazoe, *J. Phys.: Cond. Matter* **7**, 47 (1995); B. Piveteau, M-C. Desjonquères, A. M. Olés and D. Spanjaard, *Phys. Rev. B* **53**, 9251 (1996).

[57] M. A. Ojeda-López, J. Dorantes-Dávila and G. M. Pastor, *J. Appl. Phys.* **81**, 4170 (1997); also *Phys. Rev. B*, submitted (1999).

[58] A. Vega, J. Dorantes-Dávila, G. M. Pastor and L. C. Balbás, *Z. Phys. D* **19**, 263 (1991).

[59] P. Grünberg, R. Schreiber, Y. Pang, M. B. Brodsky and H. Sowers, *Phys. Rev. Lett.* **57**, 2442 (1986); C. Carbone and S. F. Alvarado, *Phys Rev. B* **36**, 2433 (1987); M. N. Baibich, J. M. Broto, A. Fert, F. Nguyen Van Dau, F. Petroff, P. Eitenne, G. Creuzet, A. Friederich and J. Chazelas, *Phys. Rev. Lett.* **61**, 2472 (1988); J. J. Krebs, P. Lubitz, A. Chaiken and G. Prinz, *Phys. Rev. Lett.* **63**, 4828 (1989); S. S. P. Parkin, N. More and K. P. Roche, *Phys. Rev. Lett.* **64**, 2304 (1990); S. Demokritov, J. A. Wolf and P. Grünberg, *Europhys. Lett.* **15**, 881 (1991).

[60] P. M. Levy, K. Ounadjela, S. Zhang, Y. Wang, C. B. Sommers and A. Fert, *J. Appl. Phys.* **67**, 5914 (1990); F. Herman, J. Sticht and M. van Schilfgaarde, *J. Appl. Phys.* **69**, 4783 (1991); K. Ounadjela, C. B. Sommers, A. Fert, D. Stoeffler, F. Gautier and V. L. Moruzzi, *Europhys. Lett.* **15**, 875 (1991); D. Stoeffler and F. Gautier, *Phys. Rev. B* **44**, 10389 (1991); Z.-P. Shi, P. M. Levy and J. L. Fry, *Phys. Rev. Lett.* **69**, 3678 (1992); J-h. Xu and A. J. Freeman, *Phys Rev. B* **47**, 165 (1993).

[61] The blocking temperature characterizes the energy barrier resulting from magnetic anisotropy and that is in the way of the magnetic moment of the cluster turning into the direction of the magnetic field.

[62] C. P. Schichter, *Principles of Magnetic Resonances*, Harper and Row, New York, 1963.

[63] J. A. Becker and W. A. de Heer, *Ber. Bunsenges. Phys. Chem.* **96**, 1237 (1992).

[64] P. Jensen and K. H. Bennemann, *Z. Phys. D* **35**, 273 (1995).

[65] F. Bødker, S. Mørup and S. Linderoth, *Phys. Rev. Lett.* **72**, 282 (1994).

[66] P. G. Watson, in *Phase Transitions and Critical Phenomena, Vol. 2*, ed. C. Domb and M. S. Green (Academic Press, London, 1972), p. 101; and M. N. Barber, in Vol. 8, ed. C. Domb and J. L. Lebowitz (Academic Press, London, 1983), p. 145.

[67] J. P. Malrieu, D. Maynau and J. P. Daudey, *Phys. Rev. B* **30**, 1817 (1984); F. Spiegelmann, P. Blaise, J. P. Malrieu and D. Maynau, *Z. Phys. D* **12**, 341 (1989); *Phys. Rev. B* **41**, 5566 (1990); F. Illas, J. Casanovas, M. A. García-Bach, R. Caballol and O. Castell, *Phys. Rev. Lett.* **71**, 3549 (1993).

[68] For $\nu > N$ it is often convenient to consider the holes as carriers ($\nu_h = 2N - \nu$). The electron–hole transformation $h_{i\sigma}^+ = c_{i\bar{\sigma}}$ leaves the Hubbard Hamiltonian unchanged, except for an additive constant and a change of sign of the hopping integrals.

[69] For $\nu = 1$ and $\nu = 2N - 1$ the optimal structures are trivially independent of U/t and coincide, respectively. In the uncorrelated limit ($U = 0$) one has the optimal structure for $\nu = 2$ and $\nu = 2N - 2$.

[70] Y. Nagaoka, *Solid State Commun.* **3**, 409 (1965); D. J. Thouless, *Proc. Phys. Soc. London* **86**, 893 (1965); Y. Nagaoka, *Phys. Rev.* **147**, 392 (1966); H. Tasaki, *Phys. Rev. B* **40**, 9192 (1989).

[71] F. López-Urías and G. M. Pastor, *Phys. Rev. B* **59**, 5223 (1999).

7   Comparison of Resonance Dynamics in Metal Clusters and Nuclei

P.-G. REINHARD

Universität Erlangen

and

E. SURAUD[†]

Université Paul Sabatier, Toulouse

7.1   INTRODUCTION

The scope of this contribution is the comparison of metals clusters and nuclei. These systems have much in common as their structure and dynamics are dominated by the behavior of fermion liquids, the protons and neutrons in nuclei and the dense electron cloud in clusters. This gives rise to shell effects and a corresponding deformation pattern as well as pronounced resonance excitations related to zero sound in homogeneous matter, the giant resonances in nuclei and the surface plasmon in clusters. The structural aspects have already been much discussed in the past and are well documented in several review articles, see e.g. [1, 2]. A prominent feature here was the appearance of supershells which are only accessible in metal clusters with their unlimited pool of system sizes [3] and which have a particularly transparent explanation in the framework of semiclassical

---

[†] Membre de l'Institut Universitaire de France

Metal Clusters. Edited by W. Ekardt

periodic-orbit theory [4]. A detailed comparison of deformation properties is found in [5]. Fission is also a feature common to the two systems. This is reviewed extensively in [6].

For reasons of space, we confine the present comparison to dynamical features. Most prominent and much studied are the resonance modes. This concerns the Mie plasmon in the case of clusters which is a key issue in several other articles in this book. These modes are preferably accessed by photoabsorption measurements and they belong to small-amplitude oscillations (for a general overview on small oscillations in fermion systems see [7]). A survey treating resonance modes in nuclei, metal clusters, and drops of liquid ^{3}He all on the same footing was given in [8–11]. We will give here in Section 7.3 a short account of the dynamics in the linear regime with emphasis on direct comparison between nuclei and clusters, incorporating most recent results, and skipping (for reasons of space) the discussion of liquid ^{3}He for which we refer the reader to [11].

The more recent developments in cluster dynamics proceed into the regime of large amplitudes and nonlinear dynamics as accessed e.g. by femtoseconds laser pulses or energetic collisions. This regime was intensively studied long ago in nuclear physics in connection with heavy-ion reactions [12–14]. This opens up a rich field of various phenomena and comparison between nuclei and clusters is particularly interesting there. We will discuss the dynamics in the nonlinear regime in Section 7.4.

The theoretical framework for all regimes is the (time-dependent) local-density approximation (TDLDA) which is much discussed also in Chapter 1 of this book. We thus will present here only a short discussion of the essential ingredients and compare it with the analog mean-field models in nuclear physics. This is done as a starter in the next section.

7.2   THEORETICAL FRAMEWORK

7.2.1   Energy Functional for Clusters

Most practical (and all large-scale) calculations in cluster and nuclear physics are based on effective energy-density functionals; for reviews concerning electronic systems see [15, 16] and Chapters 1 and 3 in this book. The enormously complicated exchange term and correlation effects are simplified by packing them into an energy functional of the local density alone. The existence of such a functional is guaranteed by basic theorems. The practical determination of a functional, however, invokes further approximations and is usually done in the local density approximation (LDA) where results from the homogeneous fermion gas are translated into a local energy-density functional by assuming that the actual system behaves piecewise like infinite matter. The starting point for the LDA description of metal clusters is thus

$$E = E_{\text{kin}}(\varphi_\alpha) + E_{\text{Coul}}(\rho(r)) + E_{\text{xc}}(\rho(r)) + E_{\text{el,Ion}} + E_{\text{Ion,Ion}}. \tag{1}$$

The kinetic energy E_{kin} still carries the full information in the occupied single-electron wave functions φ_α. This guarantees that all important shell effects are maintained at that level of approximation. The direct part of the Coulomb energy E_{Coul} is already given as a functional of the local density. The exchange–correlation part E_{xc} is approximated as a local functional by means of the LDA. There exist various proposals for such exchange–correlation functionals. We will employ in the examples of the later sections the (somewhat old-fashioned) energy-density functional from [17]. The ion–ion energy $E_{\text{Ion,Ion}}$ embodies simply the Coulomb interaction between the ions, usually treated here as point particles. There is finally the interaction between the electrons and the ionic background $E_{\mathrm{el,Ion}}$ which deserves a few more explanations.

If the ions are treated explicitly, the ion–electron interaction is described by pseudopotentials. This allows us to eliminate all core electrons from the actual calculation and to deal only with the valence electrons. There exists a wide variety of pseudopotentials. The most elaborate of them are nonlocal operators because they project out of the occupied core electron states, see e.g. [18]. Separable approximations to projection can simplify the handling [19]. The simplest to use are, of course, local pseudopotentials, as e.g. the old empty-core potential of [20] or the more recent and extensive adjustment of [21]. It is a welcome feature that most simple metals, except for Li, can be treated fairly well with local pseudopotentials [22]. It is to be remarked that all these choices have been optimized with respect to structural properties. The performance concerning dynamical features has not yet been explored systematically. In fact, most of the available pseudopotentials tend to produce a blue-shifted plasmon position. An optimization of pseudopotentials for simple metals with simultaneous adjustment of static and dynamic properties is presently under way [23]. For noble metals, it is known that pseudopotentials alone (even when nonlocal) cannot reproduce the proper plasmon position. One needs to explicitly take into account the considerable polarizability of the ionic core, in particular of the rather soft last d shell [24].

The valence electrons in metals extend over the whole cluster and effects like shell structure or plasmon response are determined by long-range structure of the electron cloud. The details of the ionic background play a minor role for many of these global features. It is thus very convenient, and often practiced, to approximate the ionic background within the jellium approach by a smooth, positively charged distribution. Finite clusters require a decision on how to shape the jellium surface. Earlier applications (as e.g. the first models for plasmon response in metal clusters [25, 26]) employed a steep jellium surface where the background was taken as a mere step function. This approach, however, gives a blue-shifted plasmon position. A much better plasmon position is achieved when using a smooth transition at the jellium surface [27, 28] and a justification comes from folding of the steep jellium distribution with appropriate pseudopotentials [29, 22, 30]. The crucial parameter for the plasmon position is the width of the surface and the detailed shape is not so important. Thus one can employ the convenient Woods–Saxon profile

$$\rho_{\mathrm{J}} = \frac{\rho_0}{1 + \exp((r - R)/d)}, \quad \rho_0 = \frac{3}{4\pi r_{\mathrm{s}}^3}, \qquad (2)$$

where, as a rule of thumb, the surface width d can be expressed in term of the Ashcroft empty-core radius r_{c} [20] as $d = r_{\mathrm{c}}/\sqrt{3}$. The radius of the background is $R \simeq_{\mathrm{s}} N^{1/3}$ (tuned to provide the right number of ions N). Thus far the profile (2) stands for the spherical jellium model. Deformed shapes can be modeled by modifying R to depend on the orientation: $R = R(\vartheta, \varphi) = r_{\mathrm{s}} N^{1/3} (1 + \sum_{lm} \beta_{lm} Y_{lm}(\vartheta, \varphi))$, where Y_{lm} are the spherical harmonics. Axially symmetric deformations use only the terms with $m = 0$ and the full expansion will produce generally triaxial deformation [31]. The soft jellium background is the appropriate model for simple metal clusters. One more ingredient is needed to describe plasmon response in noble metals. The polarizability of the core electrons can be taken into account, in the spirit of a jellium model, as a smoothly distributed dielectric constant of the ionic background [32]. Thus far the soft (and possibly dielectric) jellium model suffices for the description of electronic dynamics. Static properties, however, miss crucial contributions to the energy from the ionic structure. These can be complemented by adding an average potential from these structure effects, leading to the stabilized jellium model [33]. One needs to add, furthermore, the ionic contribution to the surface tension which then leads to the structure-averaged jellium model and which gives a reliable description also of the global observables of the cluster ground states [34].

The same energy-density functionals are often also used for the dynamical evolution of the system in the spirit of a time-dependent local-density approximation (TDLDA). This step implies one more assumption, namely that exchange and correlation effects can also be approximated as staying local in time, i.e. instantaneous. It is justified as long as the leading frequencies of the correlation modes are far above the frequencies in the actual dynamics, and the success of many TDLDA calculations for simple metal clusters since the earliest attempts [25, 26] provides a phenomenological justification for that usage somewhat beyond the strict limits. All examples shown in the following revert to the simple TDLDA approach. The question of truly dynamical energy-density functionals is, nonetheless, actively investigated in several publications in various fields of electronic dynamics. We refer e.g. to the recent review articles [35–37].

7.2.2 Energy Functional for Nuclei

Effective mean-field models are also widely used for the microscopic description of nuclear structure and dynamics. The most prominent energy-density functionals are: (i) the Skyrme–Hartree–Fock approach as proposed first in [38] and brought to a quantitative description in [39], (ii) the Gogny force [40], and (iii) the relativistic mean-field model whose basic idea was proposed first in [41] and was developed to its present form in [42, 43]. All three approaches have been further developed since, have reached a high level of descriptive power, and are widely used in calculations of the nuclear ground state, low-energy excitations and heavy-ion reactions. There are several recent reviews on the relativistic mean-field model [44–46]. For a recent publication on the Skyrme–Hartree–Fock model, we refer to [47] where a typical step of further phenomenological development is exemplified. This hints already at the most prominent difference from electronic systems. The energy-density functionals for electrons can be derived systematically from fully microscopic calculations of electronic correlations [15, 16] and they work extremely well. The microscopic basis for nuclei, however, is not yet precise enough to allow an *ab initio* derivation of energy-density functionals; for a review see [48]. The standard strategy is to deduce the form of the functional from theoretical reasoning and then to adjust the parameters phenomenologically. With experience gathered over the decades, the results are nuclear energy-density functionals that are as reliable and precise as those for the case of electrons. It remains for us to discuss similarities and differences as seen from the practioner's side.

We consider here the example of the Skyrme energy functional where the potential energy is modeled as

$$
\mathcal{E}_{\text{Sk}} =
\begin{array}{ll}
T=0 & T=1 \\
+B_0 \rho^2 & +B_0' \delta \rho^2 \\
+B_3 \rho^{2+\alpha} & +B_3' \delta \rho^2 \rho^\alpha \\
+B_2 \frac{1}{2}(\vec{\nabla}\rho)^2 & +B_2' \frac{1}{2}(\delta\vec{\nabla}\rho)^2 \\
+B_1(\rho\tau - \vec{j}^2) & -B_1'(\delta\rho\delta\tau - (\delta\vec{j})^2) \\
-B_4(\rho\vec{\nabla}\vec{J} + \vec{\sigma}(\vec{\nabla}\times\vec{j})) & -B_4(\delta\rho\vec{\nabla}\delta\vec{J} + \delta\vec{\sigma}(\vec{\nabla}\times\delta\vec{j}))
\end{array}
\tag{3}
$$

where ρ is the density, τ the kinetic-energy density, $\vec{\sigma}$ the spin density, $\vec{j}$ the current, and $\vec{J}$ the spin–orbit current. The $\rho, \tau, \ldots$ stand for the total densities and currents, related to isoscalar contributions ($T = 0$), and the corresponding $\delta\rho$, $\delta\tau$, etc. for the difference between proton and neutron densities and currents which constitute the isovector contributions ($T = 1$). First, we want to point out the similarity to electron systems which consists in the existence and success of such an energy-density functional. A closer inspection of the Skyrme functional (3), however, shows several differences. There is a term employing the kinetic-energy density τ which is needed to model the effective mass $m^*/m \sim 0.8$ of the nucleons in the nuclear medium. It appears in the combination $\rho\tau - \vec{j}^2$ to guarantee Galilean invariance of the functional [49]. There is, furthermore, a gradient correction $\propto (\vec{\nabla}\rho)^2$ which carries some information on the range of the nuclear interaction. Gradient-corrected energy functionals are also much discussed in electronic systems, particularly for those with more inhomogeneous density distributions as e.g. atoms [50]. And finally, there comes the spin–orbit term $\propto \rho\vec{\nabla}\vec{J}$ which is a crucial ingredient in the nuclear mean field. In all these aspects, the nuclear energy-density functional has to be more elaborate to cope with the more complicated nuclear mean field. On the other hand, electronic energy-density functionals carry a much more detailed density dependence which can be deduced directly from the microscopic calculations of the correlated electron cloud. Very little is known about the appropriate density dependences in the nuclear case because we have not yet sufficiently reliable microscopic calculations. As a minimal solution, all density dependence is parameterized in the terms $\propto \rho^\alpha$. This suffices for normal nuclei under normal conditions because most densities stay close to the symmetric nuclear-matter equilibrium density ρ_{nm}. Recent experimental development in nuclear physics allows us to step deeper into the regime of exotic nuclei which probe the nuclear structure much more critically than the stable nuclei known hitherto. It is one of the most exciting tasks nowadays to investigate the predictive value of the established mean-field models with respect to these new nuclei.

One more difference from the electronic case needs finally to be remarked. The functional (3) parameterizes the mean-field for two sorts of fermions, namely neutrons and protons. This introduces one more distinctive quantum number, the isospin. Correspondingly, we see isoscalar ($T = 0$) terms $\rho = \rho_p + \rho_n$, and isovector ($T = 1$) terms, $\delta\rho = \rho_n - \rho_p$. Both together provide a complete description of nuclei. In metals, on the other hand, we have only the Fermi liquid of (valence) electrons. But there come in addition the ions as partner constituents that need to be modeled additionally, e.g. by pseudo-potentials or by the jellium approach. And there is a large variety of ionic backgrounds corresponding to the overwhelming variety of materials.

7.2.3 Equations of Motion and their Solution

The static and dynamic mean-field equations are derived, for a given energy-density functional, by variation with respect to the single-particle wave functions φ_α^+. The Ritz variational principle of minimal energy yields the static mean-field equations,

$$h[\rho]\varphi_\alpha = \varepsilon_\alpha\varphi_\alpha, \quad h[\rho] = \frac{\hat{p}^2}{2m} + \frac{\partial E}{\partial\rho(r)}, \tag{4}$$

usually called the Kohn–Sham equations [51]. The mean-field Hamiltonian h depends implicitly on the occupied wave functions φ_α via its functional dependence on the local density ρ. It emerges from the variation by the identification $\partial E/\partial\varphi_\alpha^+ = h[\rho]\varphi_\alpha$. This identification holds also for the nuclear case, where, however, the mean field becomes more involved due to effective mass and spin–orbit force. The time-dependent Kohn–Sham equations are derived from the principle of stationary action and become

$$h[\rho]\varphi_\alpha = i\partial_t\varphi_\alpha, \tag{5}$$

where the functional dependence of the mean-field Hamiltonian $h[\rho]$ is formally the same as in the stationary Kohn–Sham equations (4). These equations are often named the time-dependent LDA (TDLDA) to point out more clearly the LDA which is involved in the construction of the energy-density functional and which is probably the most critical approximation in the scheme.

There are various ways to solve the (TD)LDA equations numerically. The most efficient for saturating fermion systems (such as nuclei or the electron cloud in metal clusters) is a representation of wave functions and fields on an equidistant grid in coordinate space. Depending on the symmetry, this may be a radial 1D grid, an axially symmetric 2D grid, or a Cartesian 3D grid. The kinetic energy can be represented by a finite-difference formula (recommended in 1D) or by Fourier transformation (preferable in 2D and 3D). A convenient and efficient way to solve the stationary Kohn–Sham equations is provided by the damped gradient iterations. Time-stepping is best done by schemes that consist of a manifestly unitary step, such as the Crank–Nicholson step and approximations thereof. A detailed discussion of all these aspects is given in [52]. A promising variant of the time step for 3D calculations consists in alternating exponential propagation of kinetic and potential energy, so to say a quantum leap-frog method. It combines robustness, efficiency and simple implementation [53].

Once a full TDLDA code is available, there comes the question of how to analyze the results. This will be discussed later on at the beginning of sections discussing various investigations in the linear domain (Section 7.3.1) and in the nonlinear regime (Section 7.4.2).

7.2.4　Linearized Equations of Motion

The full TDLDA equations (5) are conceptually simple and rather straightforward to code. They have the further advantage that they carry no restriction on the amplitude of the motion so that they can be used deep into the regime of non-linear effects. This has been exploited extensively in nuclear TDLDA calculations (called there Skyrme-TDHF) to describe the dramatic dynamical changes in fission, fusion and other reactions [14]. It is only recently that comparable studies of large-amplitude motion in metal clusters have been undertaken, see Section 7.4. By far, most investigations on electronic dynamics in clusters have been concerned with the regime of small-amplitude motion. The enormous expense of solving the full TDLDA equations can then be much reduced by considering linearized TDLDA. But the gain in speed requires first considerable analytical work to express TDLDA in the linear domain. There exist several formulations for that purpose. One widely used scheme starts from a density matrix formulation and derives the integral equations for the response function $\chi(\omega)$ by linearization, see e.g. Chapter 1 in this book. Expanding the time-dependent wave functions in terms of particle–hole (1ph) states leads to a matrix formulation [54]. Another technique accesses immediately the dipole response function from computing the resolvent operator (i.e. inversion of the linear-response Hamiltonian) on the grid [55], a method that is feasible for radial 1D calculations. As a further possibility, an operator calculus was proposed in [8–10]. It is based on the representation of neighboring Slater states as $|\Phi(t)\rangle \sim [1 + i\hat{Q}(t)]|\Phi_0\rangle$, ends up with an obvious algebra of coupled harmonic oscillators for the normal modes $\hat{Q}_N$, and allows a convenient coding in terms of operator wave functions. The representation also allows us to formulate various levels of further approximations within one and the same scheme. Most important here are the local Random Phase Approximation (RPA) [56, 57] which concentrates on optimizing the collective flow (a sort of variational fluid-dynamical approach) and the sum rule approach which gives a first estimate of the resonance mode from simple double commutators [58]. Finally, all these technical simplifications achieved by linearization still leave sizable work such that applications are limited to smaller and/or symmetric systems (unless one involes local RPA or sum rules as further approximation). For spherically symmetric clusters, one may reach (with compromises) the size $N_{el} \sim 1000$. Much more becomes feasible if one allows one more approximation, namely to expand the residual interaction into a sum of separable terms. With an expansion up to 12 such terms one obtains reliable results which can be carried forth easily to much larger or nonsymmetric systems [59, 60]. There is no space to discuss the details of these various techniques here. They are well documented in the literature cited and they deliver at the end all the same results. We are most interested here in comparing and discussing the results. This will be done now in the subsequent two sections.

7.3 DYNAMICS IN THE LINEAR RESPONSE DOMAIN

7.3.1 Typical Observables

Every many-body dynamics maps into a system of coupled harmonic oscillators when considered in the limit of small amplitudes. The obvious feature is then the spectrum of eigenmodes, its frequencies ω_N and amplitudes, where the latter can be characterized by the transition moments, e.g. for a dipole mode the dipole strength $\langle N|\hat{D}|0\rangle$, where $|0\rangle$ stands for the ground state. Furthermore, the modes will often acquire a width Γ_N due to correlation effects and coupling to the particle continuum. A most convenient and illustrative way to present the spectrum is the corresponding strength function, e.g. for dipole modes the dipole strength

$$S_D(\omega) = \sum_N |\langle N|\hat{D}|0\rangle|^2 \delta_{\Gamma_N}(\omega - \omega_N), \quad \delta_{\Gamma_N/\pi}(\omega - \omega_N) = \frac{\Gamma_N}{(\omega - \omega_N)^2 + \Gamma_N^2}, \quad (6)$$

where the Lorentzian line shape applies for coherent broadening mechanisms. One often uses a Gaussian shape instead, if one expects incoherent line-broadening (for a discussion of widths, see Section 7.3.5). This dipole strength is also closely related to the photoabsorption strength $\propto \omega S_D(\omega)$. The explicit computation of all spectral states and subsequent summation as done in Eq. (6) often becomes inconvenient if the spectra become very dense or even continuous. A more direct access to the strength is then the expression in terms of the resolvent operator, i.e.

$$S_D(\omega) = \left\langle 0 \left| \hat{D} \frac{1}{\mathcal{L} - \omega + i\Gamma_a} \hat{D} \right| 0 \right\rangle, \quad \mathcal{L}\hat{A} = [\hat{H}, \hat{A}]_{ph}, \quad (7)$$

where $\mathcal{L}$ is the Liouvillian for linear response, $\hat{A}$ can be any one-body operator, and the suffix ph stands for projection onto the one-particle-hole (1ph) part. The width Γ_a is an artificial width that needs to be added in practice to avoid singularities of the resolvent. It can be as small as manageable. But a reasonable finite value can be employed as a convenient means to account for broadening processes not included in the linear response dynamics. Note that some artificial width is also needed when trying to plot the strength functions as defined in Eq. (6) because the singular δ peaks in the domain of well-bound states would not fit any frame.

The full-strength functions carry a lot of information and constitute at the same time a helpful illustration of the spectral features. Nonetheless, one would often like to characterize the spectra by a few key quantities in order to investigate trends over a large sample of systems. We will see that the spectra for nuclei as well as for clusters are dominated by one large resonance peak in whose vicinity a major fraction of strength is concentrated (typically more than 90%). This resonance peak can be characterized by an average peak position and width. There are two ways to determine the average position of the resonance peak. For the given spectral strength, one can take an average over the strength within the range of the width of the peak (see e.g. [60]). A much simpler approach is to use directly the 'local RPA' which reduces the small-amplitude dynamics to a variational space of collective flow only [8–10, 56, 57] and thus averages naturally over all details of fragmentation into nearby 1ph states. The results of the two methods are similar such that one can choose whichever is more convenient in a given situation. Similarly, there are two ways to determine the width of the resonance peak. The full width at half maximum (FWHM) is the most widely used quantity in connection with given strength functions, particularly in experimental studies. The local RPA, on the other hand, immediately provides a simple estimate of the width by comparing higher-energy weighted moments [57].

Thus far, the strengths $S(\omega)$ or detailed energies ω_N with transition moment $\langle N|\hat{D}|0\rangle$ describe the gross features of the spectra as they are accessible, e.g. by photoabsorption measurements. There is much more information carried in the normal modes of a many-body system. One can thus look one step further into it by considering the whole transition density

$$\rho_N(r) = \langle N|\hat{\rho}(r)|0\rangle \tag{8}$$

or its Fourier transform, the transition form factor $F_N(q) = \int d^3 r \exp(iqr)\rho_N(r)$. This quantity carries worthwhile information on the internal structure of the modes which allows us e.g. to distinguish collective flow from single-particle transitions. This quantity cannot be measured by real photons because the fixed relation between frequency and momentum, $\omega = cq$, inhibits an exploration of the q-dependence. But it is easily accessible with virtual photons which appear e.g. in electron-scattering experiments. In fact, the transition form factors have been measured extensively for nuclei in electron-scattering experiments with high energy and momentum resolution, see e.g. [61]. Similar experiments on metal clusters are yet to come.

7.3.2 General Spectral Properties

The most dominant resonances appear in the dipole modes which therefore will be at the center of the discussion in the following. As argued in the previous subsection, an

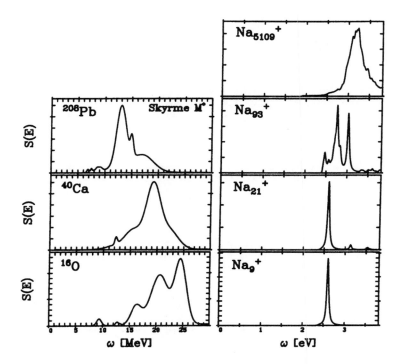

Figure 7.1 Dipole strength S for the $L = 1$ modes in a selection of spherical nuclei (left) and clusters (right)

illustrative view of the spectral properties is given by the strength function. Figure 7.1 shows the dipole strength functions for a selection of nuclei and singly charged Na clusters in jellium approximation. All systems here have magic fermion numbers and are thus spherical. In spite of the dramatically different energy scales, one sees at first glance very similar patterns. There are well-developed resonances in both systems and the underlying picture of these collective modes is comparable. The giant dipole resonance (GDR) in nuclei is predominantly a center-of-mass vibration of the protons against the neutrons, the Goldhaber–Teller mode [62]. The Mie surface plasmon in metal clusters can be understood as a vibration of the center of mass of the electron cloud against the ionic background [1, 2]. Looking longer at the picture, one also spots some differences. The supply of nuclei is limited to about $A < 300$ whereas clusters can be produced at any size. This is demonstrated here by showing an example as large as $N \sim 5000$. Furthermore, the nuclear resonance spectra are more fragmented than those of the clusters. And, finally, the trends of peak positions and widths with mass number behave basically differently.

In order to display systematically the trends with mass number, we need to compress the spectral information to a few key features. The most important is the peak position which can be computed from local RPA or direct averaging of the strength as discussed in the previous Subsection 7.3.1. The Landau fragmentation of the strength distribution can be understood from the interplay with the 1ph spectrum in the vicinity of the resonance peak. Both features, peak position and the relevant 1ph energies, are easy to compute and easy to visualize on a one-dimensional plot. This allows us to estimate in a simple fashion the various trends, e.g. with the size of the cluster. We show in Figure 7.2 these spectral

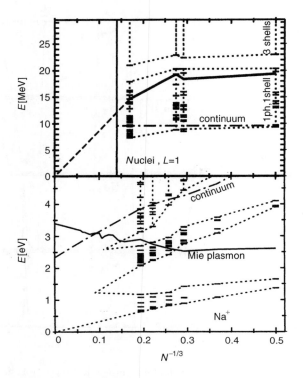

Figure 7.2 Spectral properties versus $N^{-1/3}$, or $A^{-1/3}$ respectively. Shown are the position of the dipole resonance peak (full line) in relation to the 1ph states (horizontal bars, bands of 1ph states are embraced by dashed lines) and to the continuum threshold (dashed–dotted line)

properties plotted against $N_{el}^{-1/3}$ for positively charged Na clusters (lower part), and for nuclei (upper part), again considering spherical systems throughout. The $N_{el}^{-1/3}$ is the most appropriate scale because it allows us to embrace the bulk limit simply as $N_{el}^{-1/3} \to 0$ and because most deviations from the bulk values can be expanded systematically in orders of $N_{el}^{-1/3}$. Figure 7.2 also shows the continuum threshold for neutrons (in the case of nuclei) and for electrons (in the case of clusters) to show how far the resonance reaches into the continuum.

One sees from Figure 7.2 that the frequency of the nuclear dipole giant resonance decreases with increasing particle number. This is due to the short-range character of the nuclear forces which leads to a dispersion relation $\omega = v_s q$. The finite size sets a lower limit for the momentum as $q \propto 1/R \propto N_{el}^{-1/3}$, which, in turn, predicts a trend $\omega \propto N_{el}^{-1/3}$. This asymptotic trend is indicated by the dotted continuation of the peak frequency. It becomes more than obvious that the limitation of nuclear sizes inhibits a clear empirical confirmation of this trend. The Mie resonance in the clusters, on the other hand, approaches a finite value in the bulk limit. This is due to the long-range Coulomb force, which generally produces a plasmon dispersion $\omega \propto q^0$. The overall trends for very large clusters complies with a linear growth of the surface plasmon frequency,

$$\omega(N) = \omega_{\text{Mie}} - \gamma N^{-1/3}, \qquad \omega_{\text{Mie}} = \frac{\omega_{\text{pl}}}{\sqrt{3}} = 3.4 \text{ eV (for sodium)}, \qquad (9)$$

where ω_{pl} is the bulk plasmon frequency. The trend with $N^{-1/3}$ is due to surface polarization effects and can be derived from a liquid drop expansion in terms of volume and surface effects in a simple model of the surface plasmon [56, 32]. The actual slope γ depends on the softness of the ionic background (for a detailed discussion see [60]). The more realistic soft jellium model fits the experimental peak positions very well [63] which hints that its slope is the more correct extrapolation to very large clusters. It is noteworthy that the trend (9) applies only for very large clusters with $N > 1000$. This complies with a critical number of similar size which was deduced much earlier in a nuclear physics study of the approach to bulk properties [64]. The decrease of ω is stopped when going towards smaller clusters and the trend is even slightly reversed below Na_{41}^{+}. All these features are also found in experimental data [63].

Comparison of peak position and 1ph spectra in Figure 7.2 explains the large spectral fragmentation seen in the nuclear dipole strengths. The spin−orbit force leads to a comparatively large spectral diffusion even for small nuclei, such that the giant resonance always lies in a swamp of 1ph states. Moreover, the nuclear resonance is always high above the continuum threshold such that it encounters in addition some escape width. In clusters, on the other hand, we see a large variation of spectral relations with mass numbers. The resonance is in a large gap of 1ph states for small clusters, which promises small widths. The interference with 1ph comes into play only for $N > 40$ and one should obtain growing widths there. This feature is qualitatively correct. But the spherical jellium model grossly underestimates the widths for small clusters $N \leq 40$. A closer inspection shows that there are, in fact, sizable shape fluctuations about the spherical reference point which makes the spectral distribution of 1ph states more diffuse than shown in Figure 7.2, which leads, in turn, to somewhat more fragmentation width also in small clusters. This will be discussed further in Section 7.3.5. Where the contribution from the escape width is concerned, we see that the continuum threshold for singly charged clusters stays above resonance up to $N \sim 1000$ such that resonances in these clusters are emission-stable, again very different from the nuclear case.

There are some further differences which are not shown in Figure 7.2. These are related to the fact that clusters provide far more chances for variation than nuclei. One can vary the charge state of the clusters and the collectivity of the Mie surface plasmon seems to be enhanced with increasing net charge. One can also vary the material for the clusters which amounts in the jellium model mainly to a variation of the Wigner−Seitz radius r_s. For reasons of space, we refer the reader to [10] where these variations are discussed in detail.

7.3.3 Higher Multipolarities

Photoabsorption spectra are dominated by the dipole resonances which are in any case the prevailing resonance modes. Nonetheless, there are similar resonance modes for other multipolarities. Figure 7.3 shows the strength distributions for $L = 0, 1$ and 2 in nuclei and clusters. The first feature to be remarked is that nuclei have, in fact, one more distinction besides angular momentum L. There are two sorts of fermions, protons and neutrons, which can vibrate together or against each other. This leads to the isoscalar ($T = 0$) and isovector ($T = 1$) modes which obviously behave very differently. The residual interaction is repulsive in the $T = 1$ channel which shifts the resonances up in frequency and leads them into a regime with more spectral fragmentation. The interaction is attractive in the $T = 0$ channel which gives a downshift of the resonances and subsequently less

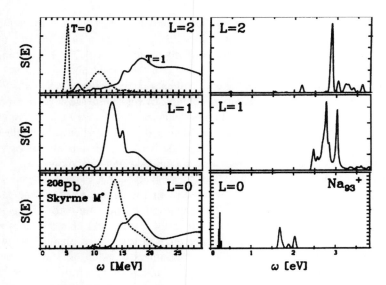

Figure 7.3 Photoabsorption strength for the $L = 0$, 1 and 2 modes in the nucleus ^{208}Pb (left) and in the cluster Na$_{93}^+$ (right). Note the distinction between isovector modes (full lines) and isoscalar modes (dotted lines) in the nuclear case

fragmentation, particularly for the $L = 2$ mode. The dipole channel is a special case in that the $T = 0$ resonance mode there corresponds to pure center-of-mass motion with the exact result $\omega = 0$. The $L = 0$ modes are both less well concentrated. The strength distribution for clusters again shows the dominant $L = 1$. But it also hints that one can expect an interesting collective resonance for $L = 2$.

The experimental access to all these other modes has to use methods beyond simple photoabsorption measurements. They have been studied extensively in the case of nuclei by various sorts of scattering experiments (with electrons, protons, ions), which could even provide the full transition formfactors $F_N(q)$. Detailed scattering experiments in the case of clusters have yet to come.

There exists a further family of modes which is distinguished by a counter-vibration of spin-up and spin-down particles. These are, for example, the Gamow−Teller modes in the nuclear case and spin excitations in clusters. There is a widespread literature on Gamow−Teller resonances which we cannot report here; for a review see [65]. Spin modes in clusters are much less studied because they are not easily accessible in measurements. But there are several theoretical explorations of these modes [66−70]. The residual interaction in the spin channel is attractive and less strong. The spin excitations thus gather at the lower end of the 1ph modes, for small clusters around 1 eV. A particularly interesting feature is the appearance of spontaneously polarized isomers [71, 72], in which the dipole and the spin-dipole modes are mixed [70]. The polarized isomers give, furthermore, rise to an anomalous behavior of the magnetic susceptibility [72].

7.3.4 Dipole Resonance and Deformation

Spherical clusters and nuclei, as discussed in the above subsections, exist mainly for fermionic shell closures. The strong pairing interaction in nuclei restores spherical symmetry also in a vicinity of the magic shells. But there is no comparable restoring force in the case of metal clusters, such that nonmagic clusters are always deformed. This deformation is the immediate result of the electronic shell structure and it has consequences for the plasmon resonances. The resonance modes are split: those along the shorter axes are blue-shifted and those along the longer axes red-shifted. This splitting of the resonances was proposed long ago as a signal of nuclear deformation [73]. But it is not so thouroughly applicable in nuclei because nuclear deformation is often accompanied by large-amplitude shape vibrations which wipes out the desired signal. The situation is much more favorable for metal clusters and the splitting of the Mie plasmon resonance has been used extensively to establish a systematics of cluster deformations [74, 75]. Theoretical predictions of the splitting were worked out much earlier [76, 77]. We report here on a more recent survey on the basis of the soft jellium, which aims to provide not only the splitting but also the appropriate overall position of the resonance peaks [78]. The cluster shape is optimized allowing for fully triaxial deformations. As the interest concentrates now on the collective splitting, the average peak positions are calculated in local RPA. Figure 7.4 shows the resonance peaks along the three principal axes (lower part) and a comparison with experimental peak positions [74] drawn as the ratio of theoretical

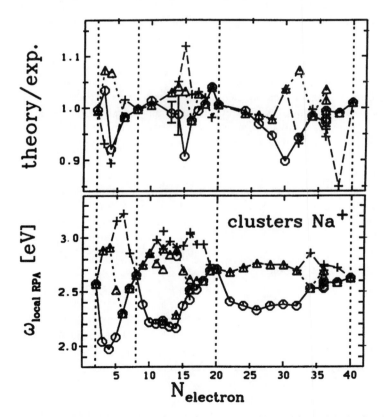

Figure 7.4 Ratio of theoretical to experimental plasmon peak positions for singly charged (and deformed) Na clusters, where the plasmon modes in x, y and z directions are considered separately (upper panel) and theoretical plasmon peak positions for the modes in x, y and z directions (lower panel)

versus experimental frequencies to fit better into the scale (upper part). Fully triaxial systems are distinguished by having three different principal axes and thus a splitting into three different plasmon peaks. Axial symmetry makes two axes of the same length and thus two degenerate peaks where the degeneracy comes at the higher peak for prolate deformations and at the lower frequency for oblate deformations. Spherical symmetry then removes the splitting totally and places all three peaks at the same position. The trends in Figure 7.4 come out as expected. The clusters just above a magic shell have prolate deformation, here up to $N_{el} = 14$. This means that the z axis is stretched whereas the orthogonal direction is compressed. Consequently, the resonance in the z direction comes out lower in energy and in the other direction higher. The cluster just below the next magic shell have oblate deformations. There, the situation is reversed. The z axis is compressed and the r axis stretched and the sequence of the resonance energies is reversed as compared to the prolate shape. That is seen in the transition from $N_{el} = 14$ to $N_{el} = 16$. A regime of triaxial deformation comes in the midst between the magic shells. The above discussed trends are equally seen in the experimental data and the comparison of calculations with measurements in the upper part of Figure 7.4 shows a very satisfying agreement. The larger deviations in the regime of triaxial clusters have a simple explanation. These triaxial systems are, in fact, extremely soft (see also [71]) and the thermal shape fluctuations are huge such that one average deformation, as considered here, is an insufficient representative of that ensemble of shapes. For these very soft systems, a better analysis in terms of full ensembles is required, already for the description of the peak positions. It is, of course, necessary in any case for the description of widths, as will be discussed in Subsection 7.3.5. It is, furthermore, to be remarked that the experiments are performed here with clusters whose temperature is around $T \sim 400$ K, i.e. above the melting point of Na. Spectra for cold clusters show much more fine structure which inhibits a simple analysis in terms of only two or three collective peaks [79]. The jellium model becomes insufficient and ionic structure begins to play a role at these low temperatures, see Subsection 7.3.6.

As hinted above, a comparable systematics of giant resonance splittings cannot be produced for nuclei because too many of them are so soft that the signal is smeared out, as is the case for the few triaxially soft clusters in the sample of Figure 7.4. One thus tries to retrieve the information on nuclear deformation from the amplitude of the low-energy quadrupole modes which is closely related to the transition strength, often called the B(E2) value [65]. This access has been used in [5] for the systematic comparison of cluster and nuclear deformation.

7.3.5 Widths

The RPA spectra already show some width, as was visible from the spectral distribution in Figure 7.1. This width is mainly the spreading width due to Landau fragmentation, which appears if the collective resonance has to share its strength with the near-lying 1ph states (it is called 'Landau damping' in continuous systems with a continuous 1ph spectrum [80]). This effect is the same in nuclei and clusters, although showing up somewhat stronger for nuclei, as discussed in Subsection 7.3.2. However, there are further line-broadening mechanisms, which are of significantly different importance in the two systems.

For nuclei, we encounter substantial escape width (0.5–1 MeV) because the giant resonances stay high above the neutron emission threshold, see e.g. Figure 7.2. And there comes also into play a large fraction of spreading width, or collisional broadening, which is due to a coupling to 2ph states, a mechanism that goes truly beyond the TDLDA. The coupling to 2ph states is often formulated as a coupling to the low-lying shape vibrations of the nucleus [81]. Landau damping and collisional broadening provide comparable contributions to the width. The actual relations depend on the mode (angular momentum and isospin) and on the size of the nucleus [82]. These questions are extensively discussed in the nuclear literature, see e.g. [81, 82].

For clusters, the escape width is less important, particularly for charged ones, as seen above. The impact of a coupling to the 2ph states has not yet been conclusively explored. But the effect is expected to be small in the domain of linear response. There comes, however, a further important mechanism, the line-broadening through the thermal shape fluctuations of the ionic (or jellium) background. This is an incoherent mechanism, as opposed to the nuclear spreading width, from which a part can be viewed as broadening through coherent coupling to the nuclear shape vibrations. The impact of thermal shape fluctuations has been much discussed at various levels [83–86]. The basic mechanism is best seen in the jellium model. We present here briefly the latest analysis from [85].

Thermal shape fluctuations are determined by the interplay between available thermal energy and the energetic cost of deformation. The lower part of Figure 7.5 shows the energy curves for quadrupole (β_2) and octupole (β_3) deformation as compared to the

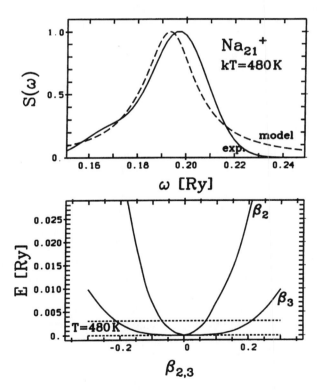

Figure 7.5 Dipole photoabsorption strength for the cluster Na_{21}^+ in the diffuse jellium model after averaging over the thermal ensemble of β_2, β_3 and β_4 shape fluctuations at $T = 480$ K. The experimental strength from [79] is also shown for comparison (upper panel). The potential energy surfaces for β_2 and β_3 deformation compared with the thermal energy at $T = 480$ K are displayed in the lower panel

typical thermal energy at room temperature for the example of the small cluster Na_{21}^+. The deformations are expressed in terms of the dimensionless deformations $\beta_L = 4\pi \langle r^L Y_{L0} \rangle / 3NR^L$ [31]. To give an example, a deformation of $\beta_2 = 0.8$ describes a prolate ellipsoid with axis ratio 2:1. The figure shows that there are non-negligible quadrupole fluctuations in the thermal ensemble. The deformation splits the peaks and averaging over many deformations (centered here around sphericity), then produces again one peak that is broadened by the incoherent superposition. The broadening due to quadrupole fluctuations alone explains about one-third of the experimentally observed width. One also sees from Figure 7.5 that the cluster Na_{21}^+ is much softer in the octupole direction. The octupole deformation breaks reflection symmetry and this symmetry-breaking gives access to a new band of 1ph transition close to the resonance peak. The thermal shape fluctuation thus gives rise to a new Landau fragmentation that was hindered by selection rules at symmetric shapes. This effect adds a dominant contribution to the width, which finally comes out very close to measured data, as can be seen in the upper part of Figure 7.5 comparing the strength distribution from a thermal ensemble of RPA calculations with experimental data. Further (successful) examples are discussed in [85].

This mechanism of thermal line-broadening acts strongest for small clusters $N_{el} \leq 40$. Shape fluctuations shrink for larger clusters, and Landau damping, on the other hand, becomes increasingly important already at the spherical shape. The trends for larger clusters are thus determined by the trends of Landau damping which have been discussed extensively in [60, 87]. The width increases steadily up to $N \approx 1000$ due to increasing level density. But then it bends over and decreases because momentum conservation is more and more restored as one approaches the infinitely large system. From general arguments of momentum coupling through the surface region one can predict a width $\propto$ 1/radius $\propto N^{-1/3}$ for very large clusters [54].

7.3.6 Effect of Ions

Thus far we have mainly discussed the soft jellium model for the ionic background. The results from that approach agree generally very well with experimental results above the melting point of Na. But the measurements for colder Na clusters show much more detailed structures which are due to ionic effects. We illustrate that for the case of Na_9^+. Figure 7.2, drawn from the jellium mode, shows that the plasmon resonance for a cluster lies in the middle of a huge shell gap, which inevitably leads to the prediction of one single strong resonance peak, see the jellium peak in Figure 7.6. But the experimental low-temperature distribution from [79] clearly shows a fragmentation into two peaks which is at variance with the jellium results. The curve denoted 'pseudopotential' in the figure is the result of a similar TDLDA calculation, where the full ionic structure has been taken into account (for details of the calculation and an extended discussion see [88]). It employs a pseudopotential composed from two error functions whose strength is adjusted to deliver an appropriate position of the plasmon peak (the question of pseudopotentials and plasmon response is presently being checked more extensively [89]). The result with ionic structure also shows this fragmented peak which is, in fact, composed from a single z peak and the fragmented x and y peaks which embrace the center peak. It is worth mentioning that such a fragmented structure has also been obtained with quantum-chemical *ab initio* methods, as reviewed in [90]. We thus find that a calculation with full ionic structure is required, if one aim at the detailed fragmentation structures, usually resolved in low-temperature experiments [79]. The computation of detailed ionic structures is very expansive and

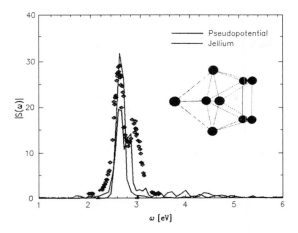

Figure 7.6 Strength functions of dipole oscillations in the linear regime for Na_9^+ with full ionic structure (solid line), on a soft jellium background (dotted), and taken from experimental data of [79] (diamonds). A pseudopotential with the strength of [18] has been employed. The ionic configuration is illustrated in the insert

technically limited to smaller clusters (today about $N \lesssim 30$). A more efficient approximate way is offered by the pseudopotential perturbation theory, which is discussed extensively in Chapter 1 of this book. After all, the soft jellium remains, nonetheless, well suited for a first survey of global features and for larger temperatures.

Ionic structures are, of course, also shaken by thermal fluctuations and one needs, in principle, to consider a thermal ensemble of configurations similar to the thermal ensemble of jellium shapes discussed in Subsection 7.3.5 above. A simple and efficient ensemble approach has been discussed by [86] where only a thermal mix of a few low-lying isomers is considered. For Na_8, the four lowest configurations are taken into account. The result at room temperature is shown and compared with experimental data in Figure 7.7. Note

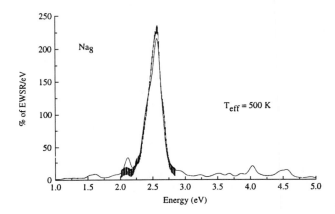

Figure 7.7 Strength functions of dipole oscillations in the linear regime for Na_9^+ computed as an ensemble average at a temperature of $T = 500$ K over the four lowest isomeric configurations in a fully three-dimensional calculation [86]. The peak position has been red-shifted to fit the experimental position

that the peak position was readjusted by a simple red-shift of the spectra, because the pseudopotential for these calculations gave too high a plasmon frequency. The agreement of the width is very satisfying. More extensive ensemble calculations have been performed recently and are reported in Chapter 1 of this book.

7.4 RESONANCE DYNAMICS BEYOND THE LINEAR REGIME

Although linear response is a more restricted limiting case of the general dynamics, it usually attracts more attention in the early studies of a system because it is the case that is more easily overseen and can be accessed in many ways, independently of the excitation mechanism. The general large-amplitude dynamics, on the other hand, is much richer in phenomena and variants. For example, the results can now depend sensitively on the excitation, i.e. on the reaction channel, and there can open up a wider choice of final states. Correspondingly, one needs also to consider more observables. It is clear that linear response theory, as such, ceases to be applicable and that one has to return to fully fledged time-dependent formalisms, e.g. the TDLDA as outlined in Section 7.2.3. Nuclear physics has a long tradition of studying truly nonlinear processes, particularly in connection with heavy-ion reactions, many of them even going beyond the capabilities of a TDLDA; for reviews see e.g. [91–94]. It is only recently that the nonlinear domain of electronic dynamics has been attacked in cluster physics and most of these phenomena can still be described by TDLDA. In the following we shall restrict our discussions to situations to which TDLDA is applicable and in which spectral properties of collective response still play a dominant role. As a new and complementary observable, we will also consider particle emission.

7.4.1 On Excitation Mechanisms

Our choice of excitation mechanisms is essentially governed by the experimental possibilities for clusters. These are short and violent excitations, such as are delivered by intense laser beams or Coulomb excitation with energetic ions. In the case of lasers, we refer here to irradiation with strong laser pulses (peak intensity I about 10^{11} to 10^{16} W/cm^2, leading to electric field amplitudes of order $E_0 \sim 0.5-15$ V/a_0) [6]. These strong laser pulses are typically much shorter than 100 femtoseconds, with frequencies ω_0 extending from the infrared to the ultraviolet. New aspects come up with those fs pulses. While a low-intensity laser beam with long coherence time very selectively excites one frequency in the spectrum of the cluster, the response to a very short and strong pulse is dominated by field effects and may access all spectral states at once.

Another line of investigation focuses on excitation with energetic and highly charged projectiles passing by at a distance larger than the cluster radius, such that only the Coulomb field of the projectile is explored. These collision processes are usually even faster than laser pulses and take place on a subfemtosecond time scale. For exploratory calculations, one may consider the limit of an arbitrarily short interaction which can easily be modeled by an instantaneous initial shift of the whole electron cloud [95]. This is, so to say, a generic excitation model. In turn, the overall strength of the excitation can be varied by varying the amplitude of the initial shift.

In both excitation mechanisms (laser or ion collision), the cluster is exposed to a short (at most 50 fs) and violent electromagnetic pulse. The first phase of the cluster's response is thus exclusively governed by the electronic response. The cluster exhibits large dipole oscillations (Mie plasmon) and it quickly loses (within about one plasmon cycle) a sizable fraction of its valence electrons, namely that part that couples directly to the open channels of electron states in the continuum. The direct electron emission is thus the dominant mechanism for energy loss. The further cluster de-excitation then proceeds by thermalization of the electron cloud and by coupling of electronic to ionic degrees of freedom. Thermalization leads, on the picosecond time scale [96], to electron evaporation through thermal fluctuations. Meanwhile, ionic motion takes the lead in the case of highly excited and/or highly charged clusters which undergo fission or fragmentation on a time scale of the order of nanoseconds.

In the nuclear case a preferred tool for large-amplitude excitations is nuclear collisions, which, as a function of the beam energy, allow us to access a wealth of different scenarios. Multiple giant resonances, for example, are excited in distant collisions with highly charged heavy ions at beam energies in the Fermi velocity domain [97]. This process is called 'Coulomb excitation' and it can be modeled, as in the cluster case, by an instantaneous initial shift. Various sorts of heavy-ion reactions (fusion, scattering, fragmentation) can be triggered at smaller impact parameters. At higher beam energies single-particle degrees of freedom progressively overbalance collective ones and dynamics becomes more and more governed by two-body effects, which go beyond the capabilities of a mean-field description (TDLDA). This also corresponds to a disappearance of giant resonances as useful indicators of the dynamics in favor of other collective variables, or in favor of totally different observables, e.g. rare particle emission [92] or fragmentation patterns [93]. We shall hence restrict ourselves to moderate beam energies, where TDLDA calculations make sense and in which giant resonances still provide a good indicator of the nuclear response. We shall also consider neutron and proton emission, which, as in the electron case, represent the preferred de-excitation channel.

7.4.2 Typical Observables

As discussed in the previous subsection, we concentrate on excitations mechanisms that can preferably be analyzed in terms of the dipole response and particle emission, in both the cluster and nuclear cases. The spectra of the dipole response has already been the central observable in the small-amplitude regime, discussed in Section 7.3.1. We thus have here the analyzing tool that covers both the linear and the nonlinear regimes, and thus allows direct comparisons of them. We analyze the dipole response by recording the dipole moment $D(t) = \int d^3 r \, r \rho(r, t)$ in a finite analyzing box $\mathcal{V}$ covering a $2r_s$ vicinity around the system [95]. The analyzing box is introduced in order to avoid perturbation of the signal from emitted particles flowing through the whole numerical grid before they are absorbed at the boundaries. The signal $D(t)$ is then Fourier-transformed into the frequency domain yielding $\tilde{D}(\omega)$. The result can be viewed and discussed in two ways: either in terms of the dipole strength function $S_D(\omega) = \Im\{\tilde{D}(\omega)\}$, or in terms of the dipole power spectrum $\mathcal{P}(\omega) = |\tilde{D}(\omega)|^2$. The strength function is more appropriate in the domain of small-amplitude excitations and it is related to the photoabsorption cross-section, see Section 7.3.1. The power spectrum is more suitable to analyze the dynamics in the nonlinear regime and the associated dissipative processes. Both objects serve equally well to characterize the general spectral structure of the system. For a detailed discussion of the spectral analysis see [95].

The description of particle escape relies on the basic relation

$$N(t) = \int_{\mathcal{V}} \mathrm{d}^3 r \rho(\boldsymbol{r}, t), \tag{10}$$

which associates the number of bound particles $N(t)$ with the time-dependent density $\rho(\boldsymbol{r}, t)$ found within the analyzing box $\mathcal{V}$. From $N(t)$ one can calculate the total number of escaped particles as $N_{\mathrm{esc}}(t) = N(t = 0) - N(t)$. Equation (10) is based on the assumption that the particle flux crossing the boundary of the analyzing box will eventually depart to infinity and thus corresponds to that part of the total time-dependent wave function that is in the continuum. Strictly speaking, such a criterion is meaningful only after long times, when the continuum contributions have propagated very far away from the center. There is, furthermore, a fraction of Rydberg states that is thus counted to the lost particles. This fraction, however, is very small and Eq. (10) has proven to be a useful definition of the bound-state occupation probability in a number of applications involving atoms in strong fields, which are probably the most critical test cases due to the dominance of the long-range Coulomb field [98]. Nuclei are rather uncritical in that respect and a finite analyzing radius has been succesfully employed there [99]. Clusters lie somewhere in between and usage of definition (10) has also been tested there [100].

In the case of clusters, a more detailed link with experiment can be established by computing the ionization probabilities for the various charge states. This information can be reliably well extracted from the single-electron wave functions of TDLDA, as has been demonstrated for the case of atoms [101]. First applications for clusters can be found in [102, 103].

Note, finally, that the average excitation energy E^* is not an observable here but a means to characterize the amplitude of the excitation. It is to be distinguished from the frequency ω of an excitation which is an observable deduced from the dipole spectrum. TDLDA describes, in fact, a sort of classical oscillations that can persist at any amplitude or E^*. Eigenstates of well-defined excitation energy can only be obtained by requantizing TDLDA, which is a trivial quantization of the harmonic oscillator in the linear regime and would require a complicated search for periodic TDLDA solutions in the general large-amplitude case.

7.4.3 From Small- to Large-Amplitude Dynamics

Collective response and particle emission are of course intrinsically connected. However, clusters offer the opportunity to separate them by varying the charge state of the system. Highly charged clusters have a high ionization threshold, which allows a swift shift from the linear to the nonlinear regime while keeping emission low. This means, on the other hand, that one is possibly dealing with metastable systems. But fission is slow as compared to plasmon oscillations such that there remains enough time to establish spectral patterns. We show in Figure 7.8 a typical strength function for the very well bound system $\mathrm{Na_{44}}^{4+}$. The system is excited by an instantaneous dipole shift, which provides a simple but realistic model of Coulomb excitation. We consider here a small-amplitude (linear-regime) excitation with $E^* = 0.6$ eV and a large-amplitude (multiplasmon-regime) case with $E^* = 10.0$ eV. In the linear case RPA results (limiting case of infinitesimal

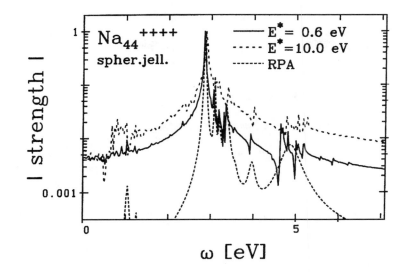

Figure 7.8 Absolute value of strength function for $Na_{44}{}^{++}$ excited with an initial dipole shift at two different amplitudes, as indicated by the excitation energies. The case $E^* = 0.6$ eV stays in the linear regime whereas $E^* = 10.0$ eV is in the multiplasmon regime (about 3–4 plasmon excitation). The strength function from linearized TDLDA (= RPA) is also drawn for comparison. The RPA peaks have been given a width of 0.04 eV for smooth representation and better comparison with the TDLDA results. All results have been normalized to the same maximum peak height for better comparison

amplitude) are also shown for comparison. We confine the presentation to the dipole strength (although the power spectrum would be more appropriate in the nonlinear case) for reasons of space.

It is gratifying to see that the small-amplitude result coincides with the RPA result, particularly in the region of the plasmon peak, $\omega = 2.8$–3.3 eV. But the TDLDA analysis is plagued by a larger background, so that the much smaller side-peaks (below 10^{-2} of the main peak) are essentially hidden, although some trace of them remains visible from the zigzag pattern in their energy region. Up to this difference, whose origin is mainly of a technical nature, TDLDA and RPA yield very similar patterns, which confirms that RPA is indeed the small-amplitude limit of TDLDA and that, in turn, the spectral analysis of (small-amplitude) TDLDA is capable of recovering the RPA results. Full TDLDA can thus provide the gross features of the strength function, i.e. the position and approximate strength of the strongest peaks. But finite sampling and numerical fluctuations yield background noise which overlays all finer details of the spectrum. This makes RPA practically superior in this context, at least for 'simple' systems, namely those that display a sufficiently high degree of symmetry to allow for exploitation of additional quantum numbers. In the general case of systems without any symmetry, full TDLDA, in turn, represents a valuable alternative to a (hardly possible) RPA calculation in full configuration space. In addition, TDLDA is easy to code and it is valuable for a first scan of the spectrum to search for the gross features and the dominant peaks with moderate resolution.

It is astonishing to see a much similar spectral distribution in the multiplasmon regime (the excitation energy of 10 eV is worth about four plasmons). The dipole resonance thus turns out to be a very robust and harmonic mode. There are, of course, some changes.

The multiplasmon case leads to broader peaks and an enhanced background, which is due to the larger fluctuations of the mean-field, and to some more electron emission. All spectral hints of the side peaks with little strength are still visible, even more pronounced at the lower-frequency side.

7.4.4 Deeper into Reaction Mechanisms

7.4.4.1 *Coulomb Excitation in Clusters and Nuclei*

In a next step, we compare nuclear and cluster response in the generic case of Coulomb excitation, as modeled by an initial shift of the electrons (respectively neutrons) with respect to ions (respectively protons). We first consider the nuclear giant dipole resonance. The lower panel of Figure 7.9 shows the power spectrum of the dipole along the z axis (symmetry axis) of ^{36}S after Coulomb excitation for several amplitudes with average excitation energies as indicated. The small-amplitude case represents the nuclear excitation spectrum in the linear regime as it is known from nuclear RPA calculations. We

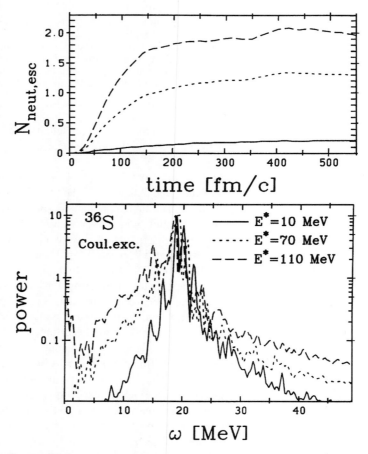

Figure 7.9 Spectra of nuclear resonances at small and large amplitudes (lower panel), as produced by Coulomb excitation. The upper panel shows the number of emitted neutrons as a function of time

see a pronounced peak corresponding to the giant dipole resonance (along the z axis). This resonance obviously survives the strong excitation associated with the larger-amplitude cases. The largest average excitation energy corresponds to as much as 4.5 resonance modes stacked on top of each other. The giant resonance thus resembles a harmonic oscillator, where coherent excitation is easily possible at any amplitude with unchanged frequency. The enormous robustness of the dipole giant resonance is supported by experiments where at least a double resonance excitation has been confirmed [97]. At second glance, one sees some modifications of the resonance peak at high energy, namely a slight broadening which is quite an expected effect because the underlying mean field undergoes stronger fluctuations. The broadening is nevertheless somewhat underestimated by pure TDLDA. Additional broadening mechanisms come from two-body collisions, but there are good reasons to expect no qualitative changes from collisions for the dipole resonance [81, 104].

For comparison we now take a 'similar' cluster case. Figure 7.10 shows results for comparable Coulomb excitations (modeled by initial shift) of the cluster Na_9^+ in the 144 Cylindrically Averaged Pseudopotential Scheme (CAPS) configuration, which consists of two subsequent rings each covering four ions and topped by one single ion [105].

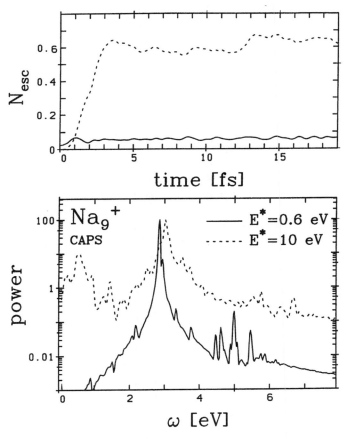

Figure 7.10 Spectra of cluster resonances at small and large amplitudes (lower panel), as produced by Coulomb excitation. The upper panel shows the number of emitted electrons as a function of time

Note that the 144 configuration is close to a spherical shape but with a small quadrupole moment and with broken reflection symmetry about the $x-y$ plane and correspondingly nonvanishing octupole moment. This *a priori* leads to a fragmented spectrum because of the large number of open 1ph transitions. Still, the Mie plasmon resonance remains a robust mode and dominates in all cases by at least two orders of magnitudes. Note that the peak position is blue-shifted for the large-amplitude case. This is due to the loss of electrons which enhances the net charge of the cluster, which, in turn, enhances the Mie plasmon energy.

In both (nuclear and cluster) cases, the initial shift leads to particle emission (upper panels of Figures 7.9 and 7.10). Note that the higher the deposited energy, the larger the number of emitted particles. Let us consider for example the highest E^* cases (which correspond to roughly 4 resonance energies in both systems), for which sizable emission takes place. Asymptotic emission values are comparable in terms of percentage of the initial particle number (10% for the nucleus, 8% for the cluster). Furthermore, the time scales on which most of the emission takes place are similar in the two systems. In both cases, it takes somewhat less than 2 collective periods to attain the asymptotic value, whereby the transition is slightly smoother in the nuclear system.

Altogether, we thus see a striking similarity, up to details, between the nuclear and the cluster responses for this 'clean' Coulomb excitation mechanism with the prominent feature that the resonance modes turn out to represent robust, harmonic oscillations. It remains to be seen to what extent this behavior persists in other experimental situations. In the following we focus on heavy-ion collisions and laser–cluster interactions as complementary and widely studied excitation mechanisms.

7.4.4.2 Heavy-Ion Collisions

Variations of the entrance channel, in fact, can easily perturb the collectivity and harmonicity of the nuclear resonance modes. Figure 7.11 shows the result from a head-on collision (zero impact parameter) of ^{12}C on ^{24}Mg leading to an excited ^{36}S as compound system. Although the average excitation is now only about 1.5 times the resonance energy, the spectrum turns out more noisy, almost filling all frequencies up to the dipole resonance (which is lowered here due to the very stretched configuration). This is the typical result for heavy-ion dynamics which suggested that nuclear collective motion is very anharmonic and tends to chaos if driven in the nonlinear regime [106]. But this anharmonicity is fed from the low-energy collective modes in nuclei associated with strong surface deformations. The nuclear collision on the way to compound nucleus formation excites a large fraction of (anharmonic) surface modes together with the dipole resonance. Nonlinear effects couple these modes and this eventually results in the noisy dipole spectrum. The emission pattern is displayed in the upper panel of Figure 7.11. One observes a steady rate of emission on a time scale of about 10 times the dipole period, at variance with the prompt emission following Coulomb excitation seen in Figure 7.9. This is due to the fact that the excitation process (formation of the compound system) is stretched over a longer time span. Still, beyond about 600 fm/c $= 1.8 \ 10^{-21}$ s emission seems to reach an asymptotic value. This example thus shows that the chaotic or harmonic pattern of a nuclear spectrum can depend sensitively on the excitation mechanism. Another 'destroyer' of collectivity is the charge asymmetry of the ground state. For example, a Coulomb excitation of the neutron-rich ^{44}S would lead to a much noisier spectrum than the one displayed

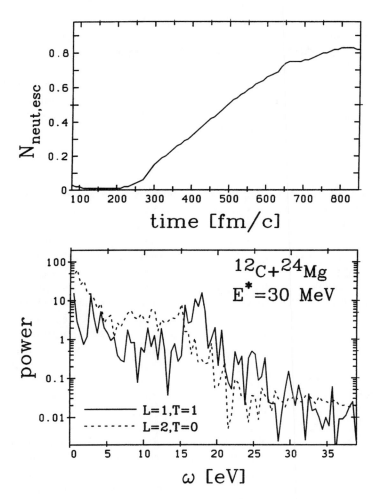

Figure 7.11 Number of escaping nucleons (upper panel) and power spectra (lower panel) for dipole and quadrupole in a heavy-ion collision of ^{12}C on ^{24}Mg

in Figure 7.9. Because of the large number of neutrons, the latter are less bound and lead to a more diffuse resonance spectrum for ^{44}S. Collectivity is nevertheless not as badly destroyed as in the case of collisional excitation. In any case, the best chances to observe clean multiple resonances exist for pure Coulomb excitation of nuclei deep in the valley of stability [107].

As seen above, the destruction of the resonance in the case of heavy-ion collisions is due to the coupling between surface and resonance modes. The cluster analog of nuclear surface vibrations is ionic motion. But this takes place at a much slower pace and cannot interfere dynamically with the electronic resonances. The plasmon is thus more robust against varying excitation mechanisms. There remains, however, the static aspect of the enormous surface distortions during nuclear fusion. Such a distortion could also be produced in the analogous process of cluster fission. And, indeed, it has been found that there is a chance to dissolve the plasmon into a broad, fragmented spectrum when going to extremely deformed shapes during cluster fission. Here one can observe a dramatic splitting and fragmentation of the resonance which leads to a noisy response in the nonlinear regime [108]. This raises immediately the further question whether other distortions could also destroy the collectivity of the plasmon response. A systematic survey of plasmon resonances in Na clusters deposited on an NaCl surface has shown that the already sizable distortions through the surface is not sufficient to destroy the pronounced collectivity of the plasmon [109]. It is thus to be concluded that one needs these extremely fuzzy shapes during fission or fragmentation to achieve sufficient spectral chaos.

7.4.4.3   Cluster Response to an Intense Femtosecond Laser

In contrast to Coulomb excitation, today's intense laser pulses still spread over at least several 10 fs and the interaction process needs to be taken into account with its detailed time profile. Systematic calculations of laser-irradiated clusters for such laser parameters have shown that the cluster response depends very much on the actual laser frequency [102]. As long as the frequency stays far away from the plasmon resonance, the dipole response closely follows the pulse profile and disappears if the laser is switched off. In turn, if the laser frequency is close to the Mie resonance, one observes damping through electron emission and a true excitation of the plasmon resonance which lasts even after the laser has been switched off. We consider here an example close to resonance, i.e. an interaction between Na_9^+ and a laser with a 20-fs Gaussian pulse of peak intensity $I = 10^{12}$ W/cm^2 and photon energy $\hbar\omega_0 = 2.86$ eV, i.e. 10% above the Mie plasmon peak at $\hbar\omega_M = 2.6$ eV. The results are shown in Figure 7.12. For the first 30 fs, the electronic dipole moment $D(t)$ follows the profile of the laser pulse. Then, however, $D(t)$ slightly falls back below the pulse envelope, actually even before the pulse reaches its peak. This reflects the fact that, at this instant, the system starts to ionize rapidly. Due to this quite abrupt loss of electrons ($N_{esc} = 4.5$ at the end of the pulse), the dipole oscillations are strongly damped. Around $t = 40$ fs, however, the rate of electron emission becomes weaker and we observe a new phenomenon: as the charge state of the cluster increases, the collective oscillation of the residual electron cloud is swept towards *higher* frequencies. As a consequence, after having gone through a minimum at $t = 35$ fs, $D(t)$ now *increases* again and reaches a maximum around $t = 50$ fs. This corresponds to the instant when the collective oscillation of the residual electron cloud comes just into resonance with the laser. As the collective oscillation frequency continues to increase, the amplitude of $D(t)$ decreases again. After the pulse is switched off and electron emission has come to an end, we find that the remaining electrons continue to perform collective oscillations at the new plasmon frequency of 2.9 eV.

7.4.5   Semiclassical Approximations to TDLDA

Semiclassical methods have been widely used for describing nuclear dynamics. In the case of intermediate-energy heavy-ion collisions semiclassical methods, based on Boltzmann-like kinetic equations, have even become a standard tool during the last decade [92, 94, 110, 111] and have been successfully used in many studies of high-energy nuclear collisions. Yet, the formal justification of these equations is not fully settled [112]. While at sufficiently high beam energy, one can hope that a sufficiently collisional regime takes place to allow us to overlook still unsatisfactory details in the description, this is no longer

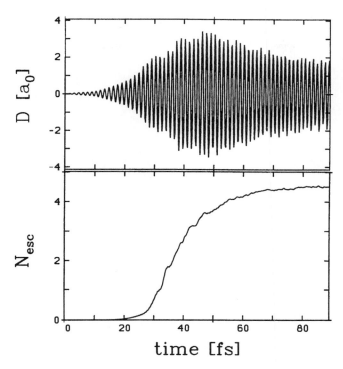

Figure 7.12 Full time-dependent dipole response of Na_9^+ to a 20-fs Gaussian laser pulse with peak at $t = 32.2$ fs, peak intensity 10^{12} W/cm^2 and photon energy 2.86 eV (upper panel), and number of escaping electrons for the same laser parameters (lower panel)

the case at moderate beam energies, such as those we have been considering here. The *a priori* appropriate Vlasov equation turns out to be poorly justified in many respects, both formally [113] and numerically [114, 115]. A major problem lies here in the insufficient treatment of the Pauli principle, which can be badly violated in long time simulations of the Vlasov equation. Although a more correct treatment is conceivable [116], the use of the Vlasov equation remains controversial for studying resonance dynamics. A way out of this problem could be offered by the time-dependent Thomas–Fermi (TDTF) approach [117], which assumes a hydrodynamical picture of the flow, employing an equation of state that properly accounts for the Pauli principle. Although such an approach can be used for the GDR, it is well known that it badly fails in the case of the other multipolarities such as the giant quadrupole resonance [118]. It is thus of limited use in a nuclear context.

In the (younger) field of cluster physics, semiclassical methods have been less systematically explored, although several applications exist [2, 119, 120]. The Vlasov equation suffers from the same formal [121] and numerical [122] deficiencies as its nuclear counterpart, and this defect can be cured in a similar (expensive) way [123]. A more promising alternative is here offered by TDTF [124]. Careful studies of the phase-space structure of the electron cloud in a cluster show that the local degree of anisotropy of the Fermi sphere remains reasonably small in the dynamical situations considered here. The situation is more gratifying than in the nuclear case, because of the overwhelming dominance of the dipole response in the cluster case, a mode for which velocity anisotropy is small.

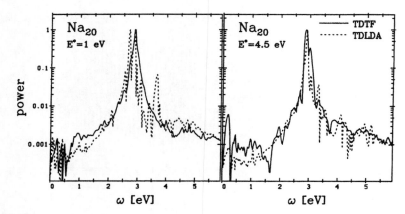

Figure 7.13 Dipole power spectra in Na_{20} in the linear (left part) or nonlinear (right part) regimes. The oscillatory motion is performed along the longest axis of the cluster

Hence, a TDTF picture of cluster dynamics seems justified. It furthermore constitutes an interesting attempt at a fully density-based description of the dynamics, thus providing a true time-dependent model of density functional theory. We illustrate the capability of TDTF in clusters in Figure 7.13, which displays the power spectra of Na_{20} (with CAPS ionic structure) in both TDTF and TDLDA [124]. Note that Na_{20} is a demanding case because of the strong interplay between collective and single-particle motion [54], which renders the comparison with a full TDLDA calculation decisive. The displayed spectra correspond to a shift along the longest cluster's axis, in both the linear ($E^* = 2.6$ eV) and in the nonlinear ($E^* = 9.7$ eV) regimes. In the linear case, the TDLDA spectrum shows a fragmented, double-peak structure. The TDTF approach, averaging over single-particle structures, smoothes the power distribution with respect to the TDLDA result and yields only one average plasmon peak. Notice, also, that it fails to reproduce the ph transition at 3.6 eV. The nonlinear TDLDA spectrum, on the contrary, hardly shows any structure. In this regime, the intrinsic deficiencies of TDTF to reproduce Landau fragmentation are of little importance and the spectra obtained in the two methods become very similar. From these computations in Na_{20} we conclude that TDTF can be reasonably trusted as far as the high-energy electron dynamics is concerned, as could be expected for the semiclassical methods at high energy.

7.5 CONCLUSION

The comparison of cluster and nuclear dynamics in this chapter has shown similarities concerning several prominent features: TDLDA is a well-established and widely used tool to describe the dynamics in the (multiple) resonance regime. There are pronounced dipole resonance modes which represent harmonic oscillations up to rather large amplitudes far beyond the linear response regime. These modes are well accessed by Coulomb excitation through a highly charged ion passing by with speeds in the Fermi velocity domain. Direct particle emission follows the excitation very quickly, within the time of about one resonance cycle.

At a second glance, however, one spots many differences in detail. The nuclear energy-density functionals need to be more elaborate, accounting for non-unit effective mass, spin–orbit force, and gradient corrections. The trend of the dipole resonance frequency with mass number is $\propto N^{-1/3}$ in the case of nuclei due to the short-range nuclear interactions whereas the long-range Coulomb force leads to a finite value in the limit $N \to \infty$. Nuclei provide a richer selection of important resonance modes with other multipolarities and due to the distinction between isospin-one and isospin-zero excitations. Nuclear dynamics is also much more sensitive to the actual excitation mechanism due to the close time scales of surface motion and resonance oscillations. Clusters, on the other hand, can be excited by laser pulses which open up a wealth of interesting experiments allowing for systematic variations of temporal profile as well as strength of the excitation. Investigations of cluster dynamics with fs laser pulses are presently a quickly growing field of research which will allow the investigations of many new aspects of electronic dynamics.

7.6 ACKNOWLEDGMENTS

The authors thank the Franco-German exchange program PROCOPE 95073 and Institut Universitaire de France for financial support during the realization of this work. They are furthermore glad to acknowledge many fruitful discussions with M. Brack, F. Calvayrac, A. Domps, C. Kohl, S. Kuemmel and C. Ullrich.

7.7 REFERENCES

[1]   W. de Heer, *Rev. Mod. Phys.* **65**, 611 (1993).
[2]   M. Brack, *Rev. Mod. Phys.* **65**, 677 (1993).
[3]   J. Pedersen, S. Bjørnholm, J. Borggreen, K. Hansen, T. P. Martin and H. D. Rasmussen, *Nature* **353**, 733 (1991); O. Genzken, *Mod. Phys. Lett. B* **4**, 197 (1993).
[4]   M. Brack and R. K. Bhaduri, *Semiclassical Physics*, Addison-Wesley, Reading, MA, 1997.
[5]   M. Koskinen, P. O. Lipas and M. Manninen, *Nucl. Phys. A* **591**, 421 (1995).
[6]   U. Näher, S. Bjørnholm, S. Frauendorf, F. Garcias and C. Guet, *Phys. Rep.* **285**, 245 (1997).
[7]   G. F. Bertsch and R. Broglia, *Oscillations in Finite Quantum Systems*, Cambridge University Press, 1994.
[8]   P.-G. Reinhard and Y. Gambhir, *Ann. Phys.* (Leipzig) **1**, 598 (1992).
[9]   P.-G. Reinhard, *Ann. Phys.* (Leipzig) **1**, 632 (1992).
[10] P.-G. Reinhard, O. Genzken and M. Brack, *Ann. Phys.* (Leipzig) **5**, 576 (1996).
[11] S. Weisgerber and P.-G. Reinhard, *Ann. Phys.* (Leipzig) **2**, 666 (1993).
[12] K. Goeke and P.-G. Reinhard, (eds) *TDHF and Beyond*, Lecture Notes in Physics Vol. 171, Springer, Berlin, 1982.
[13] J. W. Negele, *Rev. Mod. Phys.* **54**, 913 (1982).
[14] K. T. R. Davies, K. R. S. Devi, S. E. Koonin and M. R. Strayer, in *Treatise of Heavy-Ion Science, Vol. 3*, ed. D. A. Bromley (Plenum, 1985).
[15] R. M. Dreizler and E. K. U. Gross, *Density Functional Theory*, Springer, Berlin, 1990.
[16] R. O. Jones and O. Gunnarsson, *Rev. Mod. Phys.* **61**, 689 (1989).
[17] O. Gunnarsson and B. I. Lundqvist, *Phys. Rev. B* **13**, 4274 (1976).
[18] G. B. Bachelet, D. R. Hamann and M. Schlüter, *Phys. Rev B* **26**, 4199 (1982).
[19] S. Goedecker, M. Teter and J. Hutter, *Phys. Rev. B* **54**, 1703 (1996).
[20] N. W. Ashcroft and D. C. Langreth, *Phys. Rev.* **155**, 682 (1067).
[21] C. Fiolhais, J. P. Perdew, S. Q. Armster, J. M. MacLaren and M. Brajczewska, *Phys. Rev. B* **51**, 14001 (1995).
[22] S. A. Blundell and C. Guet, *Z. Phys. D* **28**, 81 (1993).
[23] S. Kümmel and M. Brack, in preparation.
[24] L. Serra and A. Rubio, *Phys. Rev. Lett.* **78**, 1428 (1997).
[25] W. Ekardt, *Phys. Rev. Lett.* **52**, 1925 (1984).

[26] D. E. Beck, *Sol. St. Comm.* **49**, 381 (1984).

[27] A. Rubio, L. C. Balbas and J. A. Alonso, *Z. Phys. D* **19**, 93 (1991).

[28] P.-G. Reinhard, S. Weisgerber, O. Genzken and M. Brack, Lecture Notes in Physics, Vol. **404**, 254 (1992).

[29] L. Serra, G. B. Bachelet, N. Van Giai and E. Lipparini, *Phys. Rev. B* **48**, 14708 (1993).

[30] P.-G. Reinhard, S. Weisgerber, O. Genzken and M. Brack, *Z. Phys. A* **349**, 219 (1994).

[31] G. Lauritsch, P.-G. Reinhard, M. Brack and J. Meyer, *Phys. Lett. A* **160**, 179 (1991).

[32] A. Liebsch, *Phys. Rev. B* **36**, 7378 (1987).

[33] J. P. Perdew, H. Q. Tran and E. D. Smith. *Phys. Rev. B* **42**, 11627 (1990); M. Brajczewska, C. Fiolhais and J. P. Perdew, *Int. J. Quant. Chem.* **S27**, 249 (1993).

[34] B. Montag, P.-G. Reinhard and J. Meyer, *Z. Phys. D* **32**, 125 (1994).

[35] E. K. U. Gross and W. Kohn, *Adv. Quant. Chem.* **21**, 255 (1990).

[36] E. K. U. Gross, C. A. Ullrich and U. J. Gossmann, in *Density Functional Theory*, ed. E. K. U. Gross and R. M. Dreizler, NATO ASI series B337 (Plenum Press, New York, 1994).

[37] E. K. U. Gross, J. F. Dobson and M. Petersilka, in *Topics in Current Chemistry*, ed. R. F. Nalewajski (Springer, 1996).

[38] T. H. R. Skyrme, *Nucl. Phys.* **9**, 615 (1959).

[39] D. Vautherin and D. M. Brink, *Phys. Rev. C* **5**, 626 (1972).

[40] J. F. Berger, M. Girod and D. Gogny, *Nucl. Phys. A* **428**, 23c (1984).

[41] H. P. Dürr, *Phys. Rev.* **103**, 469 (1956).

[42] J. D. Walecka, *Ann. Phys.* (N.Y.) **83**, 491 (1974).

[43] J. Boguta and A. R. Bodmer, *Nucl. Phys. A* **292**, 414 (1977).

[44] B. D. Serot and J. D. Walecka, *Adv. Nucl. Phys.* **16**, 1 (1986).

[45] P.-G. Reinhard, *Rep. Prog. Phys.* **52**, 439 (1989).

[46] P. Ring, *Prog. Part. Nucl. Phys.* **37**, 193 (1996).

[47] P.-G. Reinhard and H. Flocard, *Nucl. Phys. A* **584**, 467 (1995).

[48] P.-G. Reinhard and C. Toepffer, *Int. J. Mod. Phys.* **E3**, 435 (1994).

[49] Y. M. Engel, D. M. Brink, K. Goeke, S. J. Krieger and D. Vautherin, *Nucl. Phys. A* **249**, 215 (1975).

[50] J. P. Perdew, K. Barke and M. Ernzerhof, *Phys. Rev. Lett.* **77**, 3869 (1996).

[51] W. Kohn and L. J. Sham, *Phys. Rev.* **140**, A1133 (1965).

[52] V. Blum, G. Lauritsch, J. A. Maruhn and P.-G. Reinhard, *J. Comp. Phys.* **100**, 364 (1992).

[53] F. Calvayrac, PhD thesis, Toulouse, 1998, unpublished.

[54] C. Yannouleas and R. A. Broglia, *Ann. Phys.* (N.Y.) **217**, 105 (1992).

[55] G. F. Bertsch, *Comp. Phys. Comm.* **60**, 247 (1990).

[56] M. Brack, *Phys. Rev. B* **39**, 3533 (1989).

[57] P.-G. Reinhard, M. Brack and O. Genzken, *Phys. Rev. A* **41**, 5568 (1990).

[58] O. Bohigas, A. M. Lane and J. Martorell, *Phys. Rep.* **51**, 267 (1979).

[59] V. O. Nesterenko, W. Kleinig and V. V. Guzdkov. *Z. Phys. D* **34**, 271 (1995).

[60] J. Babst and P.-G. Reinhard, *Z. Phys. D* **42**, 209 (1997).

[61] R. Neuhausen, J. W. Lightbody, S. P. Fivozinsky and S. Penner, *Nucl. Phys. A* **263**, 249 (1976); A. M. Bernstein, *Adv. Nucl. Phys.* **3**, 325 (1969).

[62] M. Goldhaber and E. Teller, *Phys. Rev.* **74**, 1046 (1948).

[63] T. Reiners, C. Ellert, M. Schmidt and H. Haberland, *Phys. Rev. Lett.* **74**, 1558 (1995).

[64] J. Treiner, H. Krivine, O. Bohigas and J. Martorell, *Nucl. Phys. A* **371**, 253 (1981).

[65] A. Bohr and B. Mottelson, *Nuclear Structure*, Benjamin, New York, 1969.

[66] I. I. Geguzin, *JETP Lett.* **33**, 568 (1981).

[67] L. Serra, R. A. Broglia, M. Barranco and J. Navarro, *Phys. Rev. A* **47**, R1601 (1993).

[68] M. Manninen, J. Mansikka-aho, H. Nishioka and Y. Takahashi, *Z. Phys. D* **31**, 259 (1994).

[69] Ll. Serra, R. A. Broglia, M. Barranco and J. Navarro, *Phys. Rev. A* **47**, R1601 (1993).

[70] L. Mornas, F. Calvayrac, P.-G. Reinhard and E. Suraud, *Z. Phys. D* **38**, 73 (1996).

[71] C. Kohl, B. Montag and P.-G. Reinhard, *Z. Phys. D* **35**, 57 (1995).

[72] B. Fischer, C. Kohl and P.-G. Reinhard, *Phys. Rev. B* **56**, 11149 (1997).

[73] M. Danos, *Bull. Am. Phys. Soc.*, Ser. II, **1**, 135 (1956).

[74] J. Borggreen, P. Chowdhury, N. Kebaïli, L. Lundsberg-Nielsen, K. Lützenkirchen, M. B. Nielsen, J. Pedersen and H. D. Rasmussen, *Phys. Rev. B* **48**, 17507 (1993).

[75] Ch. Ellert, H. Haberland, Th. Reiners and M. Schmidt, *Phys. Rev. Lett.* **74**, 1558 (1994).
[76] W. Ekardt and Z. Penzar. *Phys. Rev. B* **43**, 1331 (1991).
[77] K. Selby, V. Kresin, J. Masui, M. Vollmer, W. A. de Heer, A. Scheidemann and W. D. Knight, *Phys. Rev. B* **43**, 4565 (1991).
[78] S. Kasperl, C. Kohl and P.-G. Reinhard, *Phys. Lett. A* **206**, 81 (1995).
[79] Ch. Ellert, H. Haberland, W. Orlik, Th. Reiners and M. Schmidt, *Chem. Phys. Lett.* **215**, 357 (1993).
[80] D. Pines and Ph. Nozières, *The Theory of Quantum Liquids*, Addison-Wesley, New York, 1966.
[81] G. F. Bertsch, P. F. Bortignon and R. Broglia, *Rev. Mod. Phys.* **55**, 287 (1983).
[82] P.-G. Reinhard, C. Toepffer and H. L. Yadav, *Nucl. Phys. A* **458**, 301 (1986).
[83] J. M. Pacheco and R. A. Broglia, *Phys. Rev. Lett.* **62**, 1400 (1989).
[84] C. Yannouleas, J. M. Pacheco and R. A. Broglia, *Phys. Rev. B* **41**, 6088 (1990).
[85] B. Montag and P.-G. Reinhard, *Phys. Rev. B* **51**, 14686 (1995).
[86] F. Alasia, R. A. Broglia, H. E. Roman, Le. Serra, G. Colo and J. M. Pacheco, *J. Phys. B* **27**, L643 (1994).
[87] C. Yannouleas, E. Vigezzi and R. A. Broglia, *Phys. Rev. B* **47**, 9849 (1993).
[88] F. Calvayrac, E. Suraud and P.-G. Reinhard, *J. Phys. B.* **31**, 1367 (1998).
[89] S. Kümmel, M. Brack and P.-G. Reinhard, *Phys. Rev. B* **58**, 1774 (1998).
[90] V. Bonačić-Koutecký, P. Fantucci and J. Koutecký, *Chem. Rev.* **91**, 1035 (1991).
[91] J. Cugnon, *Ann. Phys. (Fr)* **11**, 201 (1986).
[92] W. Cassing, V. Metag, U. Mosel and K. Niita, *Phys. Rep.* **188**, 363 (1990).
[93] D. H. E. Gross, *Rep. Prog. Phys.* **53**, 605 (1990), and references therein.
[94] Y. Abe, S. Ayik, P.-G. Reinhard and E. Suraud, *Phys. Rep.* **275**, 49 (1996).
[95] F. Calvayrac, P.-G. Reinhard and E. Suraud, *Ann. Phys. (N.Y.)* **255**, 125 (1997).
[96] V. Weisskopf, *Phys. Rev.* **52**, 295 (1937).
[97] K. Boretzky *et al*, *Phys. Lett. B* **384**, 30 (1996).
[98] M. Gavrila, ed., *Atoms in Intense Laser Fields*, Academic Press, New York, 1992.
[99] G. Bertsch, P.-G. Reinhard and E. Suraud, *Phys. Rev. C* **53**, 1440 (1996).
[100] C. Ullrich, P. G. Reinhard and E. Suraud, *Phys. Rev. A* **57**, 1938 (1998).
[101] C. Ullrich, and E. K. U. Gross, *Comments At. Mol. Phys.* **33**, 211 (1997).
[102] C. Ullrich, P. G. Reinhard and E. Suraud, *J. Phys. B* **30**, 5043 (1997).
[103] C. Ullrich, P.-G. Reinhard and E. Suraud, *Euro. Phys. Journ. D* **1**, 303 (1998).
[104] P.-G. Reinhard, C. Toepffer and H. L. Yadav, *Nucl. Phys. A* **458**, 301 (1986).
[105] B. Montag and P.-G. Reinhard, *Z. Phys. D* **33**, 265 (1995).
[106] A. S. Umar, M. R. Strayer, R. Y. Cusson, P.-G. Reinhard and D. A. Bromley, *Phys. Rev. C* **32**, 172 (1985).
[107] F. Calvayrac, C. Kohl, S. El-Gammal, P.-G. Reinhard and E. Suraud, to appear in *Nuov. Cim.* **110A**, 1177 (1997).
[108] P.-G. Reinhard, F. Calvayrac and E. Suraud, *Z. Phys. D* **41**, 151 (1997).
[109] C. Kohl, F. Calvayrac, P.-G. Reinhard and E. Suraud, *Surf. Science* **405**, 74 (1998).
[110] G. F. Bertsch and S. Das Gupta, *Phys. Rep.* **160**, 190 (1988).
[111] A. Bonasera, F. Gulminelli and J. Molitoris, *Phys. Rep.* **243**, 1 (1994).
[112] E. Suraud, *Ann. Phys.* (Paris) **21**, 461 (1996).
[113] P. L'Eplattenier, E. Suraud and P. G. Reinhard. *Ann. Phys. (N.Y.)* **224**, 426 (1995).
[114] P. G. Reinhard and E. Suraud, *Ann. Phys. (N.Y.)* **239**, 193 (1995).
[115] P. G. Reinhard and E. Suraud, *Ann. Phys. (N.Y.)* **239**, 216 (1995).
[116] P.-G. Reinhard and E. Suraud, *Z. Phys. A* **355**, 339 (1996).
[117] M. J. Giannoni, D. Vautherin, M. Vénéroni and D. Brink, *Phys. Lett. B* **63**, 8 (1976).
[118] P. Ring and P. Schuck, *The Nuclear Many Body Problem*, Springer, New York, 1980.
[119] M. Gross and C. Guet, *Z. Phys D* **33**, 289 (1995).
[120] L. Féret, E. Suraud, F. Calvayrac and P. G. Reinhard, *J. Phys. B* **29**, 4477 (1996).
[121] A. Domps, P. L'Eplattenier, P. G. Reinhard and E. Suraud, *Ann. Phys.* (Leipzig) **6**, 455 (1997).
[122] A. Domps, A. S. Krepper, V. Savalli, P. G. Reinhard and E. Suraud, *Ann. Phys.* (Leipzig) **6**, 468 (1997).
[123] A. Domps, P. G. Reinhard and E. Suraud, *Ann. Phys. (N.Y.)* **260**, 171 (1997).
[124] A. Domps. E. Suraud and P.-G. Reinhard, *Phys. Rev. Lett.* **80**, 5520 (1998).

Index